# Protestant Modernist Pamphlets

Medicine, Science, and Religion
in Historical Context

# Protestant Modernist Pamphlets

## Science and Religion in the Scopes Era

EDWARD B. DAVIS

Johns Hopkins University Press
*Baltimore*

© 2024 Johns Hopkins University Press
All rights reserved. Published 2024
Printed in the United States of America on acid-free paper
2 4 6 8 9 7 5 3 1

Johns Hopkins University Press
2715 North Charles Street
Baltimore, Maryland 21218
www.press.jhu.edu

Library of Congress Cataloging-in-Publication Data

Names: Davis, Edward Bradford, 1953– author.
Title: Protestant modernist pamphlets: Science and religion in the Scopes era / Edward B. Davis.
Description: Baltimore : Johns Hopkins University Press, 2024. | Series: Medicine, science, and religion in historical context | Includes bibliographical references and index.
Identifiers: LCCN 2023058945 | ISBN 9781421449821 (hardcover) | ISBN 9781421449838 (ebook)
Subjects: LCSH: Religion and science—United States—History—20th century. | United States—Intellectual life—20th century. | American Institute of Sacred Literature.
Classification: LCC BL245 .D35 2024 | DDC 261.5/5097309042—dc23/eng/20240217
LC record available at https://lccn.loc.gov/2023058945
A catalog record for this book is available from the British Library.
ISBN 978-1-4214-4982-1 (hardcover)
ISBN 978-1-4214-4983-8 (ebook)

*Special discounts are available for bulk purchases of this book. For more information, please contact Special Sales at specialsales@jh.edu.*

*For Kathy, Katie, and Julie*

CONTENTS

*Abbreviations and Archives Cited* xi
*Preface* xiii

PART ONE: PROTESTANT MODERNIST RESPONSES TO BRYAN

    Introduction    3

CHAPTER 1  "Spiking Bryan's Guns"    9
    Contested Definitions of "Science" and "Religion"

CHAPTER 2  Liberal Protestant Scientists and Clergy Join Forces    49
    The Story of the AISL Pamphlets

CHAPTER 3  Science and Religion, Chicago Style    80
    The Protestant Modernist Encounter with Science

PART TWO: THE AISL "SCIENCE AND RELIGION" PAMPHLETS

CONKLIN (1922)  *Evolution and the Bible*    107
    Edwin Grant Conklin (1863–1952)
    INTRODUCTION    107
    ANNOTATED TEXT    111

FOSDICK (1922)  *Evolution and Mr. Bryan*    123
    Harry Emerson Fosdick (1878–1969)
    INTRODUCTION    123
    ANNOTATED TEXT    125

MATHEWS (1922)  *How Science Helps Our Faith*    134
    Shailer Mathews (1863–1941)
    INTRODUCTION    134
    ANNOTATED TEXT    137

MILLIKAN (1923) *A Scientist Confesses His Faith*   143
    Robert Andrews Millikan (1856–1953)
        INTRODUCTION   143
        ANNOTATED TEXT   145

FROST (1924) *The Heavens are Telling*   160
    Edwin Brant Frost (1866–1935)
        INTRODUCTION   160
        ANNOTATED TEXT   162

SCHMUCKER (1926) *Through Science to God, The Humming Bird's Story, An Evolutionary Interpretation*   177
    Samuel Christian Schmucker (1860–1943)
        INTRODUCTION   177
        ANNOTATED TEXT   179

PUPIN (1928) *Creative Co-ordination*   190
    Michael Idvorsky Pupin (1854–1935)
        INTRODUCTION   190
        ANNOTATED TEXT   192

FOSDICK (1928) *Religion's Debt to Science*   197
    Harry Emerson Fosdick (1878–1969)
        INTRODUCTION   197
        ANNOTATED TEXT   200

COMPTON, MATHEWS, AND GILKEY (1930) *Life After Death*   210
    Arthur Holly Compton (1892–1962), Shailer Mathews (1863–1941), and Charles Whitney Gilkey (1882–1968)
        INTRODUCTION   210
        ANNOTATED TEXT   213
            FROM THE POINT OF VIEW OF A SCIENTIST   213
            FROM THE POINT OF VIEW OF A THEOLOGIAN   222
            FROM THE POINT OF VIEW OF A CHRISTIAN MINISTER   227

MATHER (1931) *The Religion of a Geologist*   235
    Kirtley Fletcher Mather (1888–1978)
        INTRODUCTION   235
        ANNOTATED TEXT   238

*Appendix One: Publication Details for AISL Pamphlet Series "Science and Religion" and Related Publications* 253

*Appendix Two: Publication Runs for AISL Pamphlets and the Millikan "Statement"* 257

*Appendix Three (A): Scientists Who Supported AISL Pamphlets, 1922–1928* 258

*Appendix Three (B): Scientists Who Supported AISL Pamphlets, 1928–1934* 266

*Notes* 273

*Index* 309

# ABBREVIATIONS AND ARCHIVES CITED

| | |
|---|---|
| AAAS | American Association for the Advancement of Science |
| AISL | American Institute of Sacred Literature |
| AISLR | American Institute of Sacred Literature. Records, [box number followed by folder number], Hanna Holborn Gray Special Collections Research Center, University of Chicago Library |
| *ANB* | *American National Biography*, ed. John A. Garraty and Mark C. Carnes, 24 vols. (New York: Oxford University Press, 1999) |
| DUA | The Papers of Kirtley F. Mather, [box number followed by folder number], Denison University Archives, Granville, Ohio |
| EBFP | Frost, Edwin B. Papers, [box number followed by folder number], Hanna Holborn Gray Special Collections Research Center, University of Chicago Library |
| EBFPY | Edwin Brant Frost Papers, Yerkes Observatory, accessed May 1999. This material was not transferred to the University of Chicago Library and can no longer be located. The author has photocopies of some items. |
| EGCP | Edwin Grant Conklin Papers, [box number followed by folder number or name of folder] or [carton number followed by folder number or name of folder], Manuscripts Division, Department of Special Collections, Princeton University Library |

| | |
|---|---|
| HEFP | Harry Emerson Fosdick Papers, Series 1A, [box number followed by folder number], The Burke Library at Union Theological Seminary, Columbia University in the City of New York |
| HEFP Evolution | Harry Emerson Fosdick Papers, Series 3C, Box 1, Folder 6 ("Evolution & W. J. Bryan—Correspondence, etc.—1922-23"), The Burke Library at Union Theological Seminary, Columbia University in the City of New York |
| KFMP Conference | Kirtley Fletcher Mather Papers, HUG 4559.500.3, Box 4, "Evolution: The Conference Method of Study," Harvard University Archives, Courtesy of the Harvard University Archives |
| KFMP Lectures | Kirtley Fletcher Mather Papers, HUG 4559.500.20, [box number followed by folder name], Harvard University Archives, Courtesy of the Harvard University Archives |
| KFMP Speeches | Kirtley Fletcher Mather Papers, HUG 4559.500.3.5, [box number followed by folder name], Harvard University Archives |
| NAS | National Academy of Sciences |
| NRC | National Research Council |
| RAC | Rockefeller Archive Center, North Tarrytown, New York, Rockefeller Family Archives, Record Group 2, Office of the Mssrs. Rockefeller, Educational Interests, Series G, box 106, folder 741, "University of Chicago—American Institute of Sacred Literature, 1921–1933" |
| RAMP | Papers of Robert Andrews Millikan, [box number followed by folder number], Archives, California Institute of Technology |
| SMP | Shailer Mathews Papers, [box number followed by folder number], Hanna Holborn Gray Special Collections Research Center, University of Chicago Library |
| *WWWA* | *Who Was Who in America* |
| YMCA | Young Men's Christian Association |

# PREFACE

Just as I was finishing doctoral work in the history and philosophy of science nearly forty years ago, I acquired a religious tract relating the spectacular story of a modern Jonah. James Bartley had supposedly spent thirty-six hours inside the belly of a sperm whale near the Falkland Islands in February 1891 and come out alive. The British Broadcasting Company based two radio programs on the resulting article. Other than Bartley, the main person in my account was Harry Rimmer, a self-educated evangelist and anti-evolutionist who made frequent use of a variant of the Bartley tale. He also relished debating all comers, especially scientists. A major rhetorical conquest (in his view) came in November 1930 when he debated naturalist Samuel Christian Schmucker, a nationally prominent writer and speaker on evolution and eugenics, in front of 2,500 people at the largest auditorium in Philadelphia.[1]

While researching that event, I found a fascinating pamphlet by Schmucker about God and evolution. Published just one year after the Scopes trial, it belonged to a series of tracts on "Science and Religion" from the American Institute of Sacred Literature, a now-defunct correspondence arm of the University of Chicago Divinity School. Six other scientists had written similar pamphlets over the next decade—two Nobel laureates, a Pulitzer Prize winner, and five presidents of the American Association for the Advancement of Science. Two more were written by one of the greatest preachers in American history, and another by the illustrious dean of the Divinity School. Even though the authors were famous and hundreds of thousands of pamphlets were distributed to tens of thousands of people, library copies are rare today and the existence of the series as a whole was virtually unknown to historians. I knew I had a historical project in the making.

At that point, I was heavily involved with other projects on early modern science, but eventually I published nine articles about aspects of the project, including studies of the religious lives and beliefs of five of the scientists. Because the content of the pamphlets is highly interesting—even more so considering the notoriety of the authors—an annotated edition is warranted. I also tell the story of how they came to be written, financed, and distributed, and what readers thought about them amid heated arguments between conservative and liberal Protestants about religion, science, and how to relate them.

My effort to tell this story and to bring the pamphlets back into view would have been fruitless without the assistance and encouragement of individuals and institutions who believe in its significance. I am pleased to acknowledge generous grants from the National Science Foundation (SES-9818198) and the John Templeton Foundation (ID#2047 and ID#12389), without which this project would have been impossible. Messiah College (now Messiah University) provided reduced teaching loads, two sabbaticals, and three student interns who assisted with various aspects of the project: James (Mac) Brodt, Elizabeth Chmielewski, and Glenn Jones.

Expert assistance from archivists has been crucial. I thank the staff of Special Collections, The College of Wooster Libraries, and University Archives, Department of Special Collections, Washington University Libraries, for the papers of Arthur Holly Compton; Manuscripts Division, Department of Rare Books and Special Collections, Princeton University Library, and Special Collections, Ohio Wesleyan University, for the papers of Edwin Grant Conklin; Archives of the Burke Library at Union Theological Seminary in the City of New York, for the papers of Harry Emerson Fosdick; Yerkes Observatory for the papers of Edwin Brant Frost; Harvard Archives and Denison University Archives for the papers of Kirtley Fletcher Mather; Hanna Holborn Gray Special Collections Research Center, University of Chicago, for the papers of Edwin Brant Frost, Shailer Mathews, and the American Institute of Sacred Literature; Oberlin College Archives and the Archives of the California Institute of Technology for the papers of Robert Millikan; Rare Book and Manuscript Library, Columbia University, for the papers of Michael Idvorsky Pupin; and Rockefeller Archive Center, North Tarrytown, New York, for the papers of John D. Rockefeller Jr.

Special thanks are owed to the late Ronald L. Numbers, who provided unrelenting encouragement and advice for decades to a scholar more familiar with seventeenth-century European natural philosophical treatises than twentieth-century American religious pamphlets. He graciously read the whole typescript, providing countless suggestions. The late Owen Gingerich helped identify certain people and provided invaluable assistance with astronomical references, especially in the notes to Frost's pamphlet. I only wish both friends had lived to see the finished product. Kyle Cudworth, Kristin Johnson, Devin Manzullo-Thomas, Jeffrey S. McDonald, Anne M. Ostendarp, Lawrence M. Principe, Jon H. Roberts, William R. Shea, the late Loyd S. Swenson Jr., and David L. Weaver-Zercher advised on certain points. Responsibility for any errors is mine alone.

Above all, I thank my family—Kathy, Katie, and Julie—who tolerated my visits to libraries far from home and who still love me despite many failings.

# PART ONE

# PROTESTANT MODERNIST RESPONSES TO BRYAN

# Introduction

What are the religious implications of scientific knowledge? What image of science and its relation to morality and religion should scientists cultivate and promote? Should scientists cooperate with clergy in educating the public about the content, scope, and limits of scientific knowledge? Do religious scientists have a special responsibility to contribute to such conversations?[1]

Americans have debated these questions since at least the early eighteenth century, when Cotton Mather's *The Christian Philosopher* (1721) brought the Scientific Revolution to America with an unmistakably pious slant. Benjamin Silliman of Yale College and Edward Hitchcock of Amherst College included theologically sophisticated discussions of Genesis and geology in antebellum textbooks on natural history. Around the end of the nineteenth century, science books for nonsectarian public schools still sometimes used religious language and discussed the moral dimensions of science from an implicitly religious framework. Even today it is unremarkable to find some attention given to various creation stories in high school biology texts. It was in the 1920s—the decade of the Scopes trial, undoubtedly the most famous American episode in the history of science and religion—that these issues were discussed in ways that still shape the conversation today.

In February 1922, William Jennings Bryan's national campaign against the teaching of evolution went upscale when the *New York Times* published his Sunday editorial, "God and Evolution." Summarizing speeches he had been giving for months, Bryan argued that evolution was "only a guess and was never anything more," though scientists might use a fancier word, "hypothesis." Evolution was religiously dangerous because it denied miracles and the supernatural, ultimately "leaving the Bible a story book without binding authority upon the conscience of man." Those who combined evolution with belief in God would follow Charles Darwin into agnosticism. Since evolution was unproved speculation that undermined Christianity, public schools should not teach it.[2] The following Sunday, the *Times* printed responses from Princeton University biologist Edwin Grant Conklin and Columbia University paleontologist Henry Fairfield Osborn, president of the American Museum of

Natural History. The popular liberal pastor of Manhattan's First Presbyterian Church, Harry Emerson Fosdick, wrote a third reply that appeared the next week, followed two days later by Bryan's reply to Conklin and Osborn.[3]

That September, the editorials by Fosdick and Conklin were issued as shirt-pocket-sized pamphlets by the American Institute of Sacred Literature (AISL), a correspondence school for Protestant pastors administered by the University of Chicago Divinity School. Founded in 1880 at Morgan Park Theological Seminary in Chicago (which later became the Divinity School) by William Rainey Harper, an expert on Semitic languages, it was originally known as the Correspondence School for Hebrew. As Harper moved first to Yale and then to the new University of Chicago in 1891 to become its first president, courses were added covering the other biblical languages, biblical literature, history, and theology, so in 1889 the name was changed to the AISL. The Institute continued its work for decades after Harper's death in 1906, providing continuing education to Protestant ministers and Sunday School teachers throughout North America, including "three hundred colored preachers in two southern states." Course materials emphasized modern biblical scholarship and modernist theology. By the turn of the century, there were 10,000 students from six continents, and by 1917 a dozen Bible courses alone enrolled 5,000 people. Publications were also sent to Young Men's Christian Associations (YMCAs), secular schools offering Bible courses, women's clubs, and parents who wanted to cooperate with churches in educating their children. The first AISL pamphlet, *Will Christ come again?* by Shailer Mathews (discussed in chapter 3), an important Baptist lay theologian who served as dean of the Divinity School from 1908 to 1933, appeared during the Great War, followed by dozens more on diverse subjects before the Great Depression took its toll and the AISL closed in 1944.[4]

The pamphlets by Conklin and Fosdick were issued simultaneously with one by Mathews. On the front cover of each, printed in small type above the title, were the words "Popular religion leaflets," while all capital letters in a larger font announced that the pamphlet was part of a series on "Science and Religion." Over the next nine years, six more pamphlets were added, including a second title by Fosdick. The others were written by prominent American scientists: Nobel laureate physicist Robert Andrews Millikan of Caltech; Yerkes Observatory astronomer Edwin Brant Frost; West Chester (Pennsylvania) State Teachers College naturalist Samuel Christian Schmucker, a popularizer of evolution and eugenics with a national following; Columbia physicist

Michael Idvorsky Pupin, the most influential Serbian immigrant of his generation and president of the American Association for the Advancement of Science (AAAS) in the year of the Scopes trial; and Harvard University geologist Kirtley Fletcher Mather. A separate pamphlet from 1930 on eternal life, with contributions by Mathews, Nobel laureate physicist Arthur Holly Compton, and their former pastor Charles Whitney Gilkey, was not actually part of the series (although originally intended as such), but its content, authorship, and attitude are so similar that it is included here. Publication details for all ten titles are given in appendix one and appendix two.

Mailed to religious workers, scientists, high school principals, lawmakers, and ordinary people, the first four pamphlets were specifically intended to counter the views of Bryan and other conservative Protestants. Together with the rest of the pamphlets, they were central to the Divinity School's effort to communicate modernist attitudes toward science to a national audience. Fosdick was modernism's leading spokesperson, Mathews its leading theological educator, and Millikan the most famous American scientist of his generation. Except for Schmucker, the other scientists were all figures of sufficient stature to have entries devoted to them in both the *Dictionary of Scientific Biography* and *American National Biography*. Furthermore, the pamphlets were funded to a significant degree by the leading modernist philanthropist, John D. Rockefeller Jr., in partnership with churches, the AAAS, various other organizations, and individuals—including more than 100 eminent scientists whose donations helped underwrite the series. Despite very wide distribution, copies today are very scarce. For three pamphlets, only one or two copies are listed in the WorldCat database, although additional uncatalogued copies of all titles are in various archives and several titles have been offered for sale. Cataloged copies are enumerated in appendix one. Just five (Conklin, both by Fosdick, Frost, and Mather) are available in some form on the internet.

Although much has been written about the Scopes trial and fundamentalist views of science, we still know very little about what leading scientists and modernist religious thinkers thought of the relationship between science and religion. The story told here fills in some of those gaps and complements work on American religion and science by Edward J. Larson and James Gilbert. In his celebrated book on the Scopes trial, Larson chronicled the efforts of clergy and scientists involved in the trial to convey the message that modern science does not contradict faith—at least for the liberal Protestants whose statements were read into the court record. This study builds on his work by showing how the trial was only the most visible part of a much larger strategy to control the religious image of science in America and by examining much more closely than hitherto liberal Protestant views of science in the early twentieth century. Gilbert took a very different approach. From a dozen case studies of religion–science engagement from the 1920s through the early 1960s, he argued that the persistence of religion in a scientific culture is possible, because "neither science nor religion has had a stable and permanent definition in American culture. They continually shift in meaning and in their relation to each other."[5] Thus when we speak of "the relationship between science and religion," as I already did in this paragraph, we must keep in mind that we are speaking about a fluid relationship that cannot be reified without doing violence to the actual historical situations. Shifting—and highly contested—definitions of both "science" and "religion" will emerge from the following pages, especially when their "relationship" is being negotiated. At such times the conversation between religion and science becomes particularly intense, and scientists themselves have been among the most outspoken voices in this conversation. Several pamphlets illustrate this. Written by prestigious scientists who sought explicitly to integrate science and theology into unified worldviews, they refute the widely repeated axiom that science and religion occupy separate, nonintersecting realms in the modern world.

The following chapters show how the pamphlets fit within the larger history of religion and science in America after the Great War. Examining diverse meanings of "science" and "religion" with particular reference to the controversy over evolution and naturalism, I analyze differences of opinion between conservative and liberal Protestants concerning the nature of science, the essence of Christian faith, and how one ought to relate science to religion. Liberals were deeply influenced on all points by the core assumption that traditional Christian theology and science have always been engaged in mortal combat, with modern science inevitably emerging triumphant. Thus, for Christian values and morality to survive in a scientific world, many key theological tenets—which they typically spoke of pejoratively as "dogma"— would need to be set aside. This "conflict" view of the history of science and religion was codified after the Civil War by historian Andrew Dickson White, the founding president of Cornell University, and chemist John William Draper of the University of New-York. Although it has been almost completely rejected by historians today, in the Progressive Era it was uncritically accepted by most scientists and scholars as the dominant leitmotif for thinking about Christianity and science. The modernists in this study wholly embraced it and advanced it in the pamphlets.

Anxious to distance themselves decisively from what they saw as sheer obscurantism coming from religious conservatives in the years surrounding *Scopes*, liberal clergy joined with religious scientists in efforts to move public opinion in more positive directions. They issued "A Joint Statement Upon the Relations of Science and Religion" in the *New York Times*, wrote books and magazine articles for general readers, and published numerous pamphlets such as those from the AISL. To a significant degree, they were responding to similar efforts on the part of Bryan and his supporters, who demonized evolution and modern biblical scholarship at every turn while enlisting state legislatures to eliminate evolution from public school curricula. From correspondence among scientists and between scientists, clergy, and others, we will see the concerns expressed by some influential scientists about the public understanding of the nature of science and its perceived religious implications. We will also learn why the pamphlets were written, how the whole series about "science and religion" was envisioned, how funds were raised to publish and distribute them, who received them, and what various readers thought.

Each pamphlet has an introduction situating its ideas within the life and thought of the author, showing how it relates to other statements about science and religion by the same person. In five instances—Compton, Mather, Millikan, Pupin, and Schmucker—these are based on separate studies of their careers as major public intellectuals that I have published elsewhere. Information gleaned from this part of the project is the basis for chapter 3, which presents and analyzes some of the ways in which these liberal Christians responded religiously and even theologically to modern science. A primary theme is the degree to which they accepted naturalism and the prestige of science as normative factors in their religious attitudes and ideas. I consider what they said about God, miracles, morality, prayer, philosophical materialism (and how they defined it), the existence of a "supernatural" realm, and the prospect of eternal life. Where appropriate, special attention is given to their views on the deity of Jesus and the Resurrection—defining beliefs not only for conservative Protestants in modern America but for nearly all Christians since the early church. The following questions emerge from this material: What fundamental assumptions and accommodations did the authors make in light of the larger history of Christianity and science? How (if at all) did they use science to support Christian beliefs? To what degree and in what ways did modernist attitudes and beliefs shape their presentation of science? How did their acceptance of the "conflict" thesis shape the ways in which modernists viewed the relationship between theology and science?

The pamphlets and their story provide a clear window on the conversation that liberal Protestants had, with and about science, in the early part of the last century. What follows is a short historical account of an episode that carries important implications for the vibrant, far more diverse, religion–science dialogue in our own time.

CHAPTER 1

# "Spiking Bryan's Guns"
## Contested Definitions of "Science" and "Religion"

> Darwinism puts God far away; the Bible brings God near and establishes the prayer-line of communication between the Heavenly Father and His children. Darwinism enthrones selfishness; the Bible crowns love as the greatest force in the world.
> —*William Jennings Bryan,* The Menace of Darwinism

The AISL pamphlets appeared in a hotly contentious context that shaped their content, motivated their authors, and influenced their readers.[1] After the Great War, American Protestants were rent by bitter disputes between two opposing factions, each tracing their intellectual and spiritual roots to the late nineteenth century. At one end was a group of radical progressives, calling themselves "modernists," who sought to transform Christianity into a new kind of faith that was much more open to modern science, higher biblical criticism, and secular education and government. Theologically, modernists emphasized the humanity of Jesus over the divinity of Jesus, divine immanence over divine transcendence, and the importance of reformulating those traditional doctrines that seemed to fly in the face of secular reason—including key elements of the Apostles' and Nicene Creeds, touchstones of orthodoxy among Christians since the fourth century. For many modernists, it was more than just a matter of emphasis: they denied the virgin birth, the Resurrection, and the deity of Christ, while baptizing the process of evolution as God's way of bringing about both the creation of the world and the gradual moral perfection of humanity through eugenics. The atoning sacrifice of Christ was either shunted off to one side or left entirely out of the picture. As Ian G. Barbour astutely observed, the modernists "emphasized God's immanence, often to the virtual exclusion of transcendence, and in some cases God was viewed as a force within a cosmic process that was itself divine."[2]

On the other hand, many traditional Protestants saw modernism as nothing more than the old infidelity of Enlightenment skeptics, foisted upon faithful but gullible believers by dangerous scholars and preachers who had placed too much faith in science and reason and too little faith in the miracle-working God of the Bible. Rather than reformulating central doctrines in the name of science, they vigorously opposed evolution and higher biblical criticism, reaffirmed traditional beliefs about God and Christ, and continued to seek the salvation of the lost by preaching the Crucifixion and Resurrection of Jesus.

After the Great War, some of the more radical conservatives began to identify under a new label. The word "fundamentalist" was defined in July 1920 by Curtis Lee Laws, editor of the *Watchman-Examiner*, a national Baptist weekly. Laws applied the term "in compliment and not in disparagement" to himself and others who "cling to the great fundamentals and who mean to do battle royal" for them. His language implies that the term was already in use verbally, but no earlier published instance is known.[3] Although differences with modernists over the Bible and theology obviously motivated fundamentalists, Laws's definition gave fundamentalism an attitude—militant rejection of modernity in the name of traditional religion. According to George M. Marsden, "until about 1960 the term 'fundamentalists' referred to those evangelical Protestants who were militantly opposed to modernist theology and to related secularizing trends in the culture."[4] That describes most of the antievolutionists in this book, but we must keep in mind that some fundamentalists were less militant than others, less inclined to sectarian division, and more willing to find common cause with other orthodox Protestants who also found modernism dangerous to the faith. Other traditional Christians declined to call themselves "fundamentalists" for various reasons, including the fact mentioned by Laws that the word was sometimes used as a pejorative hurled at anyone more conservative than the person using it, a tendency that only increased later in the 1920s. Only those who identified as "fundamentalist" are described as such in this book. Paradoxically, the modernists did not realize how modern their conservative opponents actually were in some respects. Fundamentalism was theologically innovative. Most fundamentalists embraced a type of premillennialism introduced in England in the 1830s by John Nelson Darby, a former Anglican priest who influenced the nascent Plymouth Brethren movement to embrace his eschatology. Known as dispensationalism, it divided biblical history into separate eras called "dispensations," a pattern resembling scientific classification. In addition, the fundamentalist emphasis on the very words of Scripture as divinely inspired and therefore propositionally inerrant encouraged them to interpret the Bible "as a source of data to be mined and scientifically analyzed," according to Lincoln A. Mullen.[5]

The term "fundamentalist" clearly referred to a collection of ninety articles in twelve paperbound volumes called *The Fundamentals* (1910–1915). Millions were mailed without charge to Protestant ministers and other church officers, with costs covered by California oil barons Lyman and Milton Stewart. Yet evolution was not a principal target of this prewar project. Although some authors derided evolution, most ignored it, and a few even allowed that some form(s) of evolution might be acceptable. Their main concerns lay elsewhere, especially in promoting premillennialism and defending the reliability and authority of the Bible against biblical critics. Overall, considering the radical separatism of many who later called themselves "fundamentalists," *The Fundamentals* were not nearly as fundamentalist as one might casually think.[6]

It was not until after the Great War, according to Marsden, that "opposition to evolution became an article of the faith" among conservative Protestants. Nevertheless, most essays in the prewar *Fundamentals* shared the view "that true science and rationality support traditional Christianity grounded in the supernatural and the miraculous," an attitude that would underlie much later opposition to evolution. Many of the authors believed that "the current prejudice against the miraculous, including the miraculous origins of the Bible, and the miraculous character of the fundamental Christian doctrines concerning the Incarnation, are just that, prejudices."[7] Conservative Protestants often spoke of "true science," by which they meant a Baconian science based on common sense realism, consisting only of proven "facts," in contrast to "mere theories" such as evolution—which was in their view an unwarranted, outright fanciful conclusion about the history of life. To be fair, this attitude was consistent with "the dictionary definition of the term [science]," as Jon H. Roberts has pointed out, and with scientific textbooks of the period, as Michael Newton Keas has noted.[8] Conservatives believed that evolution was plausible to the secular mind only because unbelieving scientists had *arbitrarily* ruled out miracles from further consideration. Genuine science would not inappropriately bias itself by employing the blinders of methodological naturalism but would remain open to the possibility of miracles—not only in biblical history but also in the history of nature, although the naturalism of modernist biblical scholars was an even more important target. Evolution and biblical criticism were considered "science falsely so-called," invoking Paul's warning to Timothy, "keep that which is committed to thy trust, avoiding profane and vain babblings, and oppositions of science falsely so called: Which some professing have erred concerning the faith" (1 Timothy 6:20–21 KJV). Natural history had occasionally been categorized that way for more than a century, and the practice was commonplace during the 1920s.[9]

## Bryan, the "Menace" of Evolution, and the Illegitimacy of Scientific "Hypotheses"

Opposition to evolution became a hallmark of leadership among postwar fundamentalists. Although William Jennings Bryan rejected premillennialism and did not call himself a fundamentalist, he shared their concerns about modernism and evolution. His opposition to evolution was wholly consistent with his vision of progressive politics, for it was rooted in a hatred of what he believed the teaching of evolution would lead to: the exploitation of workers, the destruction of democracy, and the moral paralysis of the populace.[10] Bryan especially opposed the application of natural selection to human behavior, calling this "the law of hate—the merciless law by which the strong crowd out and kill off the weak."[11] What ultimately led him to try to outlaw evolution in public schools was the perception that German intellectuals had used evolution to justify militarism before the Great War, as presented in works such as *Headquarters Nights* (1917) by Stanford University evolutionary biologist Vernon Kellogg. Although Kellogg believed that evolution had been misused in this way, Bryan and many other readers thought he was stating a necessary connection.[12]

Even apart from Kellogg's book, during and after the Great War conservative Protestants often associated evolution, higher biblical criticism, and liberal theology with Germany, the land of Ernst Haeckel, Julius Wellhausen, and Friedrich Schleiermacher. In their view, German intellectuals had pioneered the rejection of traditional Christian views of creation and the Bible, culminating in the moral collapse driving German militarism. Bryan was seen by many as a champion battling German enemies of Christianity (see figure 1.1).

Bryan stated his position emphatically in "The Menace of Darwinism," a speech first given in the spring of 1921 that he often repeated prior to *Scopes*.[13] Described by Lawrence Levine as "the lengthiest, the most effective, and, as the title indicates, the most hostile attack Bryan had yet delivered," it became a lengthy chapter on "The Origin of Man" in the book *In His Image* (1922), based on the James Sprunt Lectures delivered in October 1921 at Union Theological

*Figure 1.1.* "Verdun," by E. J. Pace, *Sunday School Times,* August 19, 1922, 495. This cartoon depicts Bryan as the hero of Verdun, defending "the integrity of the word of God" and stopping the German "enemies of the Bible" in their tracks. Courtesy of Special Collections, Wright Library, Princeton Theological Seminary.

Seminary in Richmond.[14] Soon it appeared separately as a small paperbound book under the original title, to which Bryan returned "to emphasize [the speech's] dominant note." It also echoes *The Menace of Modernism* (1917) by his friend William Bell Riley, a Baptist minister from Minneapolis who founded the World's Christian Fundamentals Association in 1919. For Bryan, Darwinism was a "menace to fundamental morality" because it greatly weakened belief in God, upon which "rest the influences that control life." He allowed that evolution "*permits* one to believe in a God, but puts the creative act so far away that reverence for the Creator—even belief in Him—is likely to be lost." Why "imprison such a God in an impenetrable past? This is a living world; why not a *living* God upon the throne? Why not allow Him to work *now*?"[15] This conception of divine action—that God creates only by special miracles rather than through natural processes—was contested by the pamphlet authors, especially Fosdick.

The alternative was anathema for Bryan. "Theistic evolution may be described as an anesthetic which deadens the pain while the faith is being gradually removed," making it just "a way-station on the highway that leads from Christian faith to No-God-Land."[16] The appellation "theistic evolution" has been used since at least 1877, when several American writers compared religious interpretations of evolution under various headings.[17] Congregationalist minister Joseph Cook identified "at least three schools of evolutionists," enumerating these as "the atheistic, the agnostic, and the theistic." In his view, "it is the theistic, and not the agnostic or the atheistic school of evolution which is increasing in influence among the higher authorities of science." The great Canadian geologist John W. Dawson dissented from "theistic evolution" when discussing the fifth day of creation (see Genesis 1:20–23) in *The Origin of the World, According to Revelation and Science* (1877). He did not think that the variety of verb forms—to make, form, create, or "let the waters bring forth" (Genesis 1:20)—could mean "some form of mediate creation, or of 'creation by law,' or 'theistic evolution,' as it has been termed." Decades later his reservations were cited several times in *The Fundamentals*. Evangelist Alexander Wilford Hall, a highly successful advocate of herbal medicine, vigorously attacked "theistic evolution," even "with the proviso of intelligent design in the various transmutations," in *The Problem of Human Life* (1877). Perhaps the earliest *favorable* reference to "theistic evolution" was by Harvard botanist Asa Gray, the first American proponent of Darwin's theory. He told Yale seminarians in 1880, "The specially theistic evolution referred to judges that these general causes [variability and heredity] cannot account for the whole work [of evolution], and that the unknown causes are of a more special character and higher order."[18]

By the 1920s, views such as Gray's were nearly extinct, and Bryan got to define "theistic evolution" to suit his own purposes. His first objective was to undermine the status of evolution as legitimate scientific knowledge. Wielding a sharp Baconian sword against Darwin's own frankly hypothetical method, Bryan maintained that "not a single fact in the universe" could be found "to prove that man is descended from the lower animals," so that evolution was not even a "theory," merely an unsupported "hypothesis"—which is just "a synonym used by scientists for the word guess." Had Darwin called his theory a guess, "it would not have lived a year." Guesses were no substitute for the Word of God, and Bryan concluded that belief in evolution "requires more faith in *chance* than a Christian is required to have in God."[19]

These comments evince a wide gulf separating the scientific community from the public, including religious people, regarding conceptions of scientific knowledge. Conservative Protestants and many other Americans generally accepted the older, Baconian conception of science as established matters of fact, not speculative hypotheses. This allowed them to reject evolution without seeing themselves as rejecting science. Modernists, on the other hand, fully embraced the hypothetico-deductive method as well as the evolutionary conclusions it led to. The first two pamphlets, by Conklin and Fosdick—a scientist and a minister who listened to scientists—expressed far more confidence in hypotheses than Bryan proffered.

Bryan's "chief concern," however, lay "in protecting man from the demoralization involved in accepting a brute ancestry." He went right after Darwin, quoting *The Descent of Man*, in which Darwin had lamented several of the ways in which "civilized men . . . do our utmost to check the progress of elimination" by asylums, charitable deeds, and vaccination, whereas "with savages the weak in body or mind are soon eliminated." What would Darwin think, Bryan asked, "of remedies for typhoid fever, yellow fever, and the black plague? And what would he think of saving weak babies by pasteurizing milk and of the efforts to find a specific cure for tuberculosis and cancer? Can such a barbarous doctrine be sound?" Bryan highlighted Darwin's admission that caring for weak and helpless people is bad for human evolution, but morality requires it. "Darwin rightly decided to suspend his doctrine, even at the risk of impairing the race. But some of his followers are more hardened," such as an unidentified author who "defended the use of alcohol on the ground that it rendered a service to society by killing off the degenerates."[20]

A further, and for Bryan a necessary, consequence of accepting evolution was loss of religious faith. The Christian child, Bryan noted, is taught to pray to a kindly heavenly father who made the child in the image of God, but "then he goes to college and a learned professor" shows him how he resembles the beasts and that even his "moral sense can be explained on a brute basis without any act of, or aid from, God." This is just "cold, clammy materialism!"[21] Just as Darwin recounted in his *Autobiography*, the student who accepts evolution will slip gradually from Christian theism through deism to agnosticism, with a consequent loss of belief in immortality. Bryan cited James Leuba's pioneering study in the psychology of religion, *Belief in God and Immortality* (1916), which linked higher education, especially advanced study in biology, with loss of belief in a personal God and immortality. He added anecdotes about "evolutionists" and other college professors, including seminary faculty, who made a point of attacking the Bible and urging students to discard traditional religious beliefs as outmoded and irrational.

Bryan also blamed evolution for religious modernism, a point underscored by a powerful cartoon (see figure 1.2) he conceived for his book, *Seven Questions in Dispute* (1924), that reprinted some of his articles from the *Sunday*

*Figure 1.2.* "The Descent of the Modernists," by E. J. Pace, frontispiece from William Jennings Bryan, *Seven Questions in Dispute* (New York: Revell, 1924). From the author's collection.

*School Times*, a weekly magazine edited by Charles Gallaudet Trumbull for nearly 100,000 readers in more than one hundred nations. Intended for Sunday school teachers and superintendents, the *Times* provided detailed lesson plans two weeks in advance and articles on topics of interest to conservative Protestants. For several years they had featured Christian cartoons by Ernest James Pace, a former political cartoonist from Chicago who had become a Christian and served as a missionary in the Philippines before joining Trumbull's staff. Some of Pace's cartoons also appeared in other fundamentalist periodicals, and hundreds were made into sets of glass lantern slides for churches and Bible conferences.[22] Bryan suggested the iconography in a letter to Trumbull early in 1924. "The cartoon I have in mind would represent evolution as I believe it to be, viz., the cause of modernism and the progressive elimination of the vital truths of the bible." It would have "three well-dressed modernists" descending a staircase on which "there is no stopping place," such that the stairway is effectively a slippery slope to "Atheism" at the bottom. The three modernists represent a student, a "minister with the Bible in his hands," and "a scientist stepping from Agnosticism to Atheism." A cartoon like this would show that, "the three persons who are most effected by modernism are the student, the preacher who substitutes education for religion, and the scientist who prefers guesses to the Word of God."[23]

Bryan feared that higher education had a deleterious effect on Christian faith. His recommendations included the crucial political point that, "in schools supported by taxation we should have a real neutrality wherever neutrality in religion is desired." Publicly funded schools, colleges, and universities ought not promote any particular religion, including Christianity, but at the same time they ought not promote irreligion. Evolution violated this premise, creating a situation that was profoundly unjust: "The neutrality which we now have is often but a sham; it carefully excludes the Christian religion but permits the use of the schoolrooms for the destruction of faith and for the teaching of materialistic doctrines." Academic freedom was no excuse, for "the parents who pay the salary have a right to decide what shall be taught."[24]

Ever the populist, Bryan then introduced his strongest political arguments. Starting with the frightening philosophy of Friedrich Nietzsche, "who carried Darwinism to its logical conclusion," Bryan found in Darwinism "the basis of the gigantic class struggle that is now shaking society throughout the world," as "the brute doctrine of the 'survival of the fittest'" was "transforming the industrial world into a slaughter-house." Citing the discussion of eugenics in

Benjamin Kidd's *The Science of Power* (1918), Bryan said, "Darwinism robs the reformer of hope. Its plan of operation is to improve the race by 'scientific breeding' on a purely physical basis," which could take thousands of years, whereas in "Christ's plan . . . [a] man can be born again; the springs of life can be cleansed instantly so that the heart loves the things that it formerly hated and hates the things that it once loved." Bryan concluded, "Many have tried to harmonize Darwinism with the Bible, but these efforts, while honest and sometimes even agonizing, have not been successful." This was because "the natural and inevitable tendency of Darwinism is to exalt the mind at the expense of the heart, to overestimate the reliability of the reason as compared with faith and to impair confidence in the Bible." Theistic evolution occupied an unstable middle ground, not unlike "a traveller [sic] in the mountains, who, having fallen half-way down a steep slope, catches hold of a frail bush. It takes so much of his strength to keep from going lower that he is useless as an aid to others." Bryan urged theistic evolutionists to "revise their conclusions in view of the accumulating evidence of its baneful influence."[25]

In January 1922, less than a year after Bryan had first delivered this speech, he learned of efforts by Baptists in Kentucky to ban evolution in public schools. Within a few months he and some fundamentalist leaders backed legislation in Kentucky—defeated by a single vote—and several other states. Before the end of the decade, anti-evolution laws had been introduced in more than twenty states and passed in five: Oklahoma, Florida, Tennessee, Mississippi, and Arkansas (in chronological order). An amendment banning pro-evolution radio broadcasts was debated in the United States Senate.[26] In March 1925, Tennessee governor Austin Peay signed the Butler Act, leading quickly to the carefully orchestrated show trial of John Scopes—in which Bryan played a pivotal, unsolicited role as part of the prosecution. Although the trial itself does not concern us, Bryan's anti-evolution crusade led directly to the AISL pamphlets on "Science and Religion," which debuted in September 1922.

## Responding to Ridicule: Is Evolution Atheistic?

Almost no professional scientist shared Bryan's low view of evolution. Ronald L. Numbers has identified just one biologist from this period who rejected common ancestry, the "reputable but relatively obscure" zoologist Albert Fleischmann of the University of Erlangen, but he was not an American.[27] From letters by American scientists to the AISL, just one more name can perhaps be added, that of analytical chemist Francis Perry Dunnington, recipient of the Charles Herty Medal in 1935, who spent his whole career at the University of Virginia. Describing himself as "possibly more of a fundamentalist than you dream," Dunnington told Shailer Mathews that "*all of this evolution* idea is still only 'theory' and as such should never be *taught* to youth. Teach only such science as has been proven and reserve for investigators to continue to search for the wisdom and glory which God has concealed in His works."[28]

To a considerable extent, scientists saw efforts to ridicule evolution as pure nonsense. The curator of the American Museum of Natural History, paleontologist William Diller Matthew, thought his colleagues viewed evolution "not as a theory but an obvious fact of the geologic record—a natural law, as is the law of gravitation in the physical world. Its opponents appear to us much in the same class as the believers in a flat earth—of whom there still are some, as I find occasionally."[29] Kirtley Mather responded similarly to the flood geology of creationist George McCready Price. A few months after *Scopes*, he told University of Akron zoologist Walter C. Kraatz that Price "has never tested the principles of which he himself proclaims by discovering how they work in the field." His ideas are not accepted, because "the principles which he attempts to upset are being tested repeatedly every day," by oil companies and miners, and "are found to ring true. It is, ~~however,~~ <therefore,> not a question of opinion but of fact. The geologic succession of life is absolutely trustworthy."[30] Conklin was no less annoyed. When a librarian at the University of Michigan sent him a satirical poem about Bryan, Conklin commented, "This is really the only way in which Mr. Bryan can be reached, namely by making fun of him," adding that "you have found a better way of answering him than many of us scientists who have taken him too seriously." Writing to a New York attorney on another occasion, Conklin wondered out loud, "Whether it is possible to deal with such stupidity is a very serious question. I sometimes think that our only hope is in educating coming generations and in relying upon the graveyard for those who are uneducable."[31]

Frustrated by their inability to successfully educate the public, many scientists preferred to ignore Bryan as an irrelevant intrusion on their professional activities, or at least to dodge the issue as far as possible. When the National Academy of Sciences (NAS) prepared a statement on evolution in mid-1923, the committee recommended that it be kept confidential unless they were called upon to use it. Yale University president James Rowland Angell, a psychologist, cautioned geologist John M. Clarke, chair of the committee, "that the situation is a good deal more serious than some of our scientific friends appreciate." A statement from the NAS itself (see the following section) might be useful, but Angell was "entirely unwilling to have it used unless the phraseology commended itself substantially unanimously to the personnel of the Academy. In other words, I should not want to have any back-firing which might make such a pronouncement more harmful than none at all."[32] As Baylor physiologist Fred T. Rogers told Mathews, "the attitude I find prevalent among the faculty men of the Baylor Medical College is one of indifference. The best men assume the attitude that the excitement is a temporary thing and will soon die out: that as a group it would be poor policy to take any joint action opposing the religious fanaticism represented by such men as [Fort Worth Baptist pastor] Frank Norris, etc."[33] Although Rogers considered this a poor commentary, he also thought it accurate.

Some scientists did see a particular need to respond thoughtfully and at length, especially religious scientists such as Ohio Wesleyan College biologist Edward L. Rice. His father, William North Rice, an ordained Methodist minister and recipient of the first PhD in geology granted by an American university (at Yale in 1867), had written *Christian Faith in an Age of Science* (1903), in which he cautioned readers, "The God who is seen only in the supposed gaps in the continuity of nature, is a God in whom the evolutionist can have no faith." This was an early statement of the idea now known as the "god of the gaps" to which I return in the following section. His God was "eternally immanent in an eternal universe."[34] The son held a similar attitude, and he sensed danger in letting Bryan be the only religious voice. As vice president and chair of the zoology section of the AAAS, he delivered an address in December 1924 that was published as the lead article in *Science* three months later, just a few weeks before Scopes was charged. "If I were only a scientist," he said, "I think I should pay no attention, beyond a smile, to writings like those of Mr. Bryan on evolution," for "it is of little moment" whether any given person believes in any scientific theory. "But I am not merely a scientist," he continued, "I am a teacher." Furthermore, "in common with many members of this section, I am not simply a scientist and a teacher but also a Christian," and "it is precisely as a Christian that I most resent the attitude of Mr. Bryan." Rice was content to let Bryan believe privately as he wished, but when he "uses his moral earnestness and his oratorical genius to proclaim to the world that belief in evolution precludes belief in God or, at least, is seriously hostile to religious belief, he becomes fundamentally dangerous, not to science, but to religion." Evolution is universally believed by scientists and taught to youth, who, "confronted with Mr. Bryan's alternative," might "feel themselves compelled to give up religion; this I regard as an inestimable loss to them and to the Christian church." Rice looked poorly upon "any movement which tends to split the forces of the church rather than to bring them into harmony, or upon any attempt to read essentially religious men out of the church because of nonessential differences in scientific or theological belief."[35]

These are remarkable sentiments for a scientific address carrying the imprimatur of the AAAS, but it was a remarkable moment. Rice also answered Bryan on several scientific points, including the legitimate role that hypotheses play in modern science and the alleged absence of evidence for evolution. Having read several of Bryan's books and speeches, he was not impressed by the Great Commoner's appeal to the smug certainty of biblical literalism on the one hand and ridicule of the uncertainties of genuine science on the other hand. "His method," Rice noted perceptively, "is that of the lawyer striving to win his case rather than that of the earnest seeker for truth." This was "not to the advantage of the professed defender of the faith of the Christ who characterized himself as 'the truth,'" and it contrasted with the behavior of Darwin himself, who had honestly admitted difficulties with his theory. "Sarcasm and ridicule," Rice commented, "are as conspicuous for their absence from Darwin's writings as for their presence in Bryan's." Furthermore, Bryan's belief in the "literal infallibility of the Bible" was on shaky historical ground. It "has not been universally held by the leaders of religious thought in past centuries; it was not accepted, for example, by Luther or Calvin, by Augustine or Jerome." Above all, "Jesus Christ was a Modernist, not a Fundamentalist, in the matter of Old Testament criticism; and the Gospels are full of his efforts to overcome the deadly literalism even of his own disciples."[36]

To conclude, Rice returned to the frank spiritual concerns that had prompted him to give such an address. Was evolution in fact harmful to religious belief? Where Bryan saw the evolutionary picture as "imprison[ing]" God "in an impenetrable past" and ruling out "a *living* God" who works in the present, Rice saw just the opposite. "Is this not exactly the position of the theistic evolutionist, for whom natural law is merely a human attempt to formulate the method of divine activity, and evolution a human attempt to formulate the method of divine creation?" Rice granted that "not all evolutionists are theists," and that Darwin himself had drifted "from an orthodox belief to a condition of agnosticism," but he interpreted Darwin's loss of religious belief partly as related to the declining interest in poetry, music, and art that Darwin had mentioned in his autobiography. Rice further thought that "the impossible character of the dominant orthodox theology" of Darwin's day had also contributed to this. Nevertheless, "neither in 1859 nor in 1924 can the blame for the conflict of evolution and religion be placed wholly on the theologians. There is an *odium scientificum* as well as an *odium theologicum*." Rice contrasted certain unidentified "materialistic scientists who seized eagerly upon the evolution theory as a new weapon for attacking Christian faith" with some theologians "who, from the start, recognized the truth of evolution as an aid to faith." It was fair to ask, "whether the materialistic scientist is not as responsible for the present flareup as is Mr. Bryan himself." Rice noted that "an increasing number of our leading scientists are publicly proclaiming their own theistic philosophy, and emphasizing anew the essential harmony of a progressive scientific belief with real religion," praising in this connection Conklin, Millikan, Henry Fairfield Osborn, and University of Chicago botanist John Merle Coulter. Rice hoped for the day "when a more scientific religion and a more religious science shall join in a common welcome to truth, whether revealed in nature, in human life, or in the Bible, and shall present an unbroken front in the struggle for the higher evolution of the human race."[37]

Not long after Rice's address, zoologist William Marion Goldsmith of Southwestern College, a Methodist institution in Kansas—one of a dozen scientists who went to Dayton to testify at *Scopes*—drew nearly identical conclusions in *Evolution or Christianity?* (1924), a short book of his own supplemented with the *New York Times* editorials by Fosdick and Osborn, and essays by theologian Lyman Abbott (who had recently died) and S. Parkes Cadman of Central Congregational Church in Brooklyn. Goldsmith took exception to the way in which Bryan "ofttimes substitutes ridicule and sarcasm for sound logical reasoning," using silly caricatures of evolution rather than explaining the actual theory and the evidence that supports it. This had deleterious spiritual consequences, since "thousands of our church people (not necessarily ministers) are dogmatically informing the young people that they must agree with Bryan before they are eligible for the work of the church." Taught to think for themselves, college students readily reject "the elementary, and fallacious teachings of the anti-evolutionist," whose own doctrine might weaken Christian faith "more than that of his opponents."[38]

Like Rice, Goldsmith taught biology at a small Christian college and regarded the irreligious image of evolution as the most serious obstacle to teaching it. In his own student days, Goldsmith had struggled with this himself, but his faith was strengthened by reading *The Coming of Evolution* (1910) by the English biologist John Wesley Judd, a work he encountered while teaching at Oakland City College, a Baptist school in Indiana. There he caught a glimpse of "a '**Presiding Mind**' over natural processes," an idea that "continued to grow upon me, and to grip my very life" during his doctoral studies at Indiana University. Having spent ten years working it through, Goldsmith had reached the point where he wanted to help "the young student who has only a meager knowledge of either Science or Religion" to give serious consideration to both realms. From the start, science has been "dominated by devout Christian men," and it still is. Invoking Millikan's pamphlet (in the version published in *The Christian Century*), like Millikan he recited a litany of names of eminent religious scientists, all in the hope "that this volume will stimulate hundreds of young people of America to more extensive reading and consequently bring them to an earlier solution of these vital problems of life."[39]

Goldsmith had already written a textbook, *The Laws of Life* (1922), based on his course. It is clear from the opening lines of the preface, with language about humanity "Emerging half blinded from the mist of the dark ages," and "Slowly but surely . . . freeing himself from the firm grasp of ignorance and superstition," that Goldsmith was steeped in the bogus "Conflict" thesis of Andrew Dickson White. In a chapter on "Superstition as a Retarding Factor," he repeated the standard myth about Columbus and the flat earth, blamed John Milton for expressing "primitive" views on the origin of life, and related the story (apparently autobiographical) of "a certain young man who was reared under the old school of religious faith." As a child, he had been told about Santa Claus, the Easter bunny, and various fairy tales—along with stories about "the imperceptible character called Jesus Christ," which led him to wonder what effect the false tales had "upon the reception of this, the only one with any semblance of the truth?" A few years later, he learned enough zoology to question the truth of the story of Noah's ark. As a university student, having seen the evidence for evolution, he "smile[d] at the childish inclinations of the Christian's creed," concluding that "the story of Jesus Christ must likewise fall into the same category" as other spurious tales. "Thus it is with hundreds of men whose early training was based upon error," he added. The "only hope," in his opinion, was "to overthrow Twentieth Century superstition, Twentieth Century dogmatism, and the Twentieth Century adherence to mythology" through the study of "scientific facts," especially evolution, "a science that deals with facts and nothing but facts."[40] In short, Goldsmith was Bryan's mirror image.

Goldsmith's main interest was heredity, not evolution per se, especially eugenics and euthenics seasoned with scientific racism. Mentioning a "mixed family of white and black children" in "the slums of a certain city," he noted that "the mother of this brood was said to have been a most vile prostitute and to have never been married," so her children "had inherited the lowest strains of physical, mental, and moral constitutions." If "the germ plasm of these illegitimates" were unfortunately to mix with the "pure line" of a "respectable family," then "many wrecked homes" could be foreseen to result from "the matings of such unequal young people, altho[ugh] they might be placed in an identical environment." We cannot hope to solve such problems, Goldsmith concluded, "until the principles of heredity, eugenics, and euthenics are better understood by everyone."[41]

In January 1923, Goldsmith evaluated his course by giving students a questionnaire, administered by an assistant without his presence. "What effect has the teaching of evolution in this class and reading of 'The Laws of Life' had upon your Christian faith?" Students were to indicate whether their faith had been "Strengthened," "No Effect," or "Weakened," and were invited to add comments and their signatures if they wished. Of eighty-eight students, only two thought their faith had been weakened and twenty indicated no effect. Sixty-six students, fully 75% of the class, "claimed that a knowledge of evolution strengthened their faith in Christianity," and nearly two-thirds spelled out what they meant in comments that Goldsmith printed in *Evolution or Christianity?* According to this captive audience, his approach produced fundamental changes in the attitudes of students toward evolution, especially by broadening and deepening their sense of God's power and purposeful presence. One student wrote that the course "has strengthened a conviction for **right living** and impressed me with the **responsibility of life**. It has caused a stronger feeling of reverence." Another thought that learning about evolution "has **made God seem more real and plausible to me**." A third comment reveals how Goldsmith interpreted evolution both teleologically and theologically: "Evolution is progress. **Christianity is progress or it is not Christianity**." Because Christ said, "I am the way of **Life**," and by studying the evolution of life, "we become better acquainted with the Creator of that life." Another student, apparently answering an examination question rather than the questionnaire, shows more fully what Goldsmith taught about progress: he linked evolution, morality, and divine immanence, with eugenics as the lynchpin. An evolving animal, humanity can be "bettered" and "**led higher** through the study of eugenics," becoming thereby **"more like his Creator, more capable of sensing His powers,"** and "**I believe more passionately in religion because of this evolutionary Faith**." Armed with such testimonies, Goldsmith challenged "the anti-evolution propagandists to present as convincing evidence **against** evolution." Who was better able to testify to the effect of evolution on faith than students themselves?[42]

Goldsmith was a popular teacher who also taught Sunday school at a Methodist church one block from campus. Nevertheless, his vigorous promotion of evolution got him into trouble in the winter of 1924–1925, when five church members formed a committee representing a larger group who circulated a petition in town to fire Goldsmith and a Bible professor. Fortunately for Goldsmith, the student council defended him, Southwestern president Albert East Kirk, a theologian and former biology teacher, praised Goldsmith's efforts to bring evolution to campus, and the Methodist ministers and laypeople on the board of trustees issued a strongly supportive statement. In the end, the local Methodist conference voted overwhelmingly to retain him.[43]

Goldsmith's story underscores the fact that liberal theologies and progressive types of evolution have been closely linked since the late nineteenth century, with eugenics added to the mix in the early twentieth century.[44] For a particularly stark example, we turn to the short account of "The Fall of Man," written by Goldsmith's campus pastor, P. W. Beck, and published in the book. Beck accepted the universality of the fall but not in a traditional manner. Instead of seeing humanity as originally sinless, he held that the "image of God" was somehow "breathed" into a creature whose body had been produced by evolution, and thereby "man became a dual person; he became **the image of God in a house of 'clay'**." Like a Renaissance humanist, Beck placed humans midmost in creation, with the soul lifting us toward heaven and the flesh dragging us down to earth. Our salvation lay in biology, which reveals "God's law for the salvation of the flesh," as "one by one, the evil instincts and impediments are to be eliminated," allowing the spirit to ascend. "**The laws of eugenics**," Beck concluded, "**are God's provisions for the salvation from the fall**."[45]

## Modernist Concerns about Education and Materialism: Defining "Science" and "Religion"

During the 1920s, the meaning of "science" was contested in further ways that are equally important for understanding the pamphlets. Is science simply a benign method of establishing truth about nature that is neither friendly nor hostile toward religion? Is it an enemy to all religious belief, especially belief in the biblical God? Or is it actually a welcome ally in the development of a modern religious attitude? The meaning of "religion" was no less contested than the meaning of "science," and the dangerous specter of "materialism" was never far from the door.

Some influential modernist leaders shared Bryan's concern about the irreligious image of science while disagreeing fundamentally with Bryan's proposed solution. In a rare breath of fresh air eleven years after *Scopes*, Mathews conceded that "the teachers of science themselves were partly to blame because of sometimes [holding] a 'smart Alec attitude' toward religion." Academic freedom "certainly does not mean license to insult other peoples' convictions." Christians opposed evolution not from "mere bigotry but of apprehension as to the dangers which threatened morals in the breaking down of the authority of the Bible." There was no other way to "understand the attitude of otherwise intelligent people."[46] As Fosdick wrote in his first pamphlet, "the fundamental interest" motivating Bryan was his fear that evolution "will depreciate the dignity of man." Nevertheless, for Fosdick, it did not matter whether God made us "out of the dust by sudden fiat or out of the dust by gradual process." Either way, "Here man is and what he is he is." Scientists might care about the details, "but it is not a crucially important religious problem." This is followed by a staccato flourish of eight simple words: "Origins prove nothing in the realm of values." For modernists generally, but especially for Fosdick, philosophical reductionism—not evolution—lay at the core of the religious problem posed by modern science, and this bore directly on the dangers inherent in higher education. He bluntly admitted that Bryan's "fear is well grounded, as every one closely associated with the students of our colleges and universities knows. Many of them are sadly confused, mentally in chaos, and, so far as any guiding principles of religious faith are concerned, are often without chart, compass, or anchor." In his opinion, however, the culprit was not evolution, but "types of teaching in our universities which are hostile to any confidence in the creative reality of the spiritual life—dreary philosophies which reduce everything to predetermined mechanical activity."[47]

Judging from something he told Conklin several years later, Fosdick would have connected those "dreary philosophies" with people such as University of Chicago physiologist Jacques Loeb, author of *The Mechanistic Conception of Life* (1912). Loeb held that biologists should understand living things solely as physical and chemical objects lacking freedom of action, since even our highest thoughts are merely involuntary responses to external stimuli. "Not only is the mechanistic conception of life compatible with ethics," he wrote, "it seems the only conception of life which can lead to an understanding of the source of ethics."[48] Fosdick told Conklin that he had known the recently deceased Loeb, "and honored him greatly; but he did seem to me to have an impossible philosophy of reality."[49] Nevertheless, when responding to Bryan, Fosdick said that "the real enemies of the Christian faith, so far as our students are concerned, are not the evolutionary biologists, but folk like Mr. Bryan who insist on setting up artificial adhesion between Christianity and outgrown scientific opinions, and who proclaim that we cannot have one without the other." Such dichotomous thinking, if accepted by students, might well lead them to "give up Christianity in accordance with Mr. Bryan's insistence that they must."[50] The heart of the problem was "whether we are going to think of creative reality in physical or in spiritual terms, and that question cannot be met on the lines that Mr. Bryan has laid down." In Fosdick's opinion, Bryan had drawn precisely the wrong conclusion about the implications of evolution for a theology of divine action, when he said that evolution imprisoned God in the past and prevented God from working now. In fact, "the effect of evolution upon man's thought of God, as every serious student of theology knows, has been directly the opposite of what Mr. Bryan supposes." Where eighteenth-century minds had thought of God "as the absentee landlord who had built the house and left it," in the nineteenth century "the most characteristic thought of God was in terms of immanence—God here in this world, the life of all that lives, the sustaining energy of all that lives, as our spirits are in our bodies, permeating, vitalizing, directing all." Quoting English theologian Henry Drummond, Fosdick continued, "If God appears periodically he disappears periodically. If he comes upon the scene at special crises, he is absent from the scene in the intervals. Whether is all-God or occasional-God the nobler theory? Positively the idea of an immanent God, which is the God of evolution, is infinitely grander than the occasional wonder-worker who is the God of an old theology."[51] Another English theologian, Aubrey Lackington Moore, had said very similar things, but it was probably Drummond who led others subsequently to speak explicitly of the need to avoid invoking a "god of the gaps"—that is, a God found only in putative "gaps" in our present knowledge that might later be closed as knowledge advances.[52] Just six years after Fosdick quoted this, Mather spoke of having "the encouragement that God is no longer hidden behind the gaps in our knowledge," or as he put it later, "The God who was dimly discernible through the gaps in scientific knowledge has been pushed into the discard as those gaps have been filled."[53] Fosdick clearly wanted nothing to do with such a God, which he somewhat unfairly associated with Bryan.

Another clergyman worried about the irreligious image of science was Robert Elliott Brown, pastor of the Second Congregational Church in Waterbury, Connecticut. Brown was married to Mabel Millikan, whose older brother Robert won the Nobel Prize for physics in 1923, two years after becoming de facto president of Caltech. Early in 1922, the Millikans visited the Browns and the two men discussed Bryan's attack on evolution. As moderator of the Congregationalist ministers' conference in his state, Brown sought to pass some resolutions related to the situation.[54] Writing to Millikan afterward, Brown stated that "The Darwinians and the Bryanites alike" are to blame for the cultural divide. Evolution neither "contradicts the moral idealism of the Bible," nor "in any way invalidates the fundamental grounds of a spiritual religion." Nevertheless, "too often scientific teaching is in the hands of men who are not in sympathy with religious ideals with the result that the public often confuse the findings of science with the special anti[-]religious interpretations that such men are likely to make." He wanted individual scientists and scientific societies to state publicly that evolution "is not anti-religious nor is it incompatible with a spiritual faith." Noting that Millikan had suggested that Congregational ministers might prepare an influential statement in favor of evolution, Brown urged Millikan to "make some such declaration as I suggest and have it passed at your scientific congresses." Failing this, simply compiling a list of "twenty men or so" who shared Brown's concerns "would hearten the liberal clergy every where."[55]

Repeating his request a few weeks later, Brown named Millikan, Conklin, Osborn, Einstein, and Dutch physicist Hendrik Lorentz as possible signers. Such a declaration would "do much to help steady the situation among the Baptists and encourage the liberal men to keep up the fight," and "Bryan's guns would be spiked!"[56] The Congregationalist ministers had just authorized Brown to disseminate in national publications three resolutions Brown had written. (It is unclear whether that happened.) While denouncing attacks on evolution and affirming its religious neutrality, they also demanded "that the teaching of religion and science should proceed harmoniously in our institutions of higher learning." They urged universities to evaluate candidates "not only [for] scholarship and character, but also [for] sympathy with the moral and spiritual ideals of religion." Affirming their goodwill toward "the scientific men of America," the pastors invited them "to declare with us their faith in the essential harmony of science and religion and to join with us in seeking to give moral and spiritual guidance to our common thought and life." Brown's concerns about anti-religious attitudes on the part of college professors, not high school teachers, was shared by Fosdick and several others quoted in this book. Jon H. Roberts was right that "conservative evangelicals actually devoted considerably more energy to combating offenses against the faith committed by science professors in American colleges and universities than they did to heresy within high school classrooms," but liberal Protestants were no less worried than conservatives—even if they did not think that evolution itself was part of the problem.[57]

Brown sent his resolutions to Millikan, along with two versions of "A Declaration of the Rights of Science and Religion." It depicted "man" as having "both rational and spiritual powers," with "sufficient unity" between them "to make the progressive harmony of science and religion increasingly possible." Brown repeated the importance of hiring faculty sympathetic to religion, but this time he simply recommended that "in all education secular and religious there be a generous recognition made of the rights of reason along with no less acknowledgment of the necessity of inculcating moral and spiritual ideals universally among the youth of the land." In a concluding paragraph, Brown spoke of "our God-given powers, Reason and Faith," each having a proper place and the need for their "complete cooperation in solving the mysteries of life while securing the highest good for all mankind."[58]

This is followed by two columns of names, eight clergy and ten scientists, indicating whom Brown had in mind as potential signers of his "Declaration." The clergy were Fosdick, Cadman, George Angier Gordon (a Congregationalist) of Old South Church in Boston, James Edward Freeman of Epiphany Episcopal Church in Washington (named Bishop of Washington the next year), Episcopalian bishops William Lawrence of Boston and William Thomas Manning of New York, John Kelman of Fifth Avenue Presbyterian Church in New York, and Rabbi Stephen Samuel Wise of the Free Synagogue of New York, a leading Jewish intellectual.[59] The scientists were Millikan, Osborn, Conklin, Merriam, Angell, Pupin, Chicago physicist Albert Michelson (the first American Nobel Laureate), and two of Millikan's colleagues at Caltech, astronomer George Ellery Hale and chemist Arthur Amos Noyes. Added by hand as an apparent afterthought was Smithsonian paleontologist Charles Doolittle Walcott, president of both the AAAS and the NAS in 1923.

Millikan fulfilled Brown's request about six months later, when he prepared a public statement on science and religion based on Brown's "Declaration"—a fact that has hitherto not been realized.[60] Manuscript evidence shows that Brown's statement was rewritten by Millikan and revised by Noyes before being circulated to scientists and clergy and published by the *New York Times*.

Both statements express regret for the tendency to set science and religion in opposition, both define "science" and "religion" in terms of their purposes, and both affirm the importance of science *and* religion for the future of humanity. Above all, both statements place human beings, with what Millikan called our "Godlike powers," echoing Brown's words, "God-given powers," at the pinnacle of an evolving creation.

Contrary to Brown's wishes, however, Millikan's statement made no reference to a religious test for hiring faculty—an idea he flatly rejected. Shortly before *Scopes*, Millikan helped write a statement defending academic freedom for the AAAS, and in 1929 (during his tenure as president of the AAAS) he endorsed an unpublished resolution stating that forcing science teachers to conform to any kind of dogma is "vicious in principle and harmful in practice."[61] Another committee member, Conklin, seems to have been especially keen to defend academic freedom, perhaps because his former student Henry Fox had been dismissed from Mercer University in 1924 for teaching evolution. Conklin had already worked with Osborn and eugenicist Charles Davenport to write the first such AAAS statement in 1922.[62]

Nevertheless, Millikan originally intended to include a paragraph assigning "responsibility for this deplorable situation" equally to scientists and clergy, based on Brown's version. He noted that "science is far too often misrepresented by men of little vision, of no appreciation of its limitations and of imperfect comprehension of the real role which it plays in human life," while on the other hand religion "is also continuously misrepresented by leaders of an equally narrow outlook and of a limited understanding of its true nature." Noyes nixed this. In a handwritten note that must date from early 1923, Noyes commented on Millikan's draft, a copy of which he enclosed along with a revised version that he would be willing to sign. He thought it "unnecessary" to assign blame, "because, tho there are doubtless many such scientists, they rarely if ever take any stand against religion; and it is unfair to scientists as a group; and it seldom does any good to make the charge of narrowness."[63]

Millikan's three-paragraph "Statement" appeared in the *New York Times* in May 1923, prefaced by an advertisement that its purpose was "to assist in correcting two erroneous impressions that seem to be current among certain groups of persons. The first is that religion today stands for medieval theology, the second that science is materialistic and irreligious." The signers regretted the recent "tendency to present science and religion as irreconcilable and antagonistic domains of thought," given that science is simply about knowledge of nature while "the even more important task of religion ... is to develop the consciences, the ideals, and the aspirations of mankind." Both represent "a deep and vital function of the soul of man, and both are necessary for the life, the progress and the happiness of the human race." Science provides "a sublime conception of God" that is "wholly consonant with the highest ideals of religion, when it represents Him as revealing Himself through countless ages in the development of the earth as an abode for man and in the age-long inbreathing of life into its constituent matter, culminating in man with his spiritual nature and all his Godlike powers."[64] The following month *Science* reprinted the "Statement," as did Millikan in two of his books. In July 1923, *The Literary Digest* ran an article about it, and the AISL published the "Statement" as a pamphlet (not in this volume) no later than September 1926.[65]

The forty original signers (five additional names were added to subsequent versions) were all men, representing science, religion, government, and industry, including Secretary of Commerce (later President) Herbert Hoover. In addition to Millikan, Noyes, and Brown, we find the AISL chairman—Ernest D. Burton, who had just been named president of the University of Chicago—and AISL authors Pupin and Conklin. Every scientist on Brown's list signed, except Hale and Michelson. They were joined by seven more scientists, among them New York engineer Gano Dunn, a former student of Pupin who chaired the National Research Council (NRC), the operational arm of the NAS created in 1916 by Hale to mobilize scientists to work on problems related to the military in anticipation of American involvement in the Great War. Several other NAS members also signed, and after Millikan circulated his draft in January 1923, the NAS formed a Committee on Organic Evolution headed by geologist John M. Clarke to write a statement of their own. By early summer they were finished, but the NAS Council voted to keep it confidential temporarily, and I do not know if it ever appeared. The tone is remarkably more confrontational, emphasizing how "each discovery of natural law has aroused the antagonism of the uncomprehending community and such groups within it as are more content with mystery than knowledge." Careening into arrogance and sheer speculation, it claimed, "Today probably one-half of the world's population still believes that the earth is flat," while lamenting the need to educate people about the struggle with "prepossession, tradition, ignorance and fancy against which" modern astronomy has "forced its way into recognition." Likewise, refusal to accept evolution, or "to openly lampoon it, is the expression of ignorance, involuntary or unconfessed." Efforts to block education "through prejudice, intolerance or intellectual incompetency, menace the progress of humanity and civilization."[66] *Pace* Millikan and Brown, nothing was stated or implied about a need for scientists to grant religion any legitimacy.

The process of enlisting support from individual scientists could be complicated. Osborn later said publicly that he signed "because I am thoroughly convinced that the naturalist needs a credo or profession of his faith, even if this credo is very different from that drilled into his youthful mind and memory before the world entered into universal acceptance of the law of Evolution."[67] Angell had misgivings about the precise wording, but he would sign "if you can secure the signatures of the other men upon your list with the exception of Michelson." It is hard to say exactly how to interpret this. Michelson was Jewish, and perhaps Angell objected for that reason and Millikan (who was probably mildly anti-Semitic himself) accepted the objection.[68] On the other hand, Millikan had great personal and professional respect for Michelson, who had given him his first faculty appointment at the Ryerson Laboratory of the University of Chicago.[69] Michelson was originally suggested by Brown, and Millikan obviously kept him on the list at least initially, or Angell would not have seen it. Perhaps Michelson had already been contacted but declined, and Angell knew this, or perhaps he was just reminding Millikan that Michelson was not religious—his daughter has described her father as "nihilistic" and given to "bitter outbursts," adding that he "never accompanied his family" to the Episcopal church they attended.[70] In any case, Angell signed and Michelson did not. After discussing the whole project with Merriam, Millikan added several names, including four missing from the published version: the influential Jewish educational reformer Abraham Flexner, Chicago zoologist Frank R. Lillie, Columbia geneticist Thomas Hunt Morgan, and Chicago mathematician E. H. Moore (whose student George David Birkhoff signed). The inclusion of astronomer William Wallace Campbell, director of the Lick Observatory and president-elect of the University of California, was apparently Millikan's idea, but Osborn was especially keen to have Campbell and other astronomers on the list, "because of the great prominence of astronomy in recent human thought and also in the development of religious conceptions in the past." Osborn also suggested astronomers Hale of Caltech and Henry Norris Russell of Princeton University, but neither signed. Millikan told Conklin, who thought the "Statement" was "splendid," that he hoped Thomas Edison would sign (he did not), but Edison was away and could not be reached. He also asked Conklin to approach the president of Princeton

Seminary, Ross Stevenson, but his name is also missing. As for other clergy, Millikan thought it best to leave the selection up to the religious leaders themselves, apparently working through Brown, "the hope being that the men chosen will represent those who will have the largest possible influence in the religious groups."[71] Oddly, only two of the eight clergy on Brown's original list actually signed. To the best of my knowledge, all forty signers were Christians, and all except Pupin (a Serbian Orthodox believer) were Protestants, but there is no evidence that Millikan told anyone they were signing something originally written by a Protestant minister.

Once the "Statement" appeared it elicited some vocal responses, especially from certain skeptics who thought that Millikan had played fast and loose by stating functional rather than substantive definitions of "science" and "religion," definitions specifically intended to smooth over controversy in favor of a vague harmony. One of the loudest came from J. Arthur Eddy, a Denver tour operator and rancher who sent Conklin a letter with three mimeographed documents: a paper of his own, "Are Religionists Amenable to Honesty?"; a copy of the "Statement," which Eddy called "THE DR. MILLIKAN PRONUNCIAMENTO"; and a copy of a letter from botanist Luther Burbank, a famous religious skeptic, praising some of Eddy's ideas. If he sent similar missives to other signatories, I have not found them. Eddy asked Conklin bluntly to "state whether or not you feel it encumbent [*sic*] upon you to set yourself right before the people and if not, [I] would be pleased to have you explain why." Conklin declined the opportunity to elaborate, noting that he accepted Millikan's definition of religion and saw no reason "to redefine it." This did not satisfy Eddy, who accused Conklin of "a glaring, impertinent evasion of the issue, *very plainly* set forth" in the paper he had already sent. For Eddy, the issue was "whether you can conscientiously acquiesce in a deception of the public as construed by them of the bald assertion that 'there is no conflict between science and religion', and which the public is led to believe was asserted to by yourself and your forty-four co-signers." The dictionary definition of "religion," Eddy observed, involves "the feature of belief and a relationship to God," and "such is the understanding of the term by people generally." Eddy's paper emphasized the "common acceptation" of religion as involving belief in biblical inerrancy, the deity of Jesus, the Resurrection, and other miracles. The typical person, he argued, would interpret Millikan's "Statement" to mean that science accepted these. Citing the "Statement" and the article from *Christian Century* that later became Millikan's pamphlet, Eddy claimed that Millikan had tried "to ignore, belittle, and relegate this prevailing religion to obsolete medieval memories, cherished by a small, benighted group," ignoring their ongoing significance. Eddy claimed such efforts "fly in the face of facts," such as the reaffirmation of a traditional confession of faith by the Presbyterian General Assembly the previous year. "At all events," he concluded, "the lack of plain, frank, and unequivocal language has wrought a deception and confusion on the public, involving, as seeming abettors, forty-five of our leading citizens."[72]

New York secular humanist William Floyd, pacificist publisher of the left-wing magazine *The Arbitrator* and numerous anti-religious books and pamphlets, also protested Conklin's involvement. In a letter sent to all signers, Floyd stated, if religion "goes no further than the statement specifies, we can all agree that there is no conflict" with science. "But is this all of the truth?" he asked. "Do scientists generally believe the doctrines which a minister of religion must profess on taking orders?" Would Conklin "approve of the elimination by all denominations of whatever doctrines cannot be explained to the satisfaction of science?" Conklin might actually have approved of such a step, given his very liberal religious views (see the editorial introduction to his pamphlet), but he sidestepped Floyd's question while still saying some things Floyd wanted to hear. Millikan's statement "did not imply that there is no such antagonism between science and theology," he wrote, referring Floyd to his book *The Direction of Human Evolution* (1921). He acknowledged the presence of "many conflicts between the commonly accepted views of theology and the results of science," and admitted "that theology must change in order to come into agreement with science. At the same time," he added, "I am strongly of the opinion that religion is a fundamental need in science as well as in every other human affair."[73]

Again we see the modernist distinction between *religion*, which remains crucial for the modern world, and traditional *theology*, which must be discarded. Harvey W. Wiley, a physician in charge of foods and sanitation for *Good Housekeeping* magazine, interpreted Millikan's pamphlet—which included the same definitions of religion and science used in the "Statement"—in just this way. The "whole difficulty" underlying the "battle between the Fundamentalists and the Modernists," he told Mathews, comes down to interpreting terms. "'To the Fundamentalist, religion is synonymous with theology. To the Modernist, religion is synonymous with right living," citing James 1:27. He feared "that Dr. Millikan's thesis will not appeal to the rock-ribbed Fundamentalist, especially" Bryan. "Every thinking scientific man," Wiley concluded, "realizes that theology has had its day. He also realizes that there can be no conflict between the truths of science and the Beatitudes." No "physical fact of Nature or theory of evolution can ever controvert or deny any one of these characteristics of a truly religious man."[74]

Wiley understood Millikan perfectly. At a conference at his Pasadena church in 1927, Millikan said some things his secular critics wanted to hear. Although he usually denied any conflict between science and religion, "there is an absolute clash between certain types of religious thinking and the fundamentals of scientific thinking," because science requires an "attitude of open-minded search for truth and the spread of knowledge regardless of all consequences," and "even in Protestant churches in America there are elements which are fundamentally opposed to that point of view." His was a nondenominational congregation that did not stress specific doctrines, like Fosdick's Riverside Church a few years later. As Millikan described it, the identity of his church is to "renounce entirely the validity of sectarian differences, and in so doing shake off largely the shackles of tradition and place religion upon a more idealistic basis than it has been on before." Indeed, Millikan helped orchestrate the merger between his Congregationalist church and a Unitarian church that created the Union Liberal Church (commonly known as the Neighborhood Church) in 1923.[75]

Not all scientists accepted the modernist definition of religion as readily as Millikan. University of Missouri physicist Herbert M. Reese declined to support the AISL, because "you persistently evade the real issue," he told Mathews. "It is all very well for eminent divines to protest that they have the highest regard for science, and for eminent scientists to frantically assert that they believe in God and extol the character and teachings of Jesus of Nazareth, but it all sounds like an opiate to soothe the militant fundamentalist." For Reese, "the real issue" was the atonement, that salvation is found solely in Jesus Christ, and fundamentalists had no tolerance "for hair-splitting discussions as to what constitutes belief." They affirmed "that Jesus' mother was made pregnant while remaining a virgin, that he actually turned water into wine, raised the dead to life, cured incurables, and himself experienced a bodily resurrection from the grave." If these are stripped away, the fundamentalist thinks "you might just as well deny salvation by belief altogether, and I think he is right." Earnest fundamentalists were "a great deal more honest than many of the eminent divines and scientists who—I strongly suspect—don't believe the atonement theory at all. (Neither do I.)" Reese concluded with the suggestion that the AISL issue a new tract "that states just what you do believe—that if there is a God with a personality, he must be far more interested in courage, loyalty, truth, gentleness and humility, than in adherence to a formula." Mathews replied with obvious irritation, "The great trouble with you scientists is that you take your ideas of what religious thought is *today* from what was current theology one hundred years ago," instead of staying up to date with religious literature and having serious conversations with modern clergy. Stressing the need for more interaction between theologians and scientists, Mathews noted that modern theologians "go with science as far as science is capable of going, remembering that science progresses by means of hypotheses. Religion progresses by the same method, but the investigations into religious experience have to be made in a vastly different laboratory," involving the humanities and personal experience. "So far as the doctrine of the Atonement is concerned," Mathews added, "it is one of those things which has slipped back into oblivion as a religious doctrine of the person of Christ, but as a philosophy of life, embodied in the idea of vicarious suffering it still has a place, for there is no human society in which vicarious suffering is not taking place and does not affect human life." Just as scientists expected "laymen to go to the most modern and progressive specialists in science," theologians expected scientists "to go to the most modern and progressive leaders of religious thought." The pamphlets were simply trying "to help the world to see that science and religion each taken at their best, are not antagonistic but go along together, the one dealing with the material universe and the other with that which is beyond the material."[76]

But is there anything beyond the material? Mathews fully recognized that not everyone thought so and that science was sometimes interpreted as proving that there is not. "I think we ought to distinguish sharply between the facts given by observation and the various philosophies which different men have built upon these facts," he told another correspondent from Missouri, who had asked for help preparing a local student to defend evolution in a debate. "With some of these I am in absolute disagreement. They are mechanistic and leave no place for God." Scientific facts warranted "certain large inferences ... as to the possible origin of man physically," he granted, but "to deny spiritual value to humanity or to deny the existence of God as argued by such elements of the cosmic process as can be traced, is simply to argue partisanship." He saw "reason, purpose and goodness" in the evolutionary process, and even "a new basis for the argument of design." Mathews added an interesting comment on Genesis 2:7, where God made humans "of the dust of the ground" and breathed into our "nostrils the breath of life" to make us living souls. "I don't know how any theistic evolutionist would want a better formula for the evolutionistic way in which God made man," and the language about God breathing life "is a recognition of the advance from the material into the truly human."[77]

What Mathews and other modernists meant by "materialism" was the reductionist claim that human beings are "nothing but" the product of blind material forces, entailing the claim that freedom and responsibility are illusory and talk about spiritual forces such as God and the soul are just empty words—in short, that any type of Christian faith is contradicted by science. The AISL produced at least one pamphlet on this theme, *Why I Do Not Believe in Materialism* (1927), by Union Seminary (New York) theologian William Adams Brown. In 1930 they tried unsuccessfully to get Compton to write one called, "Why I Do Not Believe in a Mechanistic Universe."[78] Of the various ways in which materialism could be defined, Brown focused on "the man who denies that there is any mind in nature but our own and hence rules purpose out altogether as a principle of explanation of the universe."[79]

Such a philosophy was advocated by Chicago real estate magnate James F. Porter, who had earned a master's degree in biology at Harvard. He wrote to Mathews, outlining his objections to the pamphlets and enclosing a copy of an open letter to Bryan that he had sent the *Chicago Tribune*. Porter criticized Mathews's pamphlet, saying, "you carefully select and quote partial truths of science," falling foul of "the criticism you yourself make of either being uninformed or dishonorably clouding vital issues. You say nothing for example of all MATTER being in a turmoil and change due to the constant collisions of electrons with atomic nuclei." Applying this atomistic picture to the whole universe, Porter put the problem directly to Mathews: "Why not take into consideration the probable fact that from chaos itself, given endless time, and the effects of innumerable collisions just such a universe as our own with all our formulas and laws might evolve. History looked squarely in the face," he added with his eyes squarely on Andrew Dickson White, "clearly shows a struggle to the death between science and faith with faith constantly giving up precious beliefs in its vain efforts to hold out." The Inquisition "recognized this fact perfectly but we are now less honest with our many smoothings over. Surely, surely," he concluded, "you cannot think there is any place for faith in any scientific mind unless that mind has compartments which the owner refuses to open to others or to honestly explore himself." In reply, Mathews described Porter's position as "both illogical and erroneous" and expounded a "line of thought you will find stated in the little tracts" and widely accepted by modernists: "As long as there are personalities resulting from evolution, there must be that within the process itself which is capable of producing it. It is quite impossible for any man to think that personality comes out from impersonality." A fully mechanistic understanding of nature "simply denies the presence of anything approaching free will." He recommended the chapter on naturalism in a book by his Divinity School colleague George Burman Foster, *The Finality of the Christian Religion* (1906).[80]

Two years later, Porter challenged the views of University of Kansas zoologist Henry Higgins Lane, a former student of Conklin and author of *Evolution and Christian Faith* (1923), a rare example then of a book by a research scientist upholding both evolution and traditional Christian doctrine. A minister's son, Lane based the book on lectures he had given at the request of his students at Phillips University, an institution related to the Christian Church (Disciples of Christ) where he had briefly taught. A few weeks before *Scopes*, Conklin told Osborn that Lane was "thoroughly orthodox in his religious faith, and at the same time, a thorough good evolutionist," and suggested that he might make a good witness for evolution at the trial. A few days later, however, he added that Lane declined because his university had threatened to fire him.[81] Although Lane did not go to Dayton, he provided Scopes's attorneys with a sworn statement that Osborn published, stating that "evolution is in *no sense* atheistic or materialistic, but strongly supports the theistic and idealistic philosophy of the universe," and that it "leaves the truths of Christianity exactly where they have always been, *i.e.*, free to stand or fall on the basis of their own intrinsic evidence and the experience of Christendom."[82] In a letter to *The New Republic* a few weeks later, Porter complained that Lane could reconcile evolution and religion only by finding purpose in natural selection. To do so transformed the struggle for existence into "the law of the destruction of those who were made unfit on purpose to be destroyed." Rejecting that view, Porter noted that "it is difficult to imagine anything more terrible than the laws of nature with purpose read into them."[83]

If fundamentalists believed that modernists had already given away the religious store by accepting biological continuity between humans and other animals, modernists believed that Christian interpretations of evolution were possible. Biblical literalism and traditional theology had to be abandoned, but a sense of purpose in the universe might still be coupled with a more general spirituality. Some scientists certainly believed that their colleagues were coming increasingly to recognize this. According to astronomer Heber Doust Curtis of Allegheny Observatory, "many men of science are gradually swinging away from the absolute materialism of forty years ago." Bryan and his supporters, however, "are doing tremendous harm by alienating our intelligent and thinking youth. Youth is too apt to put aside all belief in God because of the precise dogma of such men." Quoting a speech he had recently given at Cornell University, he confessed that "so tremendous a cosmos must have divinity in it or over it; my reason rebels at the assumption that it is purely materialistic, the result of the chance concatenation of self-created physical forces." Modern thinking "has been tempered, changed, and enriched through the knowledge that we are all members of an organization so wonderful, so unthinkably great, that we cannot go far wrong in applying to it the adjective divine."[84]

Pamphlet author Samuel Christian Schmucker of West Chester (Pennsylvania) State Teachers College, another former student of Conklin, was a nationally known popularizer of evolution and eugenics who knew Mathews from their involvement with the Chautauqua Institution. He too sensed a decline in materialism. "We have had three very distinct stages in our belief as to the relation between God and His creation," he wrote in 1908. First came the "primitive belief" that Genesis was literally true. This was followed in the second half of the nineteenth century by "an abiding sense of the existence of law," leading to the realization that the earth is extremely old and that forms of life have evolved over eons of time. Religious reactions against this had led "multitudes of great scientists" to become hostile to religion, and "there arose a school of so-called materialists to whom the earth was a great complex, self-contained and self-determined, relentlessly grinding out its own future." This materialism was the second stage of belief. Placing himself in the third stage, Schmucker confessed that "materialism died with the last century. The great scientists of the new century are to a very large degree intense spiritualists. God is now recognized in His universe as never before," but not in the same ways. "No longer is He the Creator who in the distant past created a world from which He now stands aloof, excepting as He sees it to need His interference. Now God is everywhere; now God is in everything." As a faculty member at a normal school, Schmucker urged that "the teacher who is himself filled with holy zeal, who has himself learned to find in nature the temple of the living God," would "bring his pupils into the temple and make them feel the presence there of the great immanent God."[85]

## Modernist Concerns about Education and Materialism: "Theistic" versus "Atheistic" Evolution

If scientific theories allow for multiple philosophical interpretations, including interpretations friendly to religion as modernists believed, then it might be possible to identify some scientists from the 1920s who interpreted evolution "theistically" and others who interpreted it "atheistically." One way to do this is to examine what they thought of their peers' religious views and whether they might classify them as "theistic" or "atheistic." In short, if there were scientific atheists, who were they?

Correspondence related to this very question survives among the papers of Kirtley Mather, who was consulted by three nonscientists in 1928 concerning a volume they wanted to edit with the putative title "Evolution: The Conference Method of Study."[86] Although the book was never published, correspondence between the editors and nearly two dozen Harvard scientists and other intellectuals sheds light on how scientists themselves viewed evolution in relation to religious questions. Two of the editors, Ralph Wilder Brown and Alfred Benjamin Dumm, were formally associated with the World Conference on Faith and Order, an ecumenical Christian organization begun by the American Episcopal Church that later became part of the World Council of Churches. Brown was general secretary from 1924 to 1933. Dumm earned a doctorate in philosophy from George Washington University, pastored Congregational churches in New England from 1901 to 1936, and was associate secretary from 1926 to 1927. The third editor was Louis J. A. Mercier, professor of French and education at Harvard and later head of the Graduate School at Georgetown University. Brought to America as a child by French immigrants, Mercier joined the French Army in 1914, returning in 1917 to train officers at West Point before resuming his post at Harvard after the war. Associate editor of *The New Scholasticism*, published by the American Catholic Philosophical Association, he strongly opposed naturalism, which "ignores God and makes nature self-supreme," especially as presented in *The Challenge of Humanism* (1933), a sympathetic treatment of the "New Humanism" of Harvard French professor Irving Babbitt and Platonist literary critic Paul Elmer More.[87]

The "Conference Method" was a way of promoting dialogue proposed by Mercier.[88] The editors spelled out their plans in a letter soliciting advice from at least twenty scientists and others in October 1926 and February 1927. "We have in view the publication of a book on 'Evolution,'" they said, "designed to bring between two covers not only carefully worded statements by exponents of the several representative positions which are held on the subject, but a critical analysis of the statement of each spokesman, by each of the others." The editors identified three views:

1. Rejection of all evolutionary theories.
2. Theories of evolution based on materialistic and atheistic principles.
3. Theories of evolution based on theistic principles.

> Furthermore, it seems that, within each division, the various presentations ought to be sharply differentiated (a) as scientific hypotheses, (b) as proved by facts, (c) as philosophical speculations.

Recipients were asked to state any other positions they recognized and to suggest who should represent each position.[89]

Most of the eight nonscientists who responded were highly enthusiastic, including a Jesuit priest from New York City and the editor of *Zion's Herald*, a weekly for Methodists in New England. The editors also contacted Paul Elmer More, a sophisticated Christian apologist and perceptive opponent of modernism whose ideas were later reviewed favorably in Mercier's *Challenge of Humanism*; he even contributed to the publication costs, despite rejecting Mercier's Catholic view of religious authority. More recommended that the editors contact his brother, University of Cincinnati physicist Louis Trenchard More, whose book *The Dogma of Evolution* (1925) had just appeared. His brother's position involved "antagonism to the extension of evolution to a general philosophy of life. He recognizes the solid achievements of biology, but thinks that a good deal of the theorizing in that field is not properly science." Yale paleontologist Richard Swan Lull, one of just two scientists who really supported the book, put Harvard zoologist George Howard Parker in the third category—with Conklin, Kellogg, Osborn, Dutch botanist Hugo de Vries, Scottish naturalist J. Arthur Thomson, Chicago zoologist Horatio Hackett Newman, and Yale zoologist Lorande Loss Woodruff. He did not identify anyone for the other two categories.[90]

Only two nonscientists doubted the project's merits. Henshaw Ward, a grammar teacher who wrote several popular books including *Evolution for John Doe* (1925), admired its purpose but feared that "it cannot be realized in the case of evolution." Nevertheless, he recognized a position "different from any of the three outlined," that was neutral concerning evolution's religious implications. "The fraction of scientists who are described in no. 1 is extremely small (probably not one per cent)," he noted, "and very few scientists would consider either 2 or 3 a fair category." The biological hypothesis "is not based on any principle that is atheistic or materialistic or theistic; it is not based on any principle which can be extended to the fields of reasoning about God." He urged the editors to provide precise definitions to avoid intellectual chaos.[91] Ward's position resonated with that of the modernists. "Some old-fashioned people are still afraid that the evidence from the rocks may damage the Bible," a situation that he found "comical, though disheartening." Most founders of geology "were believers in the spiritual truths of the Bible," yet "they insisted that it should not be desecrated by being dragged into the scientific arena, but that it should be elevated above and beyond science." Popular treatments of geology still needed to say this, even though it was hardly a new attitude.[92]

University of Aberdeen natural historian J. Arthur Thomson, another prolific author of popular books, echoed Ward. "If you can secure *competent* co-operators," then the conference method could work—provided the contributors "understand that science is descriptive and religion interpretative." Those who see God and natural selection "as opposed alternatives, have not begun the necessary intellectual discipline." Like Ward, Thomson could not think of "any <modern> theories of Organic Evolution ~~(since Lucretius) since~~ that are based on materialistic, atheistic, *or* theistic principles. Science has to describe and formulate in empirical terms; everything that is lugged in is irrelevant when we are keeping to science."[93] Marine biologist Joseph Thomas Cunningham of London Hospital Medical College agreed with Thomson. "I do not understand what you mean by theories of evolution based on atheistic principles, or others based on theistic principles," he said. "A theory of evolution" must be "either scientific, or philosophical." Henri Bergson represented the latter. Cunningham did not understand why any "objective and scientific" theory would be "atheistic." In Darwin's theory, "or any other scientific view it is assumed that nature works automatically, in a mechanical and intelligible way. But this is not necessarily atheistic." Objectively, "we cannot get beyond nature, so we are led to regard the universe as God, Pantheism not Atheism." Subjectively, "there is great difficulty in explaining our own consciousness by the evolution of lower forms, because what we call the facts are within the mind, & how can the mind be explained by that which is within it and is the effect of it?" Likewise, it is very difficult "to explain scientifically the existence of or evolution of ideas of right & wrong beauty & ugliness." Furthermore, the whole scheme was "too complicated" for Cunningham to help. Several other scientists also responded unenthusiastically, including Conklin, who doubted the book would "settle any important questions." He saw no way to reconcile the evolution and anti-evolution points of view. "Scientific conclusions are based upon scientific evidence," he wrote, "and not upon majority vote or a harmonizing of the views of those who rely upon scientific evidence and those who deny it." William North Rice, at eighty-three years of age still actively speaking and writing about science and religion, confessed that "I do not share your confidence that a discussion conducted on the plan which you propose would lead to any very satisfactory result."[94]

Given Mercier's Harvard affiliation, it is hardly surprising that his colleagues were so well represented, and the range of opinion displayed is informative. Arthur Stanwood Pier, editor of the *Harvard Alumni Bulletin*, suggested fundamentalist Baptist preacher John Roach Straton for the first category—Straton had debated Mather about evolution on Harvard's campus the previous year—and Fosdick for the third category, which he interpreted as a type of deism. For materialism, he suggested zoologist George Howard Parker, whose article on "The Evolution of Mind" had just appeared in his magazine. Education professor Paul Henry Hanus also mentioned Straton and Parker, while adding entomologist William Morton Wheeler to the second category. Neither Pier nor Hanus gave reasons, and their view of Parker as a materialist was contradicted by Yale paleontologist Lull (see preceding discussion). Botanist Edward Charles Jeffrey, with Lull the only other scientist favoring the project, apparently ignored the editors' categories. He thought Osborn "would on the whole be the most desirable person to state the general biological position in relation to Evolution," and that either Wheeler or Johns Hopkins zoologist Herbert Spencer Jennings would "be clear and forceful exponents" of "the philosophical aspect" of evolution, which he left ambiguous. On the "purely religious side," he suggested Harvard church historian and biblical scholar Kirsopp Lake, a modernist Anglican priest who rejected the bodily Resurrection.[95] Geneticist Edward Murray East wrote bluntly, "It is like proposing to discuss whether the world is round or flat. If our fundamentalist friends wish to inform themselves, they can find plenty of evidence." Botanist Carroll William Dodge was somewhat more sympathetic, although he noted that the disagreement boiled down to "the use of the same words in different senses by different individuals and by an attempt to take certain phrases out of their context by unscrupulous lecturers." In a rare step, Dodge essentially admitted being an atheist, placing himself "in the second group, as I think most true scientists do." Philosopher William Ernest Hocking could not think of anyone for the materialist position, but he thought Mather or Thomson would be "obvious names for the third" category. Physicist William Duane stressed leaving the conversation to the scientific experts, which the conference method did not ensure. Nevertheless, he recognized that scientific theories do not necessarily represent "ultimate truth." Most people "do not understand what a scientific theory is. A scientific theory is a tool and not a creed." Geologist Louis Caryl Graton agreed that such a book was timely and appropriate but doubted that "representatives of the fundamentalist viewpoint can be found who will fit into such a co-operative undertaking effectively and sincerely." Even if this hurdle were cleared, the editors lacked the requisite scientific expertise, which "would seem to me like appointing as officials for a football game persons who are not intimately familiar with football in order to insure that no bias may enter into their decisions." The great astronomer Harlow Shapley gave a paradigmatic ivory-tower response, admitting he had "no experience in teaching, and my position as a secluded research scientist gives me very little contact with the uninformed or with the erudite." Thus, he added a fourth category, "those who work furiously in scientific research, thoughtless of evolutionary principles, desirous of accumulating fairly solid observational data—working only for the joy of the work or in response to some half appreciated and not-at-all understood psychotropism."[96]

Perhaps the most qualified respondent, considering his experience reaching a wider audience with scientific information, was Edwin Emery Slosson. Born into an abolitionist family on the Kansas frontier in 1857, Slosson studied physical science and psychology at the University of Kansas. After marrying the prominent suffragette May Preston, the first woman to earn the PhD from Cornell, he taught chemistry at the University of Wyoming for thirteen years while completing doctoral work under Julius Steiglitz at Chicago. Doctorate in hand, he promptly left Laramie for New York City, where he was soon appointed literary editor for *The Independent*, a leading Congregationalist monthly for which he wrote ceaselessly on politics, literature, and science. This post transformed him into a prominent science journalist, and for eight years he also taught physical science at Columbia's Pulitzer School of Journalism. In 1920 he became the first editor at Science News Service, an organization founded by California newspaper magnate E. W. Scripps with support from Millikan and several other scientists, who sought to improve the popularization of science when such efforts were otherwise waning. Within a year, Slosson was named director of the renamed Science Service, an "Institution for the

Popularization of Science" (according to their letterhead) that disseminated scientific information to educators and print media.[97]

As his involvement with a Christian magazine suggests, religion was integral to Slosson's public persona. "As professor and as editor," he wrote in the preface to his aptly titled book *Sermons of a Chemist* (1925), "I have frequently been called upon for talks in college chapels, and as elder or deacon in Presbyterian and Congregational churches for the last thirty years I have occasionally had to occupy the pulpit when the preacher failed to appear or during an interregnum in the pastorate." Deeply appreciative of efforts to connect liberal Christianity with science, Slosson supported the pamphlets financially and by suggesting possible papers for inclusion in the series (see chapter 2). He also published at least one similar pamphlet himself, *The Contributions of the New Physics to Religion* (1928), based on a talk he gave at Fosdick's Park Avenue Baptist Church (now Central Presbyterian Church) inaugurating a Sunday evening series on science and religion.[98]

Like most of the other scientists, Slosson thought "it will be a difficult and delicate matter to select the proper proponents and get them to discuss the question in an amicable and cooperative spirit so as to get useful results." Similar plans had been tried before in various venues, including *The Independent*, but he doubted whether the tripartite scheme proposed by the editors would work. Scientists themselves did not agree on which of the larger generalizations associated with evolutionary theory were "proved," and which were merely "probable." Everyone agreed "that man is descended from some anthropoid, but there is the widest divergence as to the particular line of ascent and the relation of the fossils found to the genealogy of man." As for the specific religious positions defined by the editors, Slosson was sure that no "reputable scientist" would hold the first, which he associated with Alfred Watterson McCann and John Roach Straton. Louis Trenchard More "has argued against some of the philosophical implications of evolutionary theory but he admits evolution of a sort and is equally persona non grata to the fundamentalists as he is to the evolutionists." On the other hand, Slosson thought that "it would be hard, if not impossible, to find any scientific man of standing who would argue for evolution based on materialistic and atheistic principles." Here he could think of only Joseph McCabe, a former Franciscan priest from England who became rabidly anti-religious and published numerous tracts promoting *The Conflict Between Science and Religion* (1927), as one of his titles proclaims. Only for the third position, "evolution on theistic principles," could

Slosson think of any genuine scientists: Conklin, Kellogg—whose book, *Evolution: The Way of Man* (1924), Slosson regarded as "the best" on the subject—Osborn, Thomson, and Johns Hopkins zoologist Maynard Mayo Metcalf, "who gave an address on the subject at the last meeting of the American Association for the Advancement of Science." For a Catholic perspective, Slosson recommended Canon Henri de Dorlodot of Louvain. Overall, most scientists considered evolution "a scientific question and pay little attention to the theistic or materialistic implications."[99]

## Conclusion: Is Science Religiously Neutral?

Clearly, most of the scientists found no value in Mercier's scheme to debate the merits of various religious interpretations of evolution—and this was true not only for an atheist like Dodge but also for religious scientists like Conklin and Edward L. Rice. Also, it was far more difficult to identify advocates of "evolution based on materialistic and atheistic principles" than to identify advocates of "evolution based on theistic principles." Several scientists found the whole question irrelevant or even impertinent, and many (as Henshaw Ward noted) thought it reflected a fundamental misunderstanding of the nature of science itself: they viewed science as religiously neutral, whereas most of the nonscientists apparently did not. Yet some scientists *did* interpret science in ways unfriendly to religion, although they may not usually have done so publicly. Noyes alluded to them in his note to Millikan, Rice acknowledged their existence, Mercier and his friends obviously felt their presence, Fosdick and Mathews worried about them, and Bryn Mawr College psychologist James Leuba claimed to have counted them in his ground-breaking survey, *The Belief in God and Immorality* (1916).[100]

Shortly before the Scopes trial—in which a notorious agnostic who had already invoked psychological determinism to defend two vicious murderers would defend a man for teaching evolution while spouting his religious doubts to the press—Osborn found specific targets to blame for the irreligious image of science.[101] Writing in the *Forum*, a prominent national magazine, he named "[John] Dewey of Columbia and [William] McDougall of Harvard" and claimed that "psychologists have lost touch with the soul," an impression he confirmed through his friend James McKeen Cattell, the former Columbia psychologist who edited *Science*. "I can talk more intelligently about any other subject than the soul," said Cattell. "It is well known that psychology lost its soul long ago and is said now to be losing its mind. You should inquire of Descartes and the Catholic Church; it is a good subject for a paleontologist like yourself!"[102]

Osborn's indictment of psychology was echoed six years later by Frederick Lewis Allen, editor of *Harper's*, in his popular history of the 1920s, *Only Yesterday* (1931). At a time when "the prestige of science was colossal," although psychology "was the youngest and least scientific" of all the sciences, it "most captivated the general public and had the most disintegrating effect upon religious faith." The image of science presented by various "popularizers and interpreters of science" was "disquieting" to many. We reside on "an insignificant satellite of a very average star" in one of countless galaxies, we act on impulses dictated by "chromosomes and ductless glands," we pursue sex as "the most important thing in life," we believe "sin is an out-of-date term, that most untoward behavior is the result of complexes acquired at an early age, and that men and women are mere bundles of behavior-patterns, anyhow." We seem "equally ready to believe with [geneticist Edward Murray] East and [psychologist Albert Edward] Wiggam in the power of heredity and with [psychologist John B.] Watson in the power of environment."[103] Edward J. Larson has shown that Allen's account of *Scopes* is factually incorrect, his understanding of Bryan and fundamentalism "grossly oversimplified," and his suggestion that the trial weakened American churches during the decade just "sheer speculation."[104] Nevertheless, Allen's sense of the moral and religious dangers inherent in some forms of mechanistic science was widely shared at the time. For a "rediscovery of the soul and the spiritual nature of man," Osborn advised, "we must turn to physicists and physiologists." Naming Millikan his first champion, Osborn admitted that many scientists would not have signed Millikan's "Statement," because they were "still convinced of the adequacy of the mechanistic theory of philosophy and of the psychological creeds for the conduct of life, such as we have cited from Dewey and McDougall, or because they prefer to remain in the perfectly consistent and defensible fortress of agnosticism erected by my old friend and teacher [Thomas Henry] Huxley." Osborn's evolutionary theory of "aristogenesis" emphasized the role of conscious choice on the part of animals to influence the course of natural selection. In his view, "Materialism" and "pure mechanism" had failed "to give an interpretation that satisfies our reason," and "like Icarus, we have taken our flight, and the wings of reason have ceased to sustain us. If this thought impresses the physicists, it impresses the biologists still more cogently. Many biologists have abandoned mechanism entirely, and have frankly revived the old purposive interpretation of nature." Although Osborn did not take a vitalistic turn, as many of his contemporaries did, he was nevertheless confident that his one permanent contribution to biology "is the profession that living nature is purposive; . . . that Democritus was wrong in raising the hypothesis of fortuity, and that Aristotle was right in claiming that the order of living things as we know them precludes fortuity and demonstrates purpose." He was content, "with Wordsworth," to call this purpose the "Wisdom and Spirit of the Universe."[105]

As the comments of Curtis and Schmucker showed, Osborn was not alone in prophesying the end of materialism and in taking a pantheistic turn in his own efforts to unify religion and science. For Bryan and his followers, however, no form of theistic evolution was acceptable, and pantheism was a short step from atheism. Responding to such concerns, several leading scientists decided to help the clergy educate American Christians about evolution and its religious implications, as they perceived them. We turn our attention next to this very aspect of the controversy.

CHAPTER 2

# Liberal Protestant Scientists and Clergy Join Forces

## The Story of the AISL Pamphlets

> For young and old, for highbrow and lowbrow, the study of evolution makes life more significant and more beautiful. It justifies their faith and fortifies their ideals. It makes God a more imminent [sic] reality. It helps all of us to understand the purpose of life, and how to accomplish it.
> That is why I teach evolution.
> —William Patten, Why I Teach Evolution (1924)

Faced with widespread opposition to evolution, many scientists felt that a *religious* response to Bryan's rhetoric was necessary, and that political and scientific responses alone were insufficient.[1] Since many liberal clergy had similar feelings, when the opportunity to enlist support from the scientists for their modernist views appeared they took full advantage. The catalyst was Bryan's Sunday editorial for the *New York Times* in February 1922, summarizing his stump speech, "The Menace of Darwinism."[2] Replies the following two Sundays by Conklin, Osborn, and Fosdick set in motion a chain of events that led scientists and clergy to join forces, formally and informally, to dilute Bryan's influence. The pamphlets resulted from those efforts.

Fosdick's editorial brought appreciative letters from diverse parts of the country, mostly from pastors, university chaplains and scientists—not to mention the president of the University of Wisconsin, limnologist Edward Asahel Birge, who called it "the best contribution to this discussion that has appeared thus far, and it is hard to believe that there will be a better one in the future."[3] Several months later, University of Minnesota biologist Charles P. Sigerfoos actually suggested that it be reprinted as a pamphlet—he had thought of asking a local bookstore to do this, not realizing that the AISL had already done so.[4] Fosdick sent clippings to seven religious magazines, and at least one, *The*

*Christian Century*, published it within a few days.⁵ Try as he might, Fosdick realized that clergy could not do this task alone, that the problem of educating the public about evolution was at least as much a problem for the scientists. This was the central theme in an exchange with Syracuse University geologist Harry N. Eaton, who read Fosdick with alarm, because "the red flag is being waved in central New York right now, as well as in Kentucky." Eaton lamented, "The religious people and scientists of this country are too spineless—too bovine." Fosdick was glad "that you, too, are out with a stuffed club for our friend Bryan. I wish to goodness that you men who are scientifically minded would get out a simple manual massing the evidence for the evolutionary hypothesis," because "there are masses of people who not only do not understand the evidence but do not know that there is any evidence." There must be an "easily get-atable form for popular consumption." In order "to break the inflated bubble of verbose gentlemen like Mr. Bryan, I think that some scientists will have to get busy on popular propaganda distinctly distasteful though it will be to them."⁶

That is exactly what some scientists also had in mind: to produce popular propaganda, although not of the purely scientific kind that Fosdick wanted. In April 1922, the AAAS appointed Conklin chair of a committee to counter Bryan. The other two members were Osborn and Charles B. Davenport, director of the Eugenics Record Office at Cold Spring Harbor, New York. All three were liberal Christians (although Conklin had ceased attending church) who eagerly promoted eugenics and scientific racism, which were integral to what evolution meant for many eminent scientists. For example, in 1926 fundamentalist John Roach Straton of Calvary Baptist Church in New York attacked Osborn's exhibit on human evolution at the American Museum of Natural History, calling his essay in the *Forum*, "Fancies of the Evolutionists." Osborn's reply, "Facts of the Evolutionists," ended with a chart showing "Existing Facts of Human Ascent," in which Caucasians are placed at the peak of the "Ascent of Increasing Intelligence" (see figure 2.1). Likewise, the required biology textbook in Tennessee, George William Hunter's *A Civic Biology: Presented in Problems* (1914), implied its pro-eugenics slant in the title, while also teaching scientific racism. Hunter distinguished "five races or varieties of man, each very different from the other in instincts, social customs, and, to an extent, in structure." They were "the Ethiopian or negro type, originating in Africa; the Malay or brown race, from the islands of the Pacific; the American Indian; the Mongolian or yellow race, including the natives of China, Japan, and the Eskimos; and finally, the highest type of all, the Caucasians,

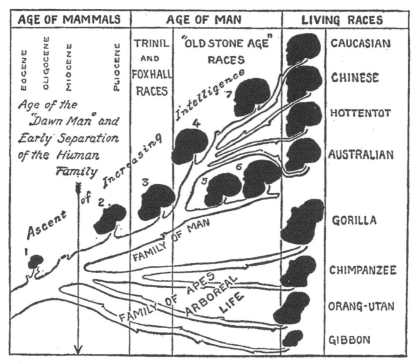

*1, 2,* Dawn stage of human prehistory. *3,* First known walking stage, the erect Trinil race of Java. *4,* Piltdown race of Sussex. *5, 6,* The low-browed Heidelberg-Neanderthal race. *7,* Crô-Magnon and related races of high intelligence. The races *3, 4, 5, 6, 7* are scattered throughout the entire period of the Age of Man, conservatively estimated at 500,000 years. Altogether, upwards of 136 skulls and skeletons of the fossil men of this period are known.

*Figure 2.1.* "Existing Facts of Human Ascent," from Henry Fairfield Osborn, "Facts of the Evolutionists," *Forum* 75.6 (June 1926), 851.

represented by the civilized white inhabitants of Europe and America."[7] Although Bryan, Straton, and other white conservative Protestants were typically no less racist than most other whites, they wholly rejected the allegedly "scientific" basis for racism and moral improvement that was dear to the liberal elites and had to be defended as such.

In connection with the AAAS committee, Osborn approached Scribner's (who printed a pamphlet based on his own *New York Times* essay the following year) about publishing a volume of essays on evolution that would reprint articles by scientists and clergy that had appeared in popular venues over the past few months, including the replies to Bryan by Osborn and Fosdick.[8] In a letter to AISL chairman Ernest D. Burton, Conklin listed the other essays the committee was considering, starting with an open letter from Birge to Edward S. Worcester, pastor of the First Congregational Church in Madison, Wisconsin, where Birge taught Sunday school.[9] Originally published by the *Wisconsin State Journal* on February 16, 1922, it answered a letter from Bryan published in Madison's *Capitol Times* on February 7, in which Bryan had attacked Birge's religious orthodoxy and his right to teach evolution at a state university. The controversy had begun in May 1921, when Bryan had spoken in Madison before Birge, Wisconsin governor John J. Blaine, and 2,500 others, and heated up again in September when Bryan revisited the state.[10] Ironically, Birge apparently screened potential faculty to ensure that the university did not hire "Freethinkers," and he was not reluctant to testify to his faith even in his official capacity.[11] His commencement address for 1925, for example, was an extended commentary on Mark 4:30–32.[12]

A second candidate was Worcester's sermon "The Fundamentals of Controversy vs. The Fundamentals of Christian Life."[13] Another was (in Conklin's words) a "strikingly fine article" by the eminent Union army surgeon William Williams Keen, titled "I Believe in Evolution and in God," undoubtedly the commencement address he had given in June 1922 at Crozer Theological Seminary (Chester), where he was the last surviving charter member of the Board of Trustees. *Science* published an excerpt the next week, and Lippincott published a book-length version in November.[14] Keen offered it to the AISL—several others had already recommended it—but they declined it. He also sent them his "brief analysis" of a "remarkable pamphlet," *The Scientific Accuracy of the Sacred Scriptures* (1920), by William Bell Riley, which he had just sent to *The Christian Century*, but that offer was also declined. He concluded that Riley "does not know anything about science and that his alleged proof is as absurd as anything can well be."[15] Keen was not only a faithful financial supporter of the AISL, but the only scientific person I am aware of who contributed before the "Science and Religion" series began publication. In a handwritten note the next year, he gave Burton a promotional statement, "I am fully persuaded that the quiet but efficient & really remarkable work of the American Institute of Sacred Literature is not only *one* of the most important agencies in teaching men to think rightly about the Bible but is *the* most important influence in this country for sane & reverent study of the Bible. ~~We deserve~~ It deserves the liberal financial & moral support of all the friends of true Religion."[16]

In addition, the committee considered William North Rice's review of Bryan's book, *In His Image*, an article by theologian Lyman Abbott, and a reply to

Bryan by one of the first national radio preachers, S. Parkes Cadman of Brooklyn's Central Congregational Church.[17] They hoped to have an article from the Presbyterian minister, poet, and diplomat Henry Van Dyke, who was in 1922 Murray Professor of English Literature at Princeton University; apparently it did not materialize.[18]

Conklin did not actually envision this anthology as an official AAAS publication. He guessed there would be "serious opposition" to involvement in a religious project, and he was sure that the AAAS "would not appropriate any funds for this purpose" (he was proven wrong, as we will see). It would be a strictly private venture on the part of the authors and the publisher, perhaps Scribner's, who had expressed interest.[19] However, his committee did issue an official statement on evolution. Apparently, that action was partly inspired by a letter to the editor of an unidentified newspaper by Robert Campbell Mac-Combie Auld, replying to an editorial by a "Mr. Cameron." A former cattle breeder who became a journalist for the *Brooklyn Daily Eagle* and editor of the *Scottish-American*, Auld wrote books and pamphlets on various literary and scientific subjects, including *The Breed That Beats the Record* (1886) and *How to Breed the Human Race* (1911). He shared Conklin's interest in eugenics, even teaching a class on that subject at Mount Morris Baptist Church in Harlem. Conklin expressed appreciation, leading Auld to propose informing Cameron "that his editorial has suggested such a course to you: to have a committee of the AAAS issue a 'Bull' (!) or encyclical on the subject. I think this is just what might be done: To have official science prepare and offer a sort of Confession of Faith or Creed on Evolution."[20] The resulting statement, endorsed by the AAAS council and issued in conjunction with the annual meeting in December 1922, strongly denied Bryan's claim that evolution is a "mere guess," questioned the wisdom of laws preventing its teaching, and asserted "the freedom of teaching and inquiry which is essential to all progress."[21]

Conklin was pleased to learn that the AISL planned to publish some essays, including his own, as pamphlets for wide distribution in September 1922, telling Burton that this would "relieve our committee of this duty and permit us to devote our attention to the more strictly scientific aspects of this controversy," adding, "the subject is of as vital importance to sane religion as it is to progressive science." As he told another correspondent, he hoped to "side step the whole matter by turning it over" to the AISL.[22] The AISL had already been thinking about responding to anti-evolutionists, well before the publication of Bryan's editorial. According to a draft list of publications apparently drawn up in 1921, a twenty-page pamphlet on "Evolution and Christianity" by an unnamed author was planned for release later that year, but this did not take place and no pamphlet with that title was ever published. The intended author may have been Lyman Abbott, who died in October 1922; the AISL did later publish a "five-minute leaflet," *The Evolutionist Believes* (ca. 1924), which reprints the final paragraphs of Abbott's book, *The Theology of an Evolutionist* (1897). In November 1921, AISL secretary Georgia L. Chamberlin told Baptist missions executive Lemuel Call Barnes that new pamphlets would include "'The Religion of an Evolutionist' or something of that sort," close to the title of Abbott's leaflet. Another possible author was Mathews, who mentioned a tract called "The Contribution of Science to Faith" in a letter to the Rockefeller Foundation. That precise title was never published, but it is very similar to the title of his own 1922 pamphlet. Yet another possibility was Chicago theologian Gerald Birney Smith, who taught a correspondence course about science and religion for the AISL (see chapter 3).[23]

Early the next year, however, Burton told the university trustees that, "The need for this literature arose acutely early in 1922 following the publication of a book by William Jennings Bryan, . . . in which the first chapters of Genesis are made the basis of scientific and historical knowledge. The book had enormous sales," and the ideas "in pamphlet form . . . are flooding the country." Pamphlets were needed to counter this effort, literature not only for "educated ministers" to give their people but also for the benefit of those "inadequately educated pastors" who were "brought face to face with the fact that they must take a position on something about which they know little." There was a similar need on the part of "students who will be reached through teachers of science, high-school principals and men of like position."[24]

Three months earlier, in March 1922, Burton's staff had mailed parcels containing five AISL pamphlets on nonscientific subjects to 30,000 ministers on their regular mailing list and to potential donors. They sought funds to support further mailings—including one planned for April, whose "chief aim will be to give accurate statements concerning the principle of *Evolution*, showing that rightly understood it is clearly an aid to *Christian Faith*."[25] At that point two pamphlets were planned, to be written by Chicago faculty, one "from the scientific point of view by Professor Coulter, and one from the religious and theological point of view of evolution, by Dr. Gerald B. Smith."[26] A theologian, Smith had written an AISL correspondence course partly about evolution, but for some reason he did not write this pamphlet.[27] Botanist John Merle Coulter, a former president of the AAAS who wrote about religion and science in a variety of popular venues, regularly taught more than

100 young men in a Bible class he had founded at Hyde Park Presbyterian Church (now part of the United Church of Hyde Park), where he had been an elder since 1898. The AISL had published one of his articles in its magazine and might have been considering reprinting it as a pamphlet, but more likely they wanted a recent article on "Evolution and Its Explanations" from *The Christian Century*—as a simple defense of the theory by a Christian biologist, it fit the situation better.[28] No such pamphlet was published.

Although the AISL canceled the April mailing, they did publish the first three "Science and Religion" pamphlets in September 1922. The initial decision concerning the texts was apparently made in early June—perhaps with an eye on the end of the fiscal year that month, in time to be included in the next year's budget. Letters to Conklin and Fosdick, seeking permission to reprint their *New York Times* essays, went out on June 8.[29] Other essays were also considered—those by Abbott, Birge, and Keen that Conklin liked—clearly the AISL wanted to cooperate with suggestions from the scientists as far as possible.[30] Chamberlin thought Birge's piece was just too long. She and Burton soon narrowed the list to Fosdick and Conklin, the latter retitled and lightly edited by Mathews to make it appear less of an attack solely on Bryan, plus a new essay by Mathews. They planned to print 30,000 of each immediately, trusting that funds to pay for them would come in over the next year.[31]

## Paying for the Pamphlets: Chasing the "Stars"

The AISL had two main sources of income—gifts and tuition from students in correspondence courses; earnings from the direct sale of pamphlets were minimal. To underwrite the evolution pamphlets, they had to raise roughly $2,000 for the first 30,000 copies of each title, and ongoing support would be needed for a sustained campaign. Fortunately, they already had a patron in mind. The first AISL pamphlets, dealing with biblical prophecy and the Great War, had been printed in 1917 and 1918. A month after the war ended, Burton wrote John D. Rockefeller Jr. to ascertain his level of interest in supporting further pamphlets opposing fundamentalism. Burton had just read Rockefeller's own pamphlet, *The Christian Church: What of Its Future?* (1918), an ecumenical reply to the wartime charge that Christianity had failed. Rockefeller's fundamentalist father had founded the University of Chicago, and the son shared the modernist views of the Divinity School. He advocated a nondogmatic, practical kind of inclusive Christianity that was highly suitable for a democracy, in which the church would be "literally establishing the Kingdom of God on earth."[32] Burton saw an opening through which he could subtly drive a truck. At the end of an appreciative and thoughtful response to Rockefeller's ideas, Burton simply mentioned the first few AISL pamphlets and the plan to publish more.[33] The following summer, Chamberlin wrote from Chautauqua, telling Burton how Chautauqua had raised $250,000 with professional help, with Rockefeller pledging $100,000 more. "I am keeping my eyes and ears open," she added, "perhaps we shall be able to work out a more extensive campaign for Institute funds next year."[34]

It took a little longer, but by the summer of 1921 Rockefeller had agreed to help. The catalyst was Rockefeller's modernist pastor, Cornelius Woelfkin, Fosdick's immediate predecessor at Park Avenue Baptist Church (now Central Presbyterian Church) and a close friend of Rockefeller, who served as a trustee. Mathews had written Woelfkin prior to the Northern Baptist Convention, apparently concerning the ways in which the AISL was combating fundamentalism and the need to get Rockefeller's support.[35] He then sent sample pamphlets to Woelfkin, who recommended the idea to Rockefeller. Mathews then sent additional pamphlets to Rockefeller Foundation staffer W. S. Richardson, with an estimate that several pamphlets could be sent to 40,000 ministers for $7,500, though the plan could be downsized depending on Rockefeller's level of interest.[36] In the meantime, Richardson sought advice from Brown University president William Herbert Perry Faunce, an early AISL donor.[37] Faunce was very enthusiastic, emphasizing the need for "millions of small leaflets, small ammunition–we need *tracts*, that will reach the multitude, that can be enclosed in letters, placed in church vestibules, distributed in Sunday schools, used in conventions, in schools and colleges, &c." Despite his enthusiasm, he had concerns about marketing. "I wish somehow it were not necessary to put the name U[niversity]. *of* C[hicago]. on each one." That "would prevent circulation just where circulation is most needed," an unambiguous reference to the very liberal reputation of the Divinity School, which would repel fundamentalist readers. Richardson agreed, adding that the pamphlets carried only the AISL name, not the university name. Rockefeller then pledged to match 40% of contributions—he thought it "unwise for him to be the only donor"—to a limit of "$7,500 from all sources."[38]

Thus, when Burton and his staff sought support specifically for the evolution pamphlets in the summer of 1922, they had a very sympathetic ear in Rockefeller—and for years to come. His support for the AISL, especially the evolution pamphlets in which he seemed particularly interested, continued at varying levels until August 1933, when his advisers recommended a "final contribution of $250." In reaching this decision, they noted the relative lack of importance of another pamphlet series, "Christianity and Modern Life," dealing with such things as appropriate forms of amusement for Christians, when compared with the "Science and Religion" series, which dealt "with questions then current and of vital interest to thousands of religiously minded people," indeed "with live matters of faith and belief."[39] Between 1922 and 1931, the period represented by the "Science and Religion" series, Rockefeller donated more than $24,200 to the AISL.[40]

The funds Rockefeller matched came from hundreds of individuals, churches, and other organizations (for a complete list of the scientists involved, see appendix three). Extrapolating from some of Conklin's comments, Chamberlin at first hoped to get as much as $2,200 from the AAAS, with Rockefeller adding $1,400, erasing the need to solicit individual donations altogether. Conklin soon disabused Burton of this idea, so Chamberlin proposed that they "go immediately to work to get $2000 from scientists and from those who are immediately benefiting [sic] from science as applied in business," as well as "leading educators." This had the advantage of not drawing on the traditional support base—in the past nearly all supporters had been religious professionals and laity. This was likely to succeed, for there was "no end of men who are making piles of money in enterprises which are possible because of modern science," and "there are also plenty of scientists who would like to appear before the world as not antagonistic to religion."[41] They initially agreed to target one hundred contributions totaling $2,000, although this was soon halved. Burton prevailed on Mathews to ask his brother, Johns Hopkins geologist Edward Bennett Mathews, to check the AAAS membership list for possible supporters.[42]

They were especially keen to enlist scientists at elite institutions, above all those who had been "starred" by placing asterisks next to their primary fields of expertise as listed in *American Men of Science*. Because this particular fact reveals wonderfully the degree to which the modernists accepted the hegemony of science and its definition of knowledge, the star system will be explained more fully before showing how the AISL used it.[43] In 1906, Columbia psychologist James McKeen Cattell, editor of the AAAS journal *Science*, issued the first edition of *American Men of Science*—now called *American Men and Women of Science*, although the original name is indicative of the misogynistic system it embodied.[44] Of the more than 4,000 scientists in this directory of biographical information, 1,000 were identified with asterisks as the "most eminent," according to a system Cattell concocted, born from his efforts to gather data for his studies of the psychological factors related to high scientific achievement. Starting in 1903, he asked ten top scientists in each of twelve disciplines, members of the NAS (itself a self-perpetuating body of elite scientists), to rank their professional colleagues in order of eminence. In the hierarchical order Cattell gave them, the fields reflected his Comptean view of the sciences, from "highest" to "lowest": mathematics, physics, chemistry, astronomy, geology, botany, zoology, physiology, anatomy, pathology, anthropology, and psychology. Other fields were not even accorded a place. From the rankings in each field, he assigned asterisks to the top 1,000 overall. He later produced a list of the 1,000 most "eminent" persons in human history—the first three were Napoleon, Shakespeare, and Mohammed—and his scheme for American scientists strikes one as comparably contrived. Once a scientist had received a star, he—and usually it was "he," since there were only nineteen "starred" women in the first edition—did not lose it until death; declining activity in later years had no effect. Furthermore, most of the starred men were of the "Puritan" type of "racial stock," a factor Cattell believed relevant to eminence. Nonetheless, as scientists were added in later editions, it became increasingly difficult to receive the coveted star, and this (among other factors) led to the system being abandoned after the edition of 1943, in which fewer than one in twenty-five scientists carried this distinction.

Those who managed to attain this sign of eminence, however, were specifically targeted by the AISL to receive sample pamphlets with a letter soliciting contributions. This is evident from the form letters apparently used in 1923–1924 that targeted the "thousand greatest scientists of U.S.,"[45] from the endorsements on certain letters,[46] and especially from the list of contributors to the pamphlet series (appendix three), which summarizes the annual fund lists for a twelve-year period overlapping with the evolution pamphlets. "Starred" scientists are omnipresent, and consequently much of the correspondence cited in this chapter is also from them. Burton and Mathews undoubtedly sensed that this would impress the Rockefeller Foundation, a leading benefactor of American science at the time, while furthering links with the scientific community that might benefit the modernist cause. Beginning in 1923–1924, the list of contributors was divided into three categories: "Scientists," "Other Individuals," and "Churches," showing the degree to which the AISL wanted to underscore the role of scientists in cooperating with this part of the larger campaign against fundamentalism.[47] For the two fiscal years overlapping with the Scopes trial (1924–1925 and 1925–1926), when national interest in these issues peaked, more than one hundred scientists gave a total of $1,325 to the pamphlet fund, including Compton (AAAS president in 1942), Davenport,

Caltech astronomer George Ellery Hale, Keen, Stanford University biologist Vernon Kellogg, Millikan (AAAS president in 1929), Princeton astronomer Henry Norris Russell (AAAS president in 1933), and paleontologist Charles Doolittle Walcott (AAAS president in 1923) of the Smithsonian. Despite Conklin's misgivings, the AAAS contributed five times: $100 in December 1922 by unanimous vote of the Executive Committee at Cattell's urging, and $60 more in January 1924, plus three additional gifts for a total of $335.[48] Kellogg apparently tried unsuccessfully to get the NRC to do the same thing; likewise for Caltech chemist Arthur A. Noyes and the NAS.[49] Further support came from wealthy individuals identified for this purpose by certain scientists, such as Conklin, who provided the names of eleven Princeton trustees whom he thought would be sympathetic to the modernist position.[50]

Support from individual scientists was solicited via a form letter, typically signed by Mathews, outlining plans and pressing the importance of keeping up the fight with the fundamentalists, while emphasizing the role of the AISL in offering an alternative religious view. Appeals were often accompanied by at least one of the "Science and Religion" pamphlets and general information about the AISL. Since "scientists recognize that we are fighting a battle for scientific truth as well as for religion, [in the past year] 25% of our distribution fund came from scientists who contributed $20 each or less," and friends of the scientists who made larger gifts.[51]

## Going beyond Evolution: More Pamphlets on "Science and Religion"

Their success at raising funds enabled the AISL to expand the "Science and Religion" series over the next several years, extending the topical scope considerably although evolution did not entirely disappear. A fourth pamphlet originated from the exchange between Millikan and Robert Brown that led to Millikan's "Joint Statement" (see chapter 1). Shortly afterward, Millikan prepared an address on "Science and Religion," to be given somewhere in Los Angeles in early March 1923 that was published right away in the Caltech alumni magazine and three months later in *Christian Century*, followed by the pamphlet in September.[52] It had come to the AISL's attention when someone sent Mathews a copy printed in a newspaper. Initially Mathews intended to publish it in his *Journal of Religion*, although he soon decided that a pamphlet was more appropriate. That decision might reflect the influence of chemist Edwin E. Slosson, director of Science Service, who had told Burton in April about Millikan's "excellent article . . . on the relations of religion and science."

Slosson asked, "Would you like to see it with the possibility of including it in your series?" He also recommended that the AISL solicit a pamphlet from University of Wisconsin zoologist Michael F. Guyer, who had been giving public talks about evolution. Burton thanked Slosson for his "most interesting and suggestive" letter, while Chamberlin said they particularly liked the way "it appeals to scientists to manifest their allegiance to religion as well as science, at this time."[53] A trustee of Science Service, Millikan had learned of the pamphlets the previous October and was among the first scientists to make a donation, so he was probably pleased to contribute an essay. He also suggested that they contact Pupin, Osborn, University of Illinois chemist William A. Noyes, and University of Iowa physicist George W. Stewart.[54] When Slosson got a copy of Millikan's pamphlet the following winter, he promised to "have a notice of them put into our Science News Letter which goes to libraries and teachers all over the country." He also suggested three more potential authors: Cattell, University of Michigan zoologist J. Playfair McMurrich (president of the AAAS in 1922), and marine biologist William Emerson Ritter of the Scripps Institution, cofounder of Science Service with Edward Willis Scripps. Ritter had already offered the AISL his article "The Religion of a Naturalist," published in Paul Carus's *Open Court* magazine, which described itself as "devoted to the science of religion [and] the religion of science"—a very liberal religious agenda that did not identify as Christian, but whose attitudes overlapped with those of the modernists. The AISL did not follow up.[55]

The fifth pamphlet resulted from another project involving Millikan, thirteen essays by leading scientists edited by Mathews, *Contributions of Science to Religion* (1924). The chapter on astronomy was written by Edwin Brant Frost, director of the Yerkes Observatory. A few months after Frost finished his essay, Mathews asked him to write "a pamphlet which gives some of the great facts, the marvels one might say, of modern astronomy," but which would also help show people "that scientific discovery increases rather than diminishes knowledge of God and the religious impulse." Frost agreed, although he was concerned that "my point of view may be so far from orthodox that some parts would perhaps better be omitted," not spelling out which parts. "I am not afraid of your unorthodoxy," Mathews replied, "but will be careful that there is nothing that will excite our fundamentalist friends unduly." His pamphlet espouses an eloquent natural theology, coupled with strong support for the relevance of Christian morality to the future of the earth and its inhabitants. Frank W. Very of the Westwood Astrophysical Observatory (Massachusetts) objected to Frost's view of the "supernatural," a word used only once, but he was a Swedenborgian and his objections seem sectarian.[56]

In choosing Millikan and Frost to write pamphlets, Mathews had picked people he knew at the University of Chicago, although Millikan had recently moved to Pasadena. For the sixth pamphlet, he looked elsewhere, to someone he knew from his work as a trustee of the Chautauqua Institution.[57] Although far from an elite scientist, Schmucker was a nationally prominent popularizer of science who drew huge crowds to his lantern-slide illustrated lectures at Chautauqua. He also wrote trade books for major publishers, including *The Meaning of Evolution* (1913), published eight times by MacMillan, and a textbook for Chautauqua courses. A former president of the American Nature Study Society, a pioneering organization in environmental education, Schmucker's father and grandfather were important Lutheran theologians, although he was an Episcopalian himself. Published in September 1926, his pamphlet displays considerable theological sophistication, using sexual selection to construct a natural theological argument and presaging process theism with its picture of uncreated, eternal laws of nature and a wholly immanent God.[58]

The central idea expressed in the title of Pupin's pamphlet, *Creative Co-ordination* (1928), is briefly mentioned in his Pulitzer Prize–winning autobiography, *From Immigrant to Inventor* (1923), read by millions of students and others in abridged forms, and more fully developed in several other publications.[59] In crucial ways Pupin was unlike the other pamphlet authors. An immigrant whose parents were illiterate Serbian farmers, he was a devout Orthodox believer, neither a Protestant nor a modernist. New York had no Serbian Orthodox church during his lifetime, so in the latter part of his life he attended the Episcopal Cathedral of St. John the Divine, where bishop William Thomas Manning held forth. By then he had risen to the pinnacle of American science and society, helping Woodrow Wilson draw the boundaries of Yugoslavia at the Paris peace talks and serving as president of the AAAS during *Scopes*. Having fled autocratic central Europe for a democracy in which he flourished, Pupin saw a close link between Christianity and American democracy, especially in the spirit of brotherhood central to both. That belief he certainly shared with Bryan, but unlike Bryan he believed evolution was good science. Pupin therefore represented an ideal political foil to Bryan's charge that evolution undermines democracy. According to Charles Harvey Arnold, it was Georgia Chamberlin who brought the AISL "such notable scholars and . . . such scientists as Michael Pupin," but nothing more is said and one can only speculate.[60] Certainly it made perfect *political* sense for the AISL to invite a publicly religious, fully "Americanized" immigrant like Pupin to join a group whose essays were being used expressly to combat fundamentalist views of science.

Nevertheless, this political dimension is only a vague hint in the final sentence of the pamphlet, which focuses on his wholesale rejection of reductionism in modern science—the same attitude Fosdick held in his 1922 pamphlet. In his autobiography, Pupin had said, "terrestrial organisms have instrumentalities with which they coordinate the non-coordinated, thus bringing . . . final cosmos out of primordial chaos."[61] He took it further in an article for *Scribner's Magazine* that became the penultimate chapter in his book *The New Reformation* (1927). Pupin told Manning he would write three articles for *Scribner's*, showing "that science, art, and religion have the same scientific basis," and that "the highest aim" was "the development of human character." "I have had it in my mind for many years," he added, "and feel confident that my philosophy is not very far from the truth and that I will not have very grave difficulty in proving that the mechanistic view of the universe is untainable [sic]." The third article, "Creative Co-ordination," appeared in August 1927, simultaneously with Mathews's statement to Rockefeller staffer Thomas Baird Appleget, that two new pamphlets were in the works, including "one from a scientist—we hope from Michael Pupin."[62]

A second pamphlet by Fosdick was published simultaneously with Pupin's. For unknown reasons, henceforth it replaced his earlier pamphlet in advertising and distribution. Even the list of titles inside the cover page omits it while naming the rest. The text came from a sermon in February 1927, a few months after he succeeded Woelfkin as Rockefeller's pastor but before Rockefeller built Riverside Church for him. Published the following spring in *Good Housekeeping* and *Reader's Digest*, the AISL issued it in May 1928.[63] The title well captures its theme: *Religion's Debt to Science*. It is not enough, Fosdick thought, simply to state "that true science cannot be anti-religious and that true religion cannot be unscientific." Rather, "religion and science are indebted to each other," because modern science enhances our appreciation for the creator, frees us from fear and fatalism, inspires sacrificial service to humanity, and helps us face facts. In turn, religion provides optimism for science.[64] The AISL apparently considered adding another pamphlet on the complementary theme, "Science's Debt to Religion," assigned to Martin Sprengling, an expert on Semitic languages at Chicago, but it never materialized.[65]

As Fosdick's pamphlet shows, Mathews and Chamberlin had moved the series in new directions since the opening salvo against Bryan. Although the issues raised by evolution were never far from the surface, they no longer dominated the agenda. The final two pamphlets continued this pattern. Technically, *Life After Death* (1930) was not published as part of the "Science and Religion" series—it lacks that heading on the cover—but it had originally been conceived as such, and the content and authorship warrant inclusion here. The AISL had already published a pamphlet called *Why I Believe in Immortality* (1925), by theologian Douglas Clyde Macintosh of Yale Divinity School, but it was never connected with the "Science and Religion" series.[66] Since at least some point in 1928, the AISL had planned to hold a symposium featuring scientists speaking about immortality.[67] It took place around Easter in 1930 as one of three informal faculty conversations held at the imposing new chapel (later named in memory of the donor, John David Rockefeller) on the campus of the University of Chicago. The other topics were "Economics and Religion" and "Does Civilization Need Religion?" Compton, Mathews, philosopher Thomas Verner Smith, and physiologist Anton Julius Carlson took part, with Smith and Carlson opposing immortality while Compton and Mathews favored it. A former Swedish Lutheran minister who had become a skeptic, Carlson later had a frank exchange with Compton about science and the "supernatural" in the pages of *The Scientific Monthly*.[68] All four talks were published by the *University of Chicago Magazine* in November 1930, two months after the AISL published only those by Compton and Mathews supplemented with an essay by Charles Gilkey, dean of the chapel and former pastor of the Hyde Park Baptist Church (now Hyde Park Union Church), where Mathews and Compton were active members.[69]

Plans for the final pamphlet were laid in 1929, when the AISL told the Rockefeller Foundation that they needed a pamphlet by a geologist. By the following summer the author had been determined: "Science and Religion from the Geologists' point of view (not an exact title) by Professor Kirtley Mather of Harvard." Around the same time, Compton agreed to write a pamphlet, "Why I Do Not Believe in a Mechanistic Universe," for a *different* series, for which they already had a pamphlet from theologian William Adams Brown of Union Theological Seminary (New York).[70] The following year, however, Compton's title was listed along with Mather's and the putative one by Sprengling, for possible inclusion in the "Science and Religion" series. Compton did not follow through, for Chamberlin was still badgering him about it in June 1931.[71] However, Mather did follow through, giving the AISL a lecture he had recently delivered at Harvard's Phillips Brooks House and published in the *Harvard Alumni Bulletin*, but clearly it had been conceived from the start as an AISL pamphlet. Excerpts soon appeared in the YMCA magazine and the Methodist *Epworth Herald*.[72]

Mather's pamphlet proved the final one in the "Science and Religion" series. Although a pamphlet on "The Religion of a Psychologist" was advertised in 1933, it was not actually printed and the putative author is unknown. Except for the first pamphlet by Fosdick and *Life After Death*, "Science and Religion" pamphlets remained in stock at least until the middle of the decade.[73] Rockefeller's interest in these issues had faded, the Great Depression was well under way, demand was clearly dwindling, and the AISL turned its attention to social ills and the dangers of Communism.

## Pamphlet Audiences and Their Responses

How important were the pamphlets? The significance of a project of this nature and magnitude should not be overlooked. In 1929–1930 the AISL mailing list contained nearly 45,000 names, including more than 1,000 scientists specifically chosen for their eminence.[74] Between 1922 and 1933 they printed perhaps as many as one million copies of the ten pamphlets, distributing these to tens of thousands of prominent Americans and a comparable number of ordinary folk, especially college students and church members.[75] High school principals, librarians, journalists, legislators at state and national levels, and Protestant minsters received pamphlets. At some research universities, scientists made them required course texts or distributed them individually to students, while chaplains on dozens of campuses also supplied them. Of course, the magnitude of a project does not necessarily betoken success. How successful were they? What did recipients think?

In September 1922 the AISL sent copies of each of the pamphlets by Fosdick, Conklin, and Mathews to 30,000 Protestant ministers, mainly in the North and on the West Coast; the South was "largely avoided, because of the limited number of pamphlets available." Earlier that year, five other pamphlets had been sent to approximately 10,000 Baptists, 6,000 Congregationalists, 4,000 Methodists, 3,500 Presbyterians, 3,000 Disciples of Christ, 1,000 Episcopalians, and 800 members of the Church of the Brethren.[76] The September mailing of evolution pamphlets almost certainly went to the same group.

Readers were divided in their assessments. An elderly Massachusetts minister recalled hearing about evolution in 1861, when he graduated from Andover Seminary. He did not believe it then, and he still rejected it. "*I think*

*Mr. Bryan is right*. Evolution has never been claimed to rest on *facts*, only on speculation. No proof exists for it." Egyptian mummies and cave men were just like modern humans. "Evolution is anti-Biblical, & anti-human, to my mind. *To me* it seems an insult to God & man to claim that Mary the mother of Jesus was descended from an *ape*!" The AISL was "far off the track in all your anti-evangelical positions." A pastor's widow said the pamphlets were "poison," lumped Fosdick with Thomas Paine and Robert Ingersoll as threats to Christianity, and said Mathews was "making more infidels & causing our young people to depart from the faith." A man from the Presbyterian Board of Foreign Missions tellingly addressed his letter to the "The American Institute against Sacred Literature." He liked Mathews's pamphlet, but those by Conklin and Fosdick were "not fit for young people." Genuine science was worthy of respect, but not "such higher critics & half baked scientists who pronounced guessing hypotheses as real truth as if they knew everything. Real scientists are very humble but these so called evolutionists are proud in their apishness & perverseness." Two recipients returned Millikan's pamphlet with added graffiti, one of them saying it was "not worth 3¢ a carload, except for kindling or the toilet." (The nominal cover price was three cents apiece.) The other wrote, "It would be nice to dream on forever. But the rich man awoke from his dream in Hell when it was to [sic] late. I hope it will not be so with you."[77]

Other ministers responded more favorably, and the AISL mailing list grew substantially during the 1920s. Some reserved judgment in an open-minded way, such as a Baptist pastor from Pontiac, Michigan, who confessed that "I have never been strong on Evolution, but am willing to 'be show'n.'" He asked for a brief statement or book "in *FAVOR* of evolution," so he could then "explain the whole thing, including Mendals [sic] Law," to his people, in order not to "let Bryan upset us."[78] Evidence is lacking, but it seems likely that many pastors on the regular mailing list followed the AISL recommendation to distribute pamphlets from racks in the vestibules of their churches, as Protestant pastors commonly did then with other tracts. Similar things could be done in clubs, schools, and colleges. That is exactly what the AISL itself did: students took 8,000 pamphlets in one year from a rack in Swift Hall (where the AISL was based) at the University of Chicago.[79] According to an internal report from 1930, "Perhaps the most marked increase in the use of the pamphlets has been in the field of colleges, of which there are now forty-one in which distribution of pamphlets is more or less active," among them the University of California, Columbia, Cornell, Indiana University, the University of Illinois, the University of Minnesota, the State University of Montana (now the University of Montana), the University of Nebraska, the University of Pennsylvania, the University of Pittsburgh, Rush Medical College, the University of Texas, the University of Utah, the University of Vermont, West Virginia University, and the University of Wisconsin. Universities in China, the Philippines, Syria, and Turkey also made them available. Distribution on campus was typically handled by religious organizations, such as the YMCA or the Wesley Foundation, but sometimes "there are men on the Science Faculties or other departments who are interested in the religious welfare of the students and keep supplies of the pamphlets on hand for their personal use." When Mathews visited Ceylon in the mid-1930s, a YMCA center there was sponsoring "a series of lectures based on our tracts."[80] Pastors used them as the basis for midweek discussion groups, and some laity bought quantities for distribution. In a particularly striking instance, a Presbyterian man sent Millikan's pamphlet to all the commissioners at his denomination's General Assembly in 1924—the year after Bryan had been an unsuccessful candidate for moderator of the same body.[81] A banker purchased 500 copies of Frost's pamphlet, one of which ironically ended up in the hands of Frost's own son, and a San Francisco businessman bought 150 of the same "work of exaltation, written in a devout spirit by an eminent astronomer." Frost reported receiving "a few letters, some of them quite commendatory," responding to his pamphlet.[82] Perhaps the strongest endorsement came from the president of Ohio State, attorney Robert E. Mathews, who gave copies of Compton's pamphlet to grieving friends. "Within the last week," he wrote, "I have heard of three cases where 'Life After Death' furnished genuine, and almost the only, consolation to friends who had just lost persons close to them. It's a great service they do."[83]

Just one month after mailing the first three pamphlets, the AISL printed another 30,000 copies of each. "We are sending at present to all the high-school principals in the United States," Chamberlin eagerly told Conklin, "for we have learned from various sources that it is among the high-school teachers that difficulty arises on account of the people who object to having their children taught on the basis of evolution." Sending pamphlets to principals "will help a little bit in the State where bills are pending," she told Ohio State zoologist Raymond Carroll Osburn. "It is the high-school men who are feeling the brunt of the difficulty because uneducated local school boards are bringing pressure to bear upon them to ignore evolution in their teaching of science."[84] A print run of that size was about right, since by the end of the decade there were nearly 24,000 public secondary schools in the United States.[85] College and university faculty, especially scientists, were also targeted at this time, and urged to enlist the support of colleagues in distributing pamphlets to their students. Pamphlets were sent to YMCA associations on college campuses and in cities, and advertisements were placed in secular and religious magazines, among them *Chautauquan, The Christian Century, The Expositor, Missions, The Nation, New Republic, Outlook, Religious Education,* and *Rural Evangel*.[86]

Efforts to distribute the pamphlets took on a crucial new dimension after November 1922, when Charles Davenport suggested that copies be sent "to each state legislator in the States of Kentucky, North and South Carolina and Minnesota. If it were feasible to send to each legislator in each state it might be a useful defense." "We really had not contemplated doing that," Burton replied, "but since you have suggested it, it seems to us a most important matter and we shall proceed at once to get hold of the list and mail pamphlets to them." Two weeks later, Burton informed Cattell that pamphlets had been sent to legislators "of five States where anti-evolution legislation has been suggested."[87] Shortly afterward, West Virginia geologist John T. Tilton suggested that lawmakers in his state—where Bryan had recently addressed the legislature—be sent first the pamphlet by Fosdick and then ten days later the one by Conklin; in his view the former was more effective. In a subsequent letter, he added, "some of the older members of the faculty here who have not had scientific training" would also benefit from reading them.[88] Texas governor Pat Neff received copies from Baylor Medical College physiologist Fred T. Rogers, who was especially concerned about academic freedom at his own university, which was under investigation by the sponsoring denomination. Texas Baptists began their inquiry into Baylor and other colleges after the publication of a textbook by Baylor sociologist Grove Samuel Dow based partly on evolutionary ideas, *Introduction to the Principles of Sociology* (1920). Rogers told Mathews that Baylor trustees ensured that "there is no liberty of biological teaching in the college branch of Baylor [in Waco]." Although "as yet there has been no interference in the teachings in the Medical School [in Dallas], . . . I do not know how long I will be permitted to follow the line of freedom in research and teaching which I have pursued." Rogers found himself somewhat out of step with his colleagues. "I can not be sure of what the outcome will be but there are some of us who as best we can are fighting for honesty in both science and religion."[89]

As AISL political efforts went nationwide in 1923–1924, according to an internal report, "Every member of a state legislature, and every national legislator received the Science and Religion series."[90] Secretary of State Charles

Evans Hughes also received copies.[91] Then in May 1925—when John Scopes was arrested for teaching evolution in compliance with his employer's wishes—Conklin's pamphlet was sent to the presidents of 281 normal schools, rising to 378 before the year ended.[92] During this critical year at least one pamphlet was sent to each of the following groups: all state legislators in Arkansas, California, Kentucky, Oklahoma, and Tennessee; editors of forty religious magazines and newspapers published in California, and sixty-one other religious magazines; 1,063 scientists; 1,826 high school principals; 3,000 librarians; 2,200 AISL members; and 355 alumni of the University of Chicago, class of 1913.[93] At the end of May, pamphlets were sent to "every minister in Tennessee," accompanied by a clipping from the *Chicago Tribune* announcing Scopes's arrest, a short commentary on this story, and information about an AISL course on "Evolution and the Christian Faith." Course texts included *Creative Evolution* (1911) by French philosopher Henri Bergson, *Readings in Evolution, Genetics, and Eugenics* (1921) by Chicago zoologist Horatio Hackett Newman, *The Direction of Human Evolution* (1921) by Conklin, *Christianity and Progress* (1922) by Fosdick, and *The Theology of an Evolutionist* (1897) by Lyman Abbott. Theology students were offered a special rate of five dollars to enroll.[94]

Responses from Tennessee pastors varied from rejection to sympathetic discernment. "When 'Evolution' has 'saved' your own damned, hell-bound Chicago and Illinois," wrote a Methodist from rural Madisonville, "then come and save Christian Tennessee! I challenge you to publish *this* to Illinois, and Chicago." A Baptist proclaimed, "my state never did a more noble thing in her history," taking the lead "in a great movement which will never stop until this damnable theory has been banished from our land." The AISL was "spreading a *theory* that is daming [sic] the lives and wrecking the faith of boys and girls. Your organization is a wolf in sheep's clothing." A Nashville pastor pointed out that the AISL promoted a type of evolution that "is most devoutly believed by all of our ministers, teachers, and Legislatures." The banned variety "excludes God from the universe," as in *Compton's Pictured Encyclopedia* (1922), which "accounts for the evolution of all material things from one original piece of protoplasm that was forever here, and this theory excludes God from the universe. Such books can no longer be taught in our public schools. Do you not think we are right?" On the other hand, Baptist evangelist J. E. Skinner thought the AISL's approach provided "the best possible means" for the "utter defeat" of evolution. Whenever evolution remained "in the realm of science there was no war upon it, and perhaps never would have been; but when it assumed a relationship in the religious realm and began its attack on Revealed religion, the war was on, and will be on till its folly is corrected."[95]

The following winter, the first woman to serve in the Mississippi state senate—the white supremacist, temperance advocate, and suffragist Carrie Belle Kearney—requested one hundred copies of each pamphlet to distribute to her colleagues in an effort to combat anti-evolution legislation.[96] Interest in the series remained high for another few years but had clearly declined by February 1932, when the annual letter to scientists went out, accompanied by the new Mather pamphlet and two other pamphlets on nonscientific subjects. "Our usual strong plea for the necessity of combating the anti-evolution legislation need hardly be made this year," the letter stated, "for with the passing away of two or three fanatical leaders the movement has died down," an apparent reference not only to Bryan's death in 1925 but also to the deaths of Reuben Archer Torrey in 1928 and Straton in 1929. "The forces of Atheism [i.e., Communism] are however gaining new strength, fed by certain foreign influences, and a fresh crop of hatreds are resulting from the attempt to create a brotherly society by force rather than by the cultivation of the spirit of brotherliness that alone can give it meaning."[97]

During the heyday of the AISL in the 1920s, however, their greatest concern was not to influence legislation but to reach young people in churches, colleges, and universities. The whole point of Bryan's crusade was to protect American youth from ideas he regarded as politically and religiously dangerous, so it was crucial to target that audience together with their pastors and teachers. The main strategy was to enlist faculty and chaplains to distribute pamphlets on their campuses. Conklin asked for 500 copies of his own pamphlet and 100 each of the Fosdick and Mathews pamphlets, saying, "I have large classes of university students who are always asking me for literature on this subject."[98] He also sent his pamphlet to various correspondents, including (among many others) a missionary to India who asked his opinion of the book, *Q.E.D.; or, New Light on the Doctrine of Creation* (1917), by prominent creationist George McCready Price; the principal of a Pittsburgh high school who requested help with a paper he was presenting to the Criterion Club; the editor of United News (New York), who wanted Conklin to write 500 words similar to his *New York Times* editorial for a series of articles on evolution; a Boston woman and a New Hampshire physician who had both attended his Lowell Lectures in 1922; and distinguished Johns Hopkins gynecologist Howard Atwood Kelly, a contributor to *The Fundamentals* who had also written several religious pamphlets.[99]

Pamphlets were also heavily used on other campuses. In the fall semester of 1922, corresponding with publication of the first three pamphlets, one or more titles were used in discussion classes run by the YMCA at Chicago, Pennsylvania, Wisconsin, and several small colleges. The following spring, Syracuse zoologist William M. Smallwood ordered fifty copies of each for a course, and Tufts College geologist Alfred C. Lane asked for one hundred copies of Mathews's pamphlet to distribute on campus. The next year, Minnesota zoologist Charles P. Sigerfoos wanted fifty copies of the first three titles and one hundred of the newly printed pamphlet by Millikan.[100] At Columbia in 1932, an adviser to student religions organizations ordered one hundred copies of six titles, in addition to other AISL materials, to fill their tract rack (apparently in Earle Hall), marking the science tracts with asterisks to indicate a heavy need.[101]

Undoubtedly the most important collegiate story involved Dartmouth College, where all freshmen took an interdisciplinary course on evolution—the first required course on evolution in the nation. Morphologist William Patten, who shared Vernon Kellogg's concerns about perceived links between German militarism and the teaching of evolution, had begun teaching it in 1919–1920, in large part to promote his alternative vision of biology as conducive to Christian ethical and moral principles: his religious goals were no less significant than his political goals. (Although this is not mentioned in their pamphlets, Conklin and Mather likewise sought to counter the prevailing image of evolution as just a deadly struggle, and they were hardly lone voices in the scientific community.) According to Gregg Mitman, Patten did his best to contribute to the war effort as a biologist, when "he began to use the metaphors of cooperation and efficiency in order to gain access to the public at large," culminating in the postwar publication of his textbook, *The Grand Strategy of Evolution* (1920).[102] The course was highly effective in challenging students' attitudes toward evolution. It also drew fire from Bryan, who spoke at Dartmouth in December 1923, with Patten responding in the *New York Times*. Several other institutions wanted to follow suit, including Chicago, which developed a famous general education course, "The Nature of the World and of Man," along similar lines in 1924.[103]

In the spring of 1923, Patten ordered 300 copies of the first three pamphlets for distribution in his course. Perhaps he learned about them from Conklin, a close friend from their involvement with the Marine Biological Laboratory at Woods Hole. Patten clearly found the pamphlets to his liking, telling Burton, "We are working along somewhat similar lines at Dartmouth." The following year he requested 700 copies of the Millikan pamphlet and promised to send Burton "the latest edition of our *Evolution* pamphlet that, if you are interested, will give you some idea of the way we teach Evolution at Dartmouth." He added, "We find it impossible to teach Evolution in a comprehensive way—without running across religious questions at about every corner—and so we have tried to help the boys with their questions as best we could—And in fact—that to my view, is the most important purpose—of this course." Dartmouth chaplain Frank Janeway and president Ernest Martin Hopkins applauded his efforts.[104]

Presumably, when Patten mentioned "our *Evolution* pamphlet," he meant *Why I Teach Evolution* (1924).[105] His motivation for teaching evolution—and for teaching it in a certain way—was explicitly religious: "the teaching of evolution is the only way to bring back a living God into those fields of human thought and experience from which the teachings of 'high-brow' philosophy and 'low-brow' religion are excluding Him with extraordinary thoroughness and rapidity." A world developing under unifying natural laws "is to me the surest and the only available evidence that there is a God-like *method* and a God-like *purpose* in it all." But, "perhaps the principal reason I teach evolution," he added, "is because the *methods of evolution exemplify the successful usage of the highest ethical and moral principles.*" Like Pupin, Conklin, Compton, and Mather, Patten saw in the continuity and overall "lawful administration" of the universe compelling evidence of "some sort of creative purpose," despite the appearance of "'chance' or accident" as the governing principle. In that part of the course dealing with the evolution of the physical world to provide the conditions for life, the argument can be seen as an early version of the anthropic principle. "For it is manifest that without these particular agencies, or even with any measurable modifications of their peculiar qualities and proportions, protoplasmic life would be impossible on this planet or any other." Given the appearance of life, "the self-sacrifice of altruistic services" becomes "the chief aim and pleasure in life." And this is "the essence of the moral and ethical teachings of Christianity as it is the essence of the moral and ethical teachings of evolution." Thus, he concluded, "there is no difference between what is vital in science and what is vital in religion. In fact, underneath, science *is* religion and religion *is* science."[106]

The importance of taking an integrative approach to religion and science was also evident in the testimony of Yale paleontologist George Reber Wieland, who responded to the first three pamphlets by crediting Hugh Miller's Victorian classic, *The Testimony of the Rocks* (1857), with helping him form a favorable attitude. "Initially there was some questioning," he reflected on his adolescent years, "suggesting, but never reaching, what is somewhat vaguely called disbelief." But if Voltaire led him in one direction, Miller "served as a balance." Although he knew that "those interested in science are widely believed to lack religion," Wieland thought "scientifically trained men have a calmer, surer belief in religion than many teachers of religion. The scientific man is not worried about creeds and dogmas, whereas in many instances the teacher of religion, taking these more seriously, comes to suffer from reactionary impulses, or is by his very ardor driven too far." The rocks teach just one lesson, "that life and the world mean something, that the processes of both the organic and the inorganic world are inseparable, and that man while he crosses, as it were, a narrow and trembling bridge between two mighty eternities has bound up within himself, and at the same time shares with all life, all of evolutionary history." Looking into ourselves, "we are near to, not remote from that which is the most sublime; and only the most sublime is God."[107]

Many of Wieland's professional colleagues also appreciated the pamphlets, even if they expressed their appreciation with less eloquence. The Astronomical Society of the Pacific sent the "valuable and interesting" tract by Frost to its members.[108] The editor of the *Journal of Heredity* was so pleased with Fosdick's pamphlet, that he asked Fosdick to write an article for his journal. When Fosdick declined, he got permission to reprint part of the pamphlet, but this did not happen.[109] When Slosson suggested that the AISL might find support from trustees of his Science Service, four of them came through: Kellogg, Hale, Robert M. Yerkes, and Millikan, who made the first of ten annual contributions. He also identified four additional scientists to contact: Osborn, University of Illinois chemist William A. Noyes, University of Iowa physicist George Walter Stewart, and Pupin.[110] After receiving the first three pamphlets from his friend Conklin, zoologist Raymond C. Osburn bought thirty-five more. Affirming that "there is no necessary conflict between science and religion and that it is perfectly possible for one to accept the teaching of science, such as evolution, without being an atheist," he offered to let the AISL reprint his own pamphlet, *Some Common Misconceptions of Evolution* (1922), an address he had given as retiring president of the Ohio Academy of Sciences.[111] Horatio Hackett Newman declined to donate, but pointed out that in the past year he had spoken for free to students, churches, YMCAs, and other groups, and thereby "reached directly not less than 10,000 persons and indirectly a much larger number with essentially the same message" as Conklin. Like Conklin, Newman no longer accepted the traditional Christianity in which he had been reared. At one point, Church of Christ theologian William Wesley Otey asked whether he believed in the divine inspiration of the Bible, the Virgin Birth, the atonement, the Resurrection, and the future general judgment. Newman wrote back, "I do not believe any of these doctrines. In this I am in accord with most of the advanced students of religion with whom I am acquainted." In his opinion, those beliefs were "not essential to true religion. They have come to be widely accepted by certain religious sects, but are not accepted by real students of religion as anything more than symbolic."[112] Iowa State geologist George F. Kay gave a similar report, telling Burton that "for fifteen years I have conducted classes on Sunday mornings for University students in the Presbyterian Church of Iowa City on the subject, 'Science and Religion.'" More than one hundred students belonged to his class, the teaching of which he regarded as "a finer service . . . than I have been able to do in connection with the many other activities in which I have participated."[113] William W. Keen bought 250 copies of Millikan's pamphlet to give to Bryan, Straton, and members of his own Baptist church.[114]

The enthusiasm expressed by many elite scientists belies the gap between them and the wide audience that the AISL hoped to reach. That gap was not always easy to cross, on both scientific and theological avenues. Some readers wanted more substance on certain theological points, such as the Colby College student who asked Conklin to spell out his beliefs on sin, righteousness, the soul and immortality, or the Presbyterian minister from Iowa who wanted the AISL to say more about sin, salvation, and biblical inerrancy. He got an evasive reply, as did Pittsburgh banker William E. Lincoln, who told Mathews that Millikan seemed "narrow and shallow both." Although Lincoln was "no 'fundamentalist'—as the word is now used," he believed the pamphlet did "more harm than good. Not what it says but what it don't say & the way it don't say it."[115] Other readers struggled to comprehend certain ideas in the pamphlets. An Indiana Baptist pastor reported on Mathews's pamphlet, which he had given to seven people from diverse educational backgrounds (including a local school superintendent with a degree in science from Chicago) to assess its effect. Most found it worthwhile but difficult to understand, and none could articulate Mathews's specific points. Some scientists echoed this conclusion, including William T. Hornaday, director of the New York Zoological Park, who thought Millikan's pamphlet was "too academic and written in much too lofty a key to impress the people who need to be impressed." Temple University psychologist Thaddeus L. Bolton expressed sympathy for the AISL but lamented that the pamphlets could interest "only the higher types of men and women in our society." The "great body" of the population were "unable to appreciate the kind of religion" advanced by the modernists, for they had "only sufficient intelligence and character to accept the very dogmatic and reactionary type of religion." Nevertheless, he saw "urgent need that the field of science and religion should be clarified and readjusted in its higher levels all the time. While this may not penetrate very far down, it is necessary that the atmosphere at the top shall be kept pure and open." Pointing to results of psychological tests given by the US Army, Harvard geneticist Edward Murray East drew a similar conclusion, noting that "we have about 25 million people in this country with a mentality of ten years or under, and it is not to be expected that they will be able to appreciate anything which makes much of a demand on intelligence." If the AISL is to be successful, he advised, "you should furnish some clever journalists with facts to put together in the ordinary newspaper style," perhaps through Slosson's Science Service—a possibility that Slosson had already suggested.[116]

A root of the problem was widespread ignorance of science, for which scientists themselves were substantially to blame. As Michigan zoologist Jacob E. Reighard told Mathews, "Teachers have too commonly assumed that evolution was fully established & have spent their energies in discussing methods. College students need, in my experience, detailed & rather elementary instruction in the evidences, but they rarely get it."[117] Patten was even more damning. In his opinion, "many scientists, as well as the vast majority of intelligent laymen, do not rightly understand what evolution is, or what the working methods of evolution are." The blame lay with scientists. "They have been too much absorbed in their own little fields of work; and few of them look beyond those fields, or take the trouble to tell in untechnical language what they have discovered, or what, if anything, they think about evolution *as a whole*." Lay people had to work it out themselves, but scientists "are now beginning to recognize their obligations to youth and to the 'man in the street,' as well as to themselves, and are trying to fulfill them."[118]

At least some scientists other than Patten tried to address this head-on. When Conklin received a request from a Missouri clergyman to identify helpful books on evolution, he recommended his own book, *The Direction of Human Evolution* (1921), and *Evolution and Christian Faith* (1923) by his former student, University of Kansas zoologist Henry Higgins Lane. He also recommended Lane's book to a Pittsburgh school principal and to an Oakland man.[119] A notable query came from a former student at Ohio Wesleyan College, J. N. Rodeheaver, a publisher of gospel music with his famous brother, fundamentalist evangelist and song leader Homer Rodeheaver, who worked with Billy Sunday. Rodeheaver was preparing a talk answering Bryan's objections to evolution and he had just written to Bryan, a good friend of his brother. Conklin sent him pamphlets and suggested that he read *The Theory of Evolution* (1917) by his Princeton colleague, paleontologist William Berryman Scott, son of a Presbyterian minister, for its short account of the evidence for evolution.[120] He also thought highly of a book by John L. Robinson, a Unitarian minister from Memphis, *Evolution and Religion* (1923), which Robinson had sent him in return for a copy of Conklin's pamphlet. Having read the book carefully enough to spot some misprints, Conklin described it as "one of the best presentations of the subject which I have ever read. It is brief and to the point. It requires no deep scientific knowledge to understand it, and yet it is really scientific. I shall be glad to recommend it to persons who are interested in this subject." Conklin's enthusiasm was shared by West Virginia botanists Robert C. Spangler, who wrote the foreword, and Perry Daniel Strausbaugh, who told Robinson that "your presentation is so clear, and easily understood that the laity can understand your meaning—a thing that is not true in so many articles dealing with the subject." The distinguished Cornell limnologist James G. Needham liked the book so much that he posted advertisements for it in his department.[121]

Some of the most interesting comments from scientists highlight the necessity and importance of the religion–science conversation. "There are unwise & intolerant minds on both sides in this discussion about the greatest things of life and faith," said University of North Carolina chemist Francis P. Venable. "Still, I believe only good will come from the discussion—Truth will prevail." There will always be mistakes, for we can have only imperfections "where finite minds are dealing with the infinite. The great value lies in the open-minded search."[122] Particularly revealing are the comments of a leading Jewish chemist and physician, Henry Leffmann of the Leffmann-Trumper Clinical Laboratory in Philadelphia. He frequented a weekly Bible reading led by the principal of Gratz College, the first independent Jewish college in North America. When first solicited for a donation, Leffmann declined, because he regarded the Bible "as a purely human document" that does not contain "a single word dictated by a power superior to humanity. It is obvious that I am wholly out of agreement with [the] basis of your propaganda." Leffmann clearly did not know the first thing about the Chicago approach to religion, so Chamberlin quickly filled him in. "We do not see how you could infer that anything coming from the University of Chicago which has been frequently referred to in the public press as a 'hot bed of heresy' could represent a method of study based on the supposition that the Bible was 'dictated by a power superior to humanity.'" Leffmann soon came around and made two donations, since he felt that "the issue between the defenders of plenary inspiration and those who wish a more liberal construction of the Bible is so sharp and I am so opposed to the former view that I can wish for success for those who take the latter view." Noting Jewish scholarship on the meaning of the Hebrew word "alma" in Isaiah 7:14 and several other examples to emphasize his point, he commented on certain "uncertainties of the text and the manner in which the standard translations ignore these difficulties." His most interesting point came as an afterthought in a postscript. "I am not unmindful of the extravagances of some of our leading scientists, especially the naturalists." His own field of chemistry "has little tendency to romancing. We can check almost everything in the laboratory. The able and justly eminent naturalists are, however, given to drawing telescopic deductions from microscopic premises, and furnish ammunition to Bryan and his ilk."[123] Similar concerns were voiced by Harvard botanist Edward Charles Jeffrey, who affirmed the need for literature like the pamphlets, "in view of the extreme prejudices which have been aroused by the mechanists in biology particularly in the southern states of the Union." Chemical engineer William DeGarmo Turner of the Missouri School of Mines was "particularly interested just now in view of the fact that the Society of Free Thinkers are apparently circularizing scientists also, and any propaganda which would present the Christian side of the relationship between science and evolution deserves the hearty support of all scientists." The Society of Free Thinkers, founded in 1915 as the Freethinkers' Society of New York, was an organization of skeptics that was led in the 1920s by publisher Joseph L. Lewis, a prominent public atheist.[124]

A few strongly anti-religious people also received pamphlets, including journalist H. L. Mencken, famous especially for his acerbic coverage of Bryan and other conservatives during the Scopes trial, whose copy of Frost's pamphlet is now in the Library of Congress. Undoubtedly the best-known skeptic in this category was the great trial lawyer Clarence Darrow, who had frustrated modernist efforts to shape perceptions of evolution during the trial by inserting himself directly into the proceedings as lead attorney for the defendant. Although he declined to contribute, noting that he had spent "about $2,000 of my own money" in Dayton, he did so cordially and invited further contact. The same cannot be said of the person using the pseudonym "Rats," who sent Conklin a postcard. "Evolution is only a theory; a presumption, but the Bible is a rotten pack of <damned> lies & bunk." Although "The Holy Pope is infallible," according to the Bible "the great Almighty God, made a series of blunders from cover to cover." Ornithologist and surgeon Robert Wilson Shufeldt, "starred" in the first edition of *American Men of Science* although he never had an academic post, was more civil but no less forceful. He objected that Millikan "has never carefully read the Bible of the Christian Church; apparently he is quite ignorant of what constitutes the modern law of organic evolution—and, under such circumstances, I fail to see any value or any truth in his comments." Shufeldt believed that "a large number of statements in the Bible of the Christians, are flatly contradicted by modern science in general, and by the law of organic evolution in particular," and that "untold tons of 'Popular Religion Leaflets' are, and will be, powerless to gainsay the statement." A Washingtonian objected to the way in which Conklin had argued for a certain compatibility between evolution and the biblical description of what Conklin called "a process in the creation of man." "But that does not answer the fact that the Bible does state a definite process, and that this process has been proved to be a myth by science," the correspondent claimed. "Science thus breaks down the very foundation of conventional religion. Those therefore that still see fit to cling to this mythology would seem to be justified in their cry that evolution tends to destroy a very important link in their faith."[125]

## Conclusion: A Successful Enterprise?

The pamphlets on "Science and Religion" were driven by a cluster of hotly contested religious and cultural issues that culminated in a sensational show trial in 1925. The concerns of Mathews and his Chicago colleagues about events in Tennessee were spelled out in a remarkable, unpublished, internally numbered four-page document, "Implication of the Scope [sic] Trial." Perhaps an early draft of something intended for public dissemination in some form, it appears to have been written while the trial was still under way.[126] As they saw it, the trial "is more than the hippodrome Mr. Bryan's convention methods have made it. It is the focus of a number of issues as widespread as our citizenship."

The document identifies six specific implications of the trial. First, it illustrated "the power of prejudice and ignorance in our legislative system. The Tennessee legislature is not a body of scientists, yet it made a scientific belief a punishable misdemeanor" for religious reasons. If found constitutional, what other laws might be passed? It was best to remove education from the realm of politics, but nothing is said about how this might be accomplished in a democratic republic. Second, the trial "sets up the dangerous doctrine that education can be absolutely controlled by a legislative majority and that the hand that holds the pocketbook has the right to control the teacher." What if the Tennessee legislature was predominantly Roman Catholic? Would that principle still hold? Teachers are not like "department store clerks," rather they are like "ministers," rendering social services rather than goods. "Would Mr. Bryan admit that his speeches must contain what the people who pay to hear him talk think he ought to say? Does the hand that buys the tickets dictate what the orator must say[?]" Surprisingly, given the intensity of religious feeling among the modernists—or, perhaps not surprisingly, given the modernist view that a reasonable use of reason had made traditional faith untenable—there is no consideration of what to do when the majority find the conclusions of an academic elite patently offensive, as the fundamentalists did in this case. Why they should be expected to pay for such ideas to be taught to their children, without a similar allowance for the teaching of their own ideas, is not discussed. Third, "The trial shows that religious prejudice is intolerant and incapable of fair play." Although Bryan had portrayed this as "a fight between evolution and the Bible," he "will not permit any discussion of the issue he himself raises. The prosecution has sought conviction and not justice." This might refer to the fact that Judge John T. Raulston did not allow the jury to hear testimony from several religious scientists and liberal theologians (including a Jewish rabbi) relative to the religious meaning of evolution—including Mather—although Mathews, Conklin, and Pupin all stayed away from Dayton. In any event, Bryan's confrontational attitude toward evolution profoundly discouraged the kind of conversation that the AISL tried to facilitate. Fourth, convicting Scopes "would mean that a legislature has the right to determine what is the teaching of the Bible," ignoring the fact that scholars differ on this. A conviction would establish a single interpretation of Genesis "for tax-supported teachers in Tennessee," making a mockery of religious liberty. Furthermore, "Few scientific laws accord with the scientific views of the biblical writers." Fifth, the trial reveals "the danger to American civilization from unintelligent voters and legislators. Constitutions cannot protect us completely from bigotry or vindictiveness any more than they can protect us from the doctrine of free silver money," a reference to Bryan's opposition to the gold standard for paper currency. Once again, we see the problem that the AISL was trying to confront—namely, how to communicate "elite" religious and scientific ideas to ordinary people. Finally, the trial "has been a school in the study of the Bible and of science. The American people will have been shown that evolution is not opposed to Genesis." Genesis teaches not science, but "the activity of God in nature and the spiritual value of man," showing "how God was involved in the creative process, and how that process culminated in man possessed of both animal and divine elements." Evolution tries to provide a detailed explanation, but "It does not take place in a vacuum, but an environment in which [there] is God. Genesis and evolution are complementary to each other, Genesis emphasizing the divine first cause and science the details of the process through which God works." The document notes that a similar view "is held by many conservative evangelical theologians," among them Baptists Augustus H. Strong and Edgar Young Mullins, Episcopalian Richard Wilde Micou, and Presbyterian Howard Agnew Johnston.[127] (I cannot confidently identify two others solely from their surnames.)

The view that evolution is just "the process through which God works" was held by several leading evangelical scientists and theologians in the decades surrounding 1900, including Asa Gray, James Dwight Dana, and Benjamin Breckinridge Warfield. The view that Genesis teaches theology rather than science is rooted in Galileo's time-honored dictum, borrowed from Vatican librarian Cardinal Cesare Baronio, that the Bible tells us how to go to heaven, not how the heavens go.[128] Predating the controversy over evolution by some 250 years, Galileo's attitude found wide support among evangelicals well before Darwin's ideas landed in their laps. It was also embodied in the pamphlets. To the extent that they popularized that attitude, they were probably successful: reaching many American Christians and their pastors with a cautionary message about misusing the Bible, showing legislators an alternative both to Darrow's agnosticism and to Bryan's bibliolatry while generating some enthusiasm in the scientific community. However, this was not the only message the pamphlets conveyed, and neither was it the only message concerning science that their authors promoted in other writings. Mathews regarded Gray's confident affirmation of evolution alongside the ecumenical creeds to be all but obsolete, and he fully realized that the pamphlets sometimes went well beyond the faith of his fathers. "It is impossible to trace the effect of such literature," he wrote a few years later, apparently overlooking hundreds of letters in his files, "but wide circulation would argue that it met the need felt by religious persons of finding something more in scientific findings than a support for romanticized orthodoxy."[129] Orthodoxy was not exactly Mathews's cup of tea. To varying degrees, Mathews and other modernists sought fundamentally to redefine the Christian understanding of God in light of what they thought modern science meant. That is the topic of the final chapter.

CHAPTER 3

# Science and Religion, Chicago Style
## The Protestant Modernist Encounter with Science

> "How much can I still believe?" is the question pathetically asked.... Beginning with two score or more doctrinal articles there ensues a process of elimination and attenuation till today, in liberal circles, the minimum creed seems to have been reduced to three tenets: belief in God, confidence in immortality, and conviction of spiritual uniqueness in Jesus of Nazareth.... Thus the pathetic game of give what must, hold what can, continues.
> —*Edwin Arthur Burtt,* Religion in an Age of Science

In the pamphlets and other writings, the AISL authors discussed core Christian beliefs in light of science, including conceptions of God, creation, miracles, and eternal life. Their views were often radically different from those of many other American Christians of their own day, to say nothing of previous generations. Their works provide a clear, if limited, window on the modernist encounter with science in the first half of the last century—especially as things looked from Chicago, the theological and geographical center of the modernist movement and the home of the seminary that published the pamphlets.

To fully appreciate the magnitude of the changes wrought by the modernists in the name of science, let us turn the clock back to February 1880, when Harvard botanist Asa Gray delivered two lectures on science and religion at the Theological School of Yale College. The first American to understand and accept Darwin's theory of evolution, Gray was also a committed orthodox Christian. As a young man of twenty-four with rationalist and materialist convictions, he had fallen under the evangelical influence of Columbia botanist John Torrey and his wife, devout New School Presbyterians in whose congenial home Gray had spent many warm hours. Soon he came around to their position, which he maintained with studied sincerity coupled with genuine irenic humility for the rest of his life, his acceptance of evolution notwithstanding. Despite marrying a Unitarian wife and spending much of his career at a Unitarian stronghold, Gray regularly attended an orthodox Congregationalist church in Cambridge and sometimes taught Sunday school. Long convinced of the necessity to relate evolution to Christian faith, for fifteen years he wrote anonymously though not secretly about it, including a review in the *Nation* of a book by Princeton theologian Charles Hodge, *What Is Darwinism?* (1874). Congregationalist pastor, theologian, and geologist George Frederick Wright, who had been defending evolution for some time, nudged Gray to take on a larger, more visible role as a proponent of what Gray himself would term "theistic evolution," specifically in response to Christian critics of evolution like Hodge and McGill geologist John W. Dawson. Gray's *Darwiniana* (1876), a collection of previously published essays augmented by a new essay on "Evolutionary Teleology," directly resulted from this.[1]

Thus, Gray agreed to speak at Yale, where the audience included philosopher Noah Porter (Yale's president), church historian George P. Fisher, and paleontologist Othniel Charles Marsh.[2] Early in his first lecture on "Scientific Beliefs," he reached back before the Civil War, when Benjamin Silliman had taught Yale students to map geology onto the Bible by interpreting the "days" in Genesis as eons of time. In those days, Gray observed, "schemes for reconciling Genesis with Geology had an importance in the churches, and among thoughtful people, which few if any would now assign to them." Such efforts "to bring the details of the two into agreement by extraneous suppositions and forced constructions of language . . . would now offend our critical and sometimes our moral sense." Gray's aim was much more modest. "Our predecessors implicitly held that Holy Scripture must somehow truly teach such natural science as it had occasion to refer to, or at least could never contradict it; while the most that is now intelligently claimed is, that the teachings of the two, properly understood, are not incompatible." Genesis "is not an original but a compiled cosmogony," intended to combat "polytheism and Nature-worship," and "its fundamental note is, the declaration of one God, maker of heaven and earth, and of all things, visible and invisible,—a declaration which, if physical [i.e., natural] science is unable to establish, it is equally unable to overthrow."[3]

Gray took this further in his second lecture, "The Relations of Scientific to Religious Belief." Instead of seeking harmonious agreement between the Bible and science, he distinguished religious truths from scientific truths while affirming that both could be held consistently. "I accept Christianity on its own evidence," Gray confessed, "and I am yet to learn how physical [i.e., natural] or any other science conflicts with it any more than it conflicts with simple theism. I take it that religion is based on the idea of a Divine Mind revealing himself to intelligent creatures for moral ends." God's revelation to humanity, "in its essence concerns things moral and spiritual," and culminated "in the advent of a Divine Person, who, being made man, manifested the Divine Nature in union with the human," which itself "constitutes Christianity." The incarnation was for Gray "the crowning miracle," attended by other miracles that he also accepted. Thereby Gray was able to consistently affirm both evolution and Christian faith, even a very traditional faith whose "essential contents" were "briefly summed up" in the Apostles' and Nicene creeds.[4]

Gray and Wright were not alone among evangelical scientists in finding ways to accept evolution alongside the ecumenical creeds. At least one more Congregationalist belongs in the same category, the distinguished Nebraska botanist Charles E. Bessey, who was president of the AAAS in 1911. In the winter of 1872–1873, Bessey studied with Gray for three months and read *Darwiniana* with apparent appreciation. He knew William Jennings Bryan well, both as a neighbor in Lincoln and from his involvement in the Round Table Club, a nonpartisan literary and debating society that Bryan and his law partner founded in 1889. Although Bessey did not write on religious topics, he lectured about geology and the Bible in churches. In 1909, he was contacted by an English clergyman who was collecting religious opinions from American scientists on behalf of the North London Christian Evidence League. Responding to specific questions, Bessey stated that the universe "most certainly" had a "first cause" possessing "personality" and "an intelligence far greater than any man." He also affirmed "that man's personality survives in a conscious state" after death, even if he could not "fully apprehend" how. He agreed "that God has revealed Himself to Man pre-eminently through Jesus Christ," adding that he found "no difficulty in such belief. Since nothing in science even so much as suggests the contrary." Thus, he was "quite willing" to think of Jesus as "the Son of God," and he had "no doubt whatsoever" that "the Bible contains a divine revelation."[5]

In addition to Gray and Bessey, it is difficult to say how many scientists or clergy active around or before 1900 embraced both evolution and orthodox Christian beliefs, but it is not hard to find some very prominent evangelical examples. Silliman's son-in-law and successor at Yale, geologist James Dwight Dana, wrote extensively about science and the Bible throughout his long career and became persuaded of evolution (except for humans) in the 1870s. Princeton theologian Benjamin Breckinridge Warfield, who contributed to the doctrine of biblical inerrancy, found a limited form of evolution theologically acceptable and even promoted it in the years leading up to the fundamentalist–modernist controversy. Baptist theologian Augustus Hopkins Strong, president of Rochester Theological Seminary, also embraced a limited form of evolution that influenced his understanding of both creation and eschatology.[6]

None of those six people were still living when Bryan began his antievolution campaign in 1922.[7] With the rise of fundamentalism after the Great War, we look almost in vain for American voices of comparable magnitude who affirmed both evolution and the ecumenical creeds. If they existed, they tended to keep their ideas under wraps as controversy raged. Perhaps the most vocal scientist to profess traditional beliefs about God and Christ while defending evolution was University of Kansas zoologist Henry Higgins Lane, a minister's son and former student of Frank R. Lillie at the University of Chicago and Conklin at Princeton University. Lane later wrote a zoology textbook and became Curator of the Natural History Museum at Kansas. He inserted himself into the controversy in 1923, when he published a course of lectures on *Evolution and Christian Faith*.[8] At a time when scientists like Compton, Conklin, Mather, Millikan, and Schmucker were defining their religious beliefs in frankly unorthodox ways, Lane defined his classically, while keeping one eye focused sharply on modern science. Like Gray, Lane rejected all-too-clever efforts to harmonize science with the Bible. "The attempt to correlate the 'days' of Genesis with the 'periods' of geological time cannot succeed," he said, for both biblical and scientific reasons. "The author of Genesis" spoke in "familiar terms," using "language suited to [the] understanding" of the audience. This did not mean that Moses was guilty "of ignorance nor deceit. The account is not *untrue*. It is simply adapted to the understanding of the kindergarten class instead of university seniors."[9]

In Lane's opinion, those Christian opponents of evolution who demanded of thinkers like Lane that evolution somehow account for Jesus's divine nature were simply barking up the wrong tree. He found that demand "preposterous. If the divinity of Christ be admitted, both He and His origin are at once removed entirely from the field of operation of evolution." The opening lines of John's gospel teach that Christ preexisted the world and helped create it. "Evolution therefore could have had no part in the production of a divine Christ. There is no precedent in nature ... for the incarnation; it can only be accepted by the believer as a unique event." Evolution could neither account for Christ's miracles, nor deny them. "Evolution does not limit the power of the Omnipotent One; it only expresses the method by which the Creator chose to work out the *creation of nature* in so far as it is manifest to finite minds." Although Lane did not mention Gray anywhere in the book, he did not hesitate to label his position "theistic evolution," the same term Gray had used and Bryan so despised, and he summed it up in language Gray would surely have endorsed. "The doctrine of evolution presents no difficulties too great to be harmonized with the gospel of Christ. It has no quarrel with His birth, life, death or resurrection."[10]

Similar views are infrequently found among the letters from scientists to the AISL, but one "starred" scientist on their mailing list, Johns Hopkins physicist William J. A. Bliss, did defend a position much like Lane's. After identifying himself as a friend of Mathews's brother (Johns Hopkins geologist Edward B. Mathews) and a member of the vestry in "one of our more ritualistic episcopal churches," Bliss said, "I have never found it necessary to strain my conscience as either a scientist or a professing Christian." Why not? "The crux of the whole matter seems to me to lie in an entire misconception of the nature and sanctions of natural law. The man of science who shares this misconception ought to be ashamed of himself, because it is his business to know better, but the layman cannot be blamed nearly so much," and the pamphlets "will acquaint the more intelligent part of the public with what science may fairly claim to know, and on what it is as ignorant as the most uneducated." It should be possible to persuade "intelligent people" that scientific evidence strongly supports evolution rather than Genesis. On the other hand, the claim "that the Virgin birth of Christ was impossible has no evidence to support it, but is a sheer case of the flagrant scientific fault of exterpolating [*sic*] beyond the limits of one's experiment." The Incarnation "and all the phenomena attending the life of Christ" are outside "the field in which deductions from natural law may safely be made." Either one believes it, or one does not, "but we cannot reject it on the ground of inco[n]sistency, because there is nothing for it to be consistent with." Bliss donated ten dollars, hoping that "such efforts as yours can lead men to discriminate as to the value of scientific evidence in relation to the Bible," thus freeing them "from the necessity of entire disbelief, because they cannot emulate the whale and swallow Jonah."[11]

## Modernist Responses to Science

Although Mathews probably appreciated Bliss's contribution, things looked almost entirely different to him. In his autobiography, Mathews reflected on his own religious journey from old faith to new, stressing the crucial role played by science. He got down to business in the fascinating chapter on "Religion and Science," which opens with characteristic candor, "It is yet to be seen how far intelligence is consonant with religion." Taking his cue from Andrew Dickson White, he wrote, "As natural forces replace Divinity and bacteria replace devils, the area of fear within which religions have had control contracts." To be credible in the modern age, religion had to recognize what science had done—and what it had done to religion. "Laboratory science," he noted, "did something more than lead to research." What follows, when compared to what Gray had said in 1880 or what Lane said in 1923, perfectly captures the profound differences between their views of science, religion, and the relationship between them. For Mathews, science had "undermined habits of thought and substituted the tentativeness of experiment for authoritative formulas," his way of referring to doctrinal statements. Gray and Lane were just red herrings on the path to modernism. "True, there were some scientists like Asa Gray who championed Darwinian evolution while holding to the Nicene Creed," but he and others "were not representative churchmen." Those "liberal preachers" who accepted evolution, such as Henry Ward Beecher, were viewed "with suspicion" by fellow evangelicals. Mathews diagnosed the illness plaguing traditional theology in stark terms. "Scientific method had not touched religious thought. It was only when educational processes had ceased to be controlled by the study of classical literature and grew more contemporary, that orthodox theology was felt to be incompatible with intellectual integrity."[12]

Mathews's negative assessment of the relationship between theology and science in late nineteenth-century education reflected a problem that had already been noticed by zoologists Edward Drinker Cope and John Sterling Kingsley, who commented on the low state of biological education in 1887. "College presidents and trustees seem to think that while some special knowledge is necessary for teaching the classics and mathematics," they said "any one is competent to give instruction in botany and zoology." Eminence in either science, they lamented, was apparently "an undesireable [sic] feature in an instructor. The teachers of biology are mostly men without biological training, men whose ideas are those of a generation ago," such that "their whole idea of botany is 'analysis,' while zoology is but cut-and-dried classification." In their opinion, too many ministers and lawyers—often alumni who "must be taken care of"—were teaching natural history. Furthermore, "the professional studies of a clergyman, instead of fitting one for a student of nature, are a positive hindrance. The whole theological training lies in the lines of faith and reverence for authority, while science demands of its devotees, if not a sceptical spirit, one of complete independence." Unlike theology, "science has no infallible gospel wherewith to settle all disputes except that presented by the book of nature, and how difficult this is of interpretation only the original investigator knows."[13] The conflict between theology and science perceived by Cope, Kingsley, and Mathews resulted to a considerable degree from the professionalization of science, especially in the context of higher education. Of course, that was no less true for White's version of the conflict thesis (as we will see), or Thomas Henry Huxley's version in Victorian Britain.[14]

To be sure, Mathews and his fellow modernists had much in common with Gray. They believed that evolution and religious faith were not contradictory, and that the world still bore evidence of purposeful intelligence. What they did not believe was the Nicene Creed. The God worshipped by the leading American evolutionist of Darwin's day was still in a meaningful sense the "maker of heaven and earth, and of all that is, seen and unseen," yet the leading American theological educator two generations later could not accept this—supposedly because of science. Authors like Gray simply had not gone far enough. In Mathews's opinion, earlier literature on religion and science dwelled in "another world" in which "men seemed to be thinking on two levels." Their assumptions "were those of contemporary orthodoxy," and their God "was an entity working in a universe which he had created and sustains. Evolution was a way in which he worked. Men could therefore believe in him without seriously affecting their theology." As "religious romanticists," they used "scientific facts in the interest of inherited belief" and could not see what was coming. Even Henry Drummond, author of "that influential book," *Natural Law in the Spiritual World* (1880), "never really abandoned a dualistic philosophy."[15] For Mathews, the fundamental problem with theistic evolution was not that it accepted evolution, but that it tried to be too theistic in doing so.

Evolution lay at the heart of Mathews's modernist theology. Nowhere is this more apparent than in the very first AISL pamphlet on any subject, *Will Christ come again?* (1917), which was issued just as the United States entered the Great War. Mathews wrote it specifically for a "campaign against the ultra-premillenialists [sic]," substantially but not entirely the same group of conservative Protestants who would start calling themselves "fundamentalists" three years later.[16] Premillennialists assumed that the world is locked in a downward spiral of moral decline, only to be saved supernaturally in the near future, when Christ returns to inaugurate a 1,000-year reign of peace. Postmillennialists, on the other hand, believed that Christians themselves can infuse the world with moral values, eventually bringing about a utopian society, symbolically seen as the millennial kingdom. Chicago theologians clearly agreed with the latter position, whether or not they used the term "postmillennialist," and they supported President Woodrow Wilson's view that the war in Europe might help ensure democracy and prevent future conflagrations.

Although Mathews's famous pamphlet was mostly about eschatology, it also reveals the central role of evolution in his approach to religion and science. Premillennialism is "a danger," because it is both "without moral emphasis" and "contrary to the recognizable facts of history and nature." Furthermore, *"premillenarians deny that Christianity is consistent with the findings of modern science particularly as regards evolution."* These denials "show that the writers know nothing about evolution or the world of science. In many cases their statements about science are untrue." An anti-scientific attitude was inseparable from this. "Such an attack upon modern science is demanded by the central principle of premillenarianism. No man can hold the premillenarian view whose mind has been really affected by the modern scientific methods and discoveries. One or the other has to be abandoned." If religion is to survive, it must recognize that "there are catastrophes in nature, but they are subject to the laws of nature and history." The "premillenarian views" were developed by "non-Christian Jews who knew nothing about the discoveries of modern science," and they can be "introduced into our own times only . . . by an attack upon modern science and an insistence that we accept the scientific conceptions of the Jewish writers." This would force people to "choose between that [premillenarian] Christianity and science," thus "injuring religious faith." Ultimately, the premillenarian position would "separate Christianity from *the growing knowledge of the universe and society given by science.*" The modernist approach, however, "sees in evolution, not a mechanical, impersonal force which replaces God, but the way in which the Spirit operates." Instead of "a trumpet calling people from a cavern under the earth and a flat earth from which the church is to be lifted into the sky during a period of tribulation, it sees a universe filled with a God of law and love." Rather than repudiating "the indubitable facts of human history and the universe, it sees the divine process working toward moral ends." Discarding eschatological diagrams and mistaken claims of fulfilled prophecies, "it undertakes to show that Christian truths are sufficiently reasonable to be worthy of acceptance by men and women who are in heartiest sympathy with our modern world." The modern Christian, Mathews urged, should look for the prophecies of the Old Testament to be fulfilled in "the discovery of God and his laws in social evolution."[17]

Reprinted at least five times within a few months, Mathews's pamphlet was very widely read and was followed the next year by a full-length book with a similar message, *The Millennial Hope* (1918), by Mathews's colleague Shirley Jackson Case. With the war now consuming American lives, Case even dragged in wholly unsubstantiated charges that German money was underwriting the spread of premillennialist ideas in America.[18] Needless to say, Mathews and Case provoked heated replies. Several premillennialist leaders responded with pamphlets of their own, including one by New York attorney Philip Mauro, who later wrote *Evolution at the Bar* (1922). Two of many angry letters to Mathews show how conservative Protestants responded to the first AISL pamphlet in ways that connect with their responses to the later "Science and Religion" series. A Canadian-born Presbyterian minister from Syracuse, John Murdoch MacInnis, opened by stating that he had been reading Mathews's writings for "nearly twenty years" and had "received considerable help from them." His response to the pamphlet on the second coming, however, was one of "keen disappointment." He accused Mathews of taking advantage of "fanatical and ridiculous" statements made by certain premillennialists and chided him for failing to engage the more thoughtful scholars who held that position. Then he attacked Mathews's claim that the person who has truly been informed by modern science and its methods could not hold the premillennialist view. "If by the 'modern scientific methods' you mean the study of ancient facts in the light of a modern theory and torturing them to fit them into the theory," MacInnis wrote, "then I grant that it would be hard for such a mind to accept the premillenarian point of view." But it was "the scientific study of the New Testament from the historical point of view" that had convinced MacInnis himself "to accept the premillenarian position, and the thing about your pamphlet that hurts me is not any fact that you present but the unchristian and intolerant spirit of it. Indeed I regard it as the rankest kind of propagandist literature." Reading it had convinced MacInnis "that Dr. [William Edwin] Orchard is right when he says, 'We have arrayed against us not so much the philosophers and the scientists but the second hand opinion of the very popular and imperfectly understood science of a generation ago.'" MacInnis turned the sensitive issue of German connections back on the modernists. "If you men in this country who stand for that rank German rationalistic interpretation of the Bible are disposed to invite a controversy by such pamphlets as your little leaflet," he said, "I am afraid that you are going to see a result that you have not counted on." The "great danger that threatens the church in this country to-day" was not "the extremes of premillenarianism," but "the blatant rationalism of Germany which is being retailed by a coterie of men who in an arrogant spirit are constantly arrogating to themselves the exclusive right to be recognized as modern scholars." MacInnis's comments are particularly interesting, given that he later taught Bible and philosophy at the Bible Institute of Los Angeles, where he succeeded Reuben A. Torrey as Dean in 1925, only to be forced out in a firestorm of criticism after the Institute published his book, *Peter the Fisherman Philosopher: A Study in Higher Fundamentalism* (1927). Having originally endorsed the book enthusiastically, the board of trustees reversed itself, condemned the book, banned it from the library, and destroyed remaining copies—only to see the highly respected G. Campbell Morgan and half the faculty resign and Harper & Brothers republish the book in 1930. Needless to say, MacInnis's vision of a more irenic, less militant, "higher" style of fundamentalism did not succeed at the Institute.[19]

Another Canadian-born evangelical, Baptist evangelist George Wilson McPherson of the Old Tent Evangel Committee in Manhattan, whose book *The Modern Conflict over the Bible* (1918) had just been published, also expressed disgust with the Chicago theologians. Recognizing that he "belongs to that ignorant class to whom you refer in your pamphlet," the writer did not "presume the right to submit even a complimentary criticism, for how can one who is unlearned criticise so learned, scriptural, [and] profound a scholar as Dr. Mathews." Indeed, he wrote sarcastically, "so noted a disciple of the German, atheistic, and pantheistic philosophies is beyond the class to which the writer belongs." Blaming that type of thinking for "so fine a product as the present war, and the present antagonistic sprit to Evangelical Christianity as manifest in the Churches everywhere," he called it "The German Apostacy" and described Mathews as a "brilliant, gifted genius of the German University." Confessing that he had once floundered "exactly where you are today," McPherson hoped that "grace and honest study of the Scriptures can do for you what it did for me. No longer am I a follower of Darwin, Spencer, and Haeckel, but of Jesus Christ, the Son of God."[20]

McPherson was a perceptive critic: Mathews and his colleagues believed that science, especially evolution, had changed almost everything. As Victor Anderson has observed, "They wanted a theology that would contribute to the advancement of learning, guiding human intelligence creatively, spiritually, and ethically through the age of positive science."[21] A perfect example is Chicago theologian Gerald Birney Smith, like Mathews and Compton a member of the highly eclectic Hyde Park Baptist Church (now Hyde Park Union Church), where devout theists such as Compton sat alongside skeptics such as historian and philosopher Edwin Arthur Burtt (who helped draft the first Humanist Manifesto).[22] Smith taught the AISL correspondence course, "Significant Movements in Modern Theology," a detailed study of changing theological attitudes in the first fifteen years of the twentieth century, focusing on theological method, the content of the Christian gospel, early Christianity, and the impact of evolution. Smith held that modern theologians "are not concerned primarily with theological *systems*." Rather, they are "concerned with the failure of traditional theology to answer satisfactorily certain pressing questions which modern men are asking. Something is wrong with the *method*." His course took the form of a commentary on twelve required books (including one of his own and one by Mathews), three in each of the four units.[23]

Smith's presentation of "The Problem of Theological Method" reveals his main objection to traditional approaches, particularly in his comments about one of the set books, *Reconstruction in Theology* (1901) by Congregationalist theologian Henry Churchill King, the president of Oberlin College who had taught mathematics, philosophy, and theology in Millikan's student days. Like many at the turn of the century, King was a liberal theologian who still upheld the classic creeds and did not believe that science made biblical miracles simply incredible. For Smith, on the other hand, "Science has cast doubt on miracles" and "has been putting a doctrine of naturalistic evolution in the place of direct divine intervention." Where King wanted "to preserve some form of appeal to authority," Smith "advocates a radical disentangling of theological inquiry from the effort to validate some superhuman authority." Only "the method of studying the facts before us," rather than "the method of appeal to authority," would put theology on a proper footing. Only that "would align theology with the great constructive forces of the modern world."[24]

Smith's final unit, "Theology and the Doctrine of Evolution," reveals a further aspect of the dynamic involving the AISL pamphlets: the Chicago theologians all but discarded divine transcendence in the name of divine immanence. Indeed, Smith chose the subject of evolution "for the very reason that we cannot here evade the ultimate question as to the actual way in which we construct our theological beliefs." At first, theologians had "generally denounced" evolution because it seemed to contradict the Bible. Then, as scientific evidence accumulated, "theologians began to recognize the facts without giving up their belief in the divine authority of scripture," mainly by appealing to the "harmony" model—the same basic approach that Asa Gray had also seen as outmoded. "Such a harmony was made possible by a frankly allegorical interpretation of the statements of the Bible, according to which the literal meaning of the text was transformed into something more in accord with the demands of scientific accuracy." Smith saw this as a "makeshift" that "could not long be satisfactory to anybody." "A more wholesome attitude," he added, "was inaugurated by two men who were thoroughly in sympathy with modern science, but who were also earnestly concerned for the welfare of religion." These were Berkeley geologist and scientific racist Joseph LeConte, author of *Evolution and Its Relation to Religious Thought* (1888), which "showed that the frank acceptance of the evolutionary position was quite compatible with a vital religious faith," and English theologian Henry Drummond, author of *Natural Law in the Spiritual World* (1880), which "took the principles of biological evolution and applied them to the elucidation of familiar Christian ideas." Smith did not explicitly mention LeConte's rejection of divine transcendence and the divinity of Jesus, but they were surely what he appreciated most.[25]

If LeConte and Drummond had enabled "a positive use of the conception of evolution in the interpretation of Christianity," the three set books for this unit went further. Lyman Abbott's *The Theology of an Evolutionist* (1897) took traditional doctrines such as sin, redemption, and immortality, and asked how these might be reinterpreted in light of evolution, ultimately showing how "we can refer all events to the providential activity" of a God who "is actually immanent in the whole world-process." Smith summarized Abbott's ideas by saying, "Sin is the survival in us of brute traits," redemption is "the emancipation of the spirit of man from the control of these lower powers," and "the end of life is to be Christlike," which entails immortality. Yet Abbott's efforts were still not sufficiently "scientific," precisely because he retained an irreducibly transcendent element—miracles, which (as Smith described it) "are possible if only we define them correctly. Miracles are simply unique events, while non-miraculous events are so often repeated as to be familiar." In this way, Smith quite accurately pointed out, "the main tenets of the Christian system seem quite compatible with the principles of evolutionary process." However, Smith was not interested in such a God, regardless of evolution. From his perspective, the problem was that Abbott's God "retains in the main the precise characteristics of the transcendent God of the pre-evolutionary theology. The essentials of the familiar 'plan of salvation' remain, modified where necessary by the logic of the evolutionary conception." Abbott's approach "is edifying and practical; but it moves easily in the literary realm of imaginative exposition rather than in the more exact pathways of accurate science."[26] Those pathways apparently bypassed entirely the fruited plain of divine transcendence.

The second set book, on the other hand, "shows the influence of more exact scientific training." This was *The Spiritual Interpretation of Nature* (1912), by Edinburgh zoologist James Young Simpson—not to be confused with his famous great-uncle, obstetrician Sir James Young Simpson. Simpson "does not attempt to withdraw the field of religion from the reach of scientific criticism." Rather, he "attempts to show how religious faith may express itself without coming into conflict with the scientific spirit." Simpson emphasized that humans are inherently religious and asked why this should be so. Because "we are conscious of exercising purposive activity, and since nature yields to our purpose," as Smith put it, "we are justified in concluding that there is in the cosmic process itself a quality which is teleological. In other words, we are justified in believing in a divine purpose which directs the cosmic process." While Smith endorsed this as a legitimate inductive inference, he took exception to Simpson's effort to "relate certain inherited Christian doctrines to the scientific interpretation of facts." First, he did not think that "the word 'creation,' which inevitably suggests a definite beginning in time, is an appropriate word to use in connection with the idea of the immanent direction of a never-ceasing process." Second, Smith objected to Simpson's acceptance of various New Testament miracles, in which he "quite fails to examine the evidence with the thoroughness which is demanded by scientific exactness," assuming instead "an attitude of positive credence which he would not assume in the case of wonders recorded in pagan literature." Smith thought Simpson, like Abbott, was simply "bringing his ready-made conclusions, and finding a way to justify them." Thus, despite Simpson's "admirable . . . use of the scientific spirit in the examination of religious problems," he "betrays . . . the pressure of the older theological ideal of 'preserving the faith once delivered.'"[27]

Only the third set book, Francis Howe Johnson's *God in Evolution* (1911), "attempts to be consistently empirical," by which Smith clearly meant that Johnson simply rejected biblical authority and the acceptance of miracles that came with it. Unlike Abbott and Simpson, Johnson "does not try to fit the picture of a transcendent God into the framework of an immanent process." Consistent with evolution, Johnson's God "is not the omnipotent Absolute familiar to us in the treatises on theology," but "definitely limited in his activity by certain circumstances which we must recognize." Like us, God also "has to make his way against obstacles." Consequently, we must work together with

God "for spiritual ends. In fact, the essence of religion may be put in the Pauline formula, 'Work out your own salvation; for it is God which worketh in you.'" Our doctrine of God must relate to our experience. If we need "to abandon some honored doctrine, like that of the omnipotence of God," we should realize that in theology as in science "the only reason for abandoning any theory is because we have found a better means of interpreting the facts." This new method does not destroy faith, Smith believed, rather "it makes possible the development of a faith suited to the precise problems of our modern life," and is therefore "perhaps more important than anything else for this generation of Christians."[28]

By offering a course giving significant attention to theology and science, Chicago was nearly unique among American seminaries at that time. They also offered a course on "The Psychology of Religion" taught by philosopher Edward Scribner Ames, the author of a book on the subject.[29] Seminary curricula were surveyed in the early 1920s by The Institute of Social and Religious Research, an arm of the Rockefeller Foundation charged with studying "the entire problem of ministerial training, not only in all the seminaries in the United States and Canada engaged in training white Protestant ministers, but in the Bible and Religious Training Schools."[30] Among other findings, the report included full curricular information from seven institutions representing specific denominations, for 1870, 1895, and 1921 or 1922. Only one, Oberlin Graduate School of Theology (Congregationalist), offered a course whose title directly refers to natural science—an elective in the 1895 catalog on "Harmony of science and revelation," presumably taught by George Frederick Wright, who at that point was Professor of the Harmony of Science and Revelation. Garrett Biblical Institute (Methodist) and Union Theological Seminary in New York (Presbyterian in 1870, but nondenominational by 1893) each had a course on "Natural Theology" in 1870, but not in 1895; Garrett offered "Archaeology" in 1895. Oberlin, Rochester Theological Seminary (Baptist), and Union offered courses on "Psychology of Religion," but not in the nineteenth century. Several schools taught apologetics and/or philosophy of religion, and evolution was perhaps included in such courses. The report includes less detailed information about curricula and other aspects at many other theological schools, but nothing is mentioned about courses on religion and science.

By the 1920s, courses related to psychology (including the psychology of religion) were offered at about one-sixth of the 162 schools whose materials were noted in an appendix, including Chicago; most met requirements for degrees in religious education. Bonebrake Theological Seminary (now United Theological Seminary), Drew Theological Seminary (now Theological School), Hartford Theological Seminary (now Hartford International University for Religion and Peace), and Vanderbilt Divinity School taught "Genetics" or "Genetic Psychology," probably reflecting perceived connections between religion, eugenics, and determinism. However, only two seminaries advertised courses about religion and science more generally. Students at Rochester could take "Science and Religion," while students at Chicago could take "Christian Theology in Relation to Modern Science," apparently a version of Smith's course. The niche occupied by Chicago is especially important. They had made a "scientific" approach to theology central to its research and teaching since the late nineteenth century, they had the second largest enrollment among Protestant seminaries in North America (Southwestern Baptist Seminary in Fort Worth was larger) by the 1920s, and no other school awarded more doctorates in theology, making Chicago the most influential institution in that regard.[31]

The significance of this is perhaps best seen in the lengthy section of the report devoted to problems identified by the study, where the question was raised, "Are the seminaries meeting their responsibility . . . [i]n interpreting science?" The illuminating answer reads as if it had been written by someone like Mathews: "Some of the seminaries are virtually untouched by the progress and method of science." They assume "that science and religion occupy mutually exclusive fields, if they are not indeed in actual conflict." Other seminaries take for granted "a scientific view of the world," but do little "to enlarge the conceptions of theology so as to include the remarkable advance of scientific knowledge and to arrive at a unified world." Nevertheless, a growing number of seminaries "are formally committed to the scientific procedure," with faculty who "know and speak the language of science, use its methods in the classroom and the laboratory, and undertake to interpret the life of the individual, the community and the world in terms of principles found in harmony with scientific theories and discoveries." In that way, "science becomes an ally of religion, deepening and clarifying insight and confirming faith." The author of the report wondered whether negative attitudes toward science might result partly from "the traditional requirement of the classical subjects for seminary admission, and whether the pre-seminary requirements in the future may not include the sciences." This led them to ask, "Is there an essential reason why the student with the scientific bent may not look forward to the ministry? Or is the reason that has kept him out merely accidental?"[32] Perhaps indeed Mathews wrote this, since one of the seven officers of the Institute was Ernest DeWitt Burton, president of the University of Chicago and former AISL chair.

In general, AISL literature placed them at one end of Protestant theological opinion, with various fundamentalist organizations occupying the other end. By the early 1920s, they found themselves in a pamphlet war with the fundamentalists, and the absence of intermediate positions was noticed by at least one discerning correspondent, Pittsburgh banker William E. Lincoln. He and other unidentified "wealthy men" were solicited for donations a few months before the "Science and Religion" series was launched, to pay for perhaps as many as half a million pamphlets countering "an amazing amount of tras[h]y and even harmful so called religious literature being circulated in enormous quantities by well meaning, but mistaken people." Ministers needed "literature as convenient in form and as easily secured with which not only to offset the influence of the injurious sort, but also to enable them to reach more people and to help them see how the gospel of Christ fits into our modern life." A few "penny leaflets" and pamphlets came with the solicitation, including a leaflet by Mathews, "Christ's Interpretation of His Second Coming," and another by Smith, "A Christian Test of Christianity," neither of which is cataloged today. "I have read [your] literature with some care," Lincoln told Burton. "I hate to knock yet some of it seems about as far off one side as the 'Moody' school [Moody Bible Institute] is on the other side." On the one hand, he agreed that certain widely circulated religious tracts were harmful, including "ideas like S[unday]. S[chool]. Times is now printing." On the other hand, he found Mathews's leaflet unconvincing, "because of truths apparently ignored which should be considered—Were the two angels at X[Christ]'s ascension ignoramuses as to eschatology?" Smith's leaflet was "very crooked reasoning & conclusions or at least general impression untrue," and he accused Smith of setting up "a man of straw" to argue against.[33]

### The Influence of Andrew Dickson White's Conflict Thesis on the Modernists

A salient fact about Smith's course is buried in a bibliography: he enthusiastically recommended Andrew Dickson White's *A History of the Warfare of Science with Theology in Christendom* (1896) as "a most readable and striking account of the gradual substitution of the empirical method for the method of conformity to authorized doctrine in various realms of thought. It reflects the scientific man's impatience with the traditional theological ideal."[34] Both the first president of Cornell and the first president of the American Historical Association, White began writing what became his magnum opus shortly after the Civil War, when he lectured at the Cooper Union on "The Battle-Fields of Science." The argument presented there and in a subsequent book, *The Warfare of Science* (1876), posited perpetual conflict between progressive forces of science and reason against obscurantist theology, for which he liked to reserve the pejorative adjective "dogmatic"—science was never "dogmatic," even when it was. Contrary to what his titles might suggest, the Episcopalian White was no zealot for atheism, although atheists ever since have used his work against Christianity. In his view, "religion pure and undefiled" remained vital to the modern age for moral reasons, but traditional Christian creeds and attitudes toward the Bible were more than obsolete and should be gradually discarded. Religion stripped of "this mass of unreason" would "flow on broad and clear, a blessing to humanity." He sought to enlist liberal clergy as allies, hoping that "as they cease to struggle against scientific methods and conclusions," they would "do work even nobler and more beautiful than anything they have heretofore done."[35]

Although his religious agenda drove him and profoundly shaped his ideas, White took such an antagonistic stance largely in response to political opposition. The denominational colleges, as all colleges in New York then were, fought his efforts in the New York State Senate to use the considerable resources of the Morrill Land Grant College Act (1862) to establish Cornell as the first nonsectarian university in New York—which would also be (as he saw it) a secular haven for science. In his view, the very notion of Christian education was a sectarian oxymoron, and his lectures and writings were his way of getting revenge. As he told his friend Ezra Cornell, he wanted to give them "a lesson which they will remember."[36]

English-born chemist John William Draper of the University of New-York advanced an overlapping view in *History of the Conflict Between Religion and Science* (1874), but White's work was much more comprehensive and probably more influential on the modernists.[37] White constructed an elaborate, seemingly authoritative, but highly misleading account of how Christian theology and science engaged inevitably in mortal combat, with science emerging triumphant. His big book is not only replete with untenable interpretations of historically significant events and texts, but it also contains numerous false "facts" of great importance. Although historians now see his scholarship as wholly unreliable and ideologically motivated, the modernists and many others accepted its contents unquestioningly, embraced its attitude uncritically, and often mimicked his over-the-top rhetoric.[38]

White influenced the pamphlet authors and their contemporaries in two principal ways, both visible in Mathews's description of the Chicago attitude. First, they assumed that the history of Christianity and science was irredeemably marred by "dogmatism," which (like White) they used as a synonym for obscurantism on the part of orthodox theologians, always to be contrasted with an open-minded, unfettered search for truth on the part of scientists. White sought to show how foolish it was for theologians ever to criticize scientific claims, a strategy he carried out even by quoting "statements" that were never uttered, such as an imagined anti-Copernican "quotation" from John Calvin that was repeated by Bertrand Russell, Thomas S. Kuhn, and countless others, before being debunked by Edward E. Rosen in 1960.[39] More often, White simply quoted authors wildly out of context, leaving his readers—including many modernist authors who relied on him—innocently to draw their own false conclusions. For example, near the end of Conklin's pamphlet, John Wesley is erroneously said to have thought that the Copernican theory "tended toward infidelity." A lapsed Methodist, Conklin lifted those three words from White, who had presented them unambiguously as an anti-Copernican quotation from Wesley, despite the fact that elsewhere in the book Wesley is seen as a Copernican. The words come from a sermon, in which Wesley spoke against the idea of multiple inhabited worlds, not the motion of the earth around the sun. Furthermore, Wesley supported his rejection of multiple worlds partly by appealing to Christiaan Huygens's conclusion that the Moon could not support life. In context, Wesley had made a good-faith effort to understand a scientific theory and its perceived implications and to argue against it on scientific grounds, but White provided not even the faintest suggestion of this—exhibiting the same obscurantism with which he implicitly charged Wesley.[40] The editorial footnotes to the pamphlets by Conklin and Fosdick identify further examples of misinformation taken from White.

A second influence of White on the modernists follows directly from the first. With White, they assumed there has never been a constructive conversation involving theology and science, and there will never be one in the future—at least not until theology has been entirely purged of "dogmatic" content, that is, of its claim to be able to state some truths about ultimate reality from higher authority. White's assumption about the ubiquity of conflict became the prevailing scholarly attitude among modernist theologians and many others—including historians of science, the professional group that ultimately debunked the Conflict thesis late in the twentieth century. The prime example is George Sarton, the positivistic Belgian scholar who founded the History of Science Society and its two journals, *Isis* and *Osiris*. He sought White's financial support when he launched *Isis*, and in the very first issue he cited White as the source of the attitude his journals would take toward the history of science and religion. Usually there has been "a real warfare," but "not a warfare between science and religion—there can be no warfare between them—but between science and theology." Sarton's student, the great sociologist Robert K. Merton, wrote a brilliant dissertation partly arguing for the formative influence of the Puritan "ethos" on the reception of science in seventeenth-century England. Sarton offered to publish it in *Osiris*, if Merton shortened that section—he did not, but Sarton published it anyway. As Merton recalled, obviously thinking of Sarton among others, scholars found his celebrated thesis "an improbable, not to say, absurd relation between religion and science," precisely because Draper and White had convinced them "that the prime historical relation between religion and science is bound to be one of conflict," such that "a state of war between the two was constrained to be continuous and inevitable."[41]

White's work found such positive reception among the modernists precisely because they recognized a fellow traveler whom they gleefully took on board. Just like White, they wanted to separate Christian morality from what White called "outworn creeds and noxious dogmas," so that it might remain relevant to the modern world. As White wrote in his memorable, inimitable preface, "I simply try to aid in letting the light of historical truth into that decaying mass of outworn thought which attaches the modern world to mediaeval conceptions of Christianity, and which still lingers among us—a most serious barrier to religion and morals, and a menace to the whole normal evolution of society." In the end, "Science, though it has evidently conquered Dogmatic Theology based on biblical texts and ancient modes of thought, will go hand in hand with Religion." This notion of religion as ethical attitudes wholly divorced from traditional doctrines matched the one found in Millikan's "Statement" that several dozen leading scientists and clergy endorsed. Borrowing from the Bible and Matthew Arnold (who likewise separated religion from dogma), White added, "theological control will continue to diminish, Religion, as seen in the recognition of 'a Power in the universe, not ourselves, which makes for righteousness,' and in the love of God and of our neighbor, will steadily grow stronger and stronger." Seeking St. James's vision of "pure religion and undefiled," most of all he wished that "the precepts and ideals of the blessed Founder of Christianity himself, be brought to bear more and more effectively on mankind."[42] This was exactly the "new" faith with which Mathews replaced the "old" faith of his youth, and which Millikan had in mind when he wrote a chapter on "New Truth and Old" in *Evolution in Science and Religion* (1927), proclaiming that "*the supreme question for all mankind is how it can best stimulate and accelerate the application of the scientific method to all departments of human life.*"[43]

It was specifically White who motivated Fosdick and Conklin also to put new faith for old. As the leading liberal preacher of his generation, Fosdick's case is uniquely important. White's book was published in 1896, the year after Fosdick's first year at Colgate. He read it that summer, and it shattered his traditional Christian faith, especially his belief in biblical inerrancy. There he found "shocking facts about the way the assumed infallibility of the Scriptures had impeded research, deepened and prolonged obscurantism, fed the mania of persecution, and held up the progress of mankind." His turn toward modernism was a direct result of reading White, as he "rose in indignant revolt" against "the old stuff I had been taught."[44] Conklin's deconversion was less dramatic but no less encompassing. Once a devout Methodist with a Local Preacher's license, through his long academic career he slid slowly but steadily toward pantheism, at least partly because Draper and White "showed the impossibility of harmonizing many traditional doctrines of theology with the demonstrations of modern science."[45] To the extent that the modernists needed White's Conflict thesis to make their narratives work, they built their castles on historical sand.

Whether or not they were directly influenced by White, most AISL authors held nontraditional, sometimes amorphous, notions of God while at the same time unambiguously rejecting philosophical materialism and underscoring the ultimate significance of human persons. It is fitting that when the only theologian in the group, Mathews, was asked whether he believed in God, he answered evasively, "That, my friend, is a question which requires an education rather than an answer."[46] While it is unclear whether he actually believed in a personal God, he nevertheless identified God in terms of "*the personality-evolving and personally responsive elements of our cosmic environment.*"[47] Mather, who attended a Sunday school class taught by Mathews, likewise spoke about "the motive power which tends to produce a fine personality in a human being."[48] His god was "a creative and regulatory power operating within the natural order," who "is immanent, permeating all of nature, unrestricted by space or time," yet "transcendent only in that His spirit transcends every human spirit, possibly the sum total of all human spirits melded together. He is not supernatural in the sense of dwelling above, apart from, or beyond nature."[49] Mather particularly liked to speak about the "administration of the universe," an uncapitalized and impersonal phrase he borrowed from another former teacher, the great geologist Thomas Chrowder Chamberlin, while stressing the importance of free persons who are more than mere machines.[50] Mather's god did no miracles, could not answer prayer by acting in nature, and offered no clear hope of immortality.[51] Compton, who prayed often, contributed to *Life After Death* (1930), and wrote an entire book on *The Freedom of Man* (1935), held a more traditional notion of God than Mathews or Mather. Yet, he refused to recite the Apostles' Creed in church because he did not believe in the Virgin Birth, the Incarnation or the Resurrection.[52] Schmucker was convinced that "materialism died" with the nineteenth century and "the great scientists of the new century are to a very large degree intense spiritualists," while in his pamphlet he announced that "the laws of nature are eternal even as God is eternal" and that God had not created them.[53] Millikan denied that materialism was "a sin of modern science," while identifying "the God of Science" minimally as "the spirit of rational order and of orderly development" and affirming that "*The most important thing in the world is a belief in the reality of moral and spiritual values.*"[54] He was also very active in a church that did not require assent to any specific creed, like Fosdick's Riverside Church. When Conklin spoke in Philadelphia about the "Religion of Science" around the time of the Scopes trial, having not attended church for many years, his talk consisted of nothing more than a litany of negatives. According to his outline (the actual text does not survive, if it ever existed on paper), there was "No Personal God, No Miracles, No Supernatural Revelation, No Personal Immortality," and "No Objective Efficacy of Prayer."[55]

Nevertheless, consistent with their emphasis on the reality of a spiritual realm within a fixed cosmic order, several AISL authors spoke without hesitation in almost traditional terms about purpose or design in the universe, even—especially—in the process of biological evolution that had produced intelligent creatures capable of free, moral actions. An obvious example is

Conklin. Although he believed that "the scientific conception of nature and of the universality of natural law" conflicted with "the usual conception of God . . . as a supernatural being . . . who created the universe out of nothing" and occasionally intervenes for various purposes, this was no warrant to "deny the existence of that which is symbolized by the word 'God.'" Rather, as in William Wordsworth's poem "Tintern Abbey," he found "God in all truth and beauty and love, in the order and constitution of the universe, in the eternal and immutable laws of nature, [and] in the mind and soul of man!" Following Harvard chemist Lawrence Henderson, a pioneering proponent of the anthropic principle, Conklin found "many remarkable fitnesses or preparations for life . . . in the lifeless world." Considering "the whole course of evolution from amoeba to man, from the simplest motor responses to the development of intelligence and reason capable of studying the universe and its origin," he was "impressed with the thought that evolution must have been guided by something other than chance. If progressive evolution is increasing the complexity of organization and increasing adaptation to the environment, it is surely no accident that organization and environment have been so correlated." Overall, "it is impossible to escape the conclusion that evolution has revealed a larger teleology than was ever dreamed of before—a teleology which takes in not only the living but also the lifeless world."[56] Compton spoke even more bluntly. "The chance of a world such as ours occurring without intelligent design becomes more and more remote as we learn of its wonders," he said at a Unitarian church in 1940. As he saw it, "our world is controlled by a supreme Intelligence, which directs evolution according to some great plan" leading to "the making of persons with free intelligence capable of glimpsing God's purpose in nature and of sharing that purpose."[57] Millikan's "Statement" saw God "revealing himself through countless ages in the development of the earth as an abode for man and in the age-long inbreathing of life into its constituent matter, culminating in man with his spiritual nature and all his Godlike powers."[58] Mathews spoke in his pamphlet of "a cosmic intelligence" that underlies the "rational and purposeful activity which in the course of evolution results in personal life."[59] Similar thoughts and attitudes are also found in other pamphlets.

Today the term "intelligent design" is often associated with Christians who accept the classical creeds and oppose evolution, not the other way around. This betokens fundamental changes in the American Protestant landscape since the *Scopes* era, but the positions represented by Bryan and Mathews have not disappeared. The founder of the Intelligent Design movement, the late Phillip E. Johnson of Berkeley Law, defended a far more sophisticated version of Bryan's old-earth creationism for identical reasons: he considered evolution scientifically unproved, he rejected theistic evolution as intellectually incoherent and religiously dangerous, he blamed evolution for many social ills, and he sought to reform science education to make it friendlier to traditional Christian beliefs. Mathews has also had descendants today, such as the late John Shelby Spong, Episcopalian Bishop of Newark, who urged Christians to abandon anything remotely resembling classical theism in the name of science.[60] At the same time, since the 1960s young-earth creationism has won the hearts and minds of tens of millions of American Christians. Ironically, the fundamentalist leaders of Bryan's day would not have embraced it—William Bell Riley, the first president of the World's Christian Fundamentals Association, once said, "There is not an intelligent fundamentalist who claims that the earth was made six thousand years ago; and the Bible never taught any such thing."[61] Proponents of young-earth creationism contest most conclusions of geology, cosmology, and evolutionary biology, deny them status as legitimate sciences, reject the ways in which most nineteenth-century Christian scientists related their findings to the Bible, and defend interpretations of Genesis that would have made Bryan blush, while sharing Bryan's acerbic attitude toward the religious and cultural dangers of evolution.[62] Meanwhile, on the opposite end of the religious spectrum, some scientists and others have emerged as aggressive, no-holds-barred spokespersons for the New Atheism. Blurring lines they do not see between science, values, and metaphysics, authors such as evolutionary biologist Jerry Coyne and cosmologist Lawrence M. Krauss proclaim that science is virtually equivalent to atheism, making belief in God all but impossible.[63] Although there is nothing new about atheism, its newly acquired high visibility in print and on the internet have made it a major player, not merely an elephant in the room, in contemporary Protestant conversations about science.

   In the middle stands another position that existed in the 1920s but lacked visibility: Protestant thinkers who hold views like those of Asa Gray. Where one would be hard pressed to identify top-drawer scientists one hundred years ago who accepted both evolution and the ecumenical creeds, it is much easier to do so now. Three examples suffice for many others. Undoubtedly the most famous is geneticist Francis Collins, who succeeded Nobel laureate James Watson as director of the Human Genome Project in 1993 and was appointed by President Barack Obama in 2009 to head the National Institutes of Health.

Converted from atheism to evangelical Christianity while studying medicine, in 2007 Collins launched BioLogos, a nonprofit organization dedicated to promoting acceptance of evolution among evangelicals. Although they are much less known than Collins outside scientific circles, physicists Charles Townes of the University of California (who died at age ninety-nine in 2015) and William D. Phillips of the National Institute of Standards and Technology both received Nobel prizes and have identified as traditional Christians.[64] As a historian, not a prophet, I cannot say whether the reemergence of views like Gray's will ultimately persuade many conservative Protestants to accept evolution, but the contemporary conversation is certainly subtler, broader, and deeper than in Bryan's time, when many American Protestants were confronted with the stark choice of rejecting modern science or ancient faith.

PART TWO

# THE AISL
# "SCIENCE AND RELIGION"
# PAMPHLETS

# *Evolution and the Bible*

Edwin Grant Conklin (1863–1952)

> What kind of a religion then does science leave to man? Not one of supernaturalism, mythology, and magic but one of nature, order, humanism. The religion of science must be based on the solid ground of reality, but it must reach up into the atmosphere of high ideals.
>
> —*Edwin Grant Conklin,* Man: Real and Ideal

## Introduction

The first three pamphlets appeared simultaneously in September 1922.[1] Those by Conklin and Fosdick were based on their essays for the *New York Times* in March 1922.[2] Because the *Times* published Conklin's essay one week earlier, it comes first in this volume. The AISL felt that the original version, "Bryan and Evolution," was too narrowly aimed at Bryan, so Conklin and Mathews gave a new title to a version that they edited to make it appear less of an attack on Bryan alone.[3] Conklin had already reused parts of the newspaper article in the preface of a new edition of his book, *The Direction of Human Evolution* (1922), augmented by some new material that in turn went into the pamphlet.[4] The AISL printed 30,000 copies each of the first three pamphlets in August (the date on the front cover is September), followed by two more printings of the same quantity in September and October. At his request, in late October Conklin received 500 copies of his own pamphlet and 100 each of the other two from a subsequent printing.[5] His pamphlet must have been popular, because another 27,000 copies were printed in September 1931, making it (with the first and only edition of Mather's pamphlet) one of the last two impressions in the "Science and Religion" series, as well as the first.

A leading public intellectual who had taught biology at Princeton University since 1908, Conklin was surely an obvious choice for this topic. His interest in the religious implications of evolution dated back to his student days at

Ohio Wesleyan College in the 1880s. Encountering faculty members who refused even to discuss the Darwinian hypothesis, which they regarded as inherently atheistic, he had chafed at their obscurantism. "This unreasoning attitude of anti-evolutionists, rather than the scientific evidences in its favor, first made me one of its advocates." Over the years he ran into the same attitude repeatedly, leading him "to devote as much time as I could spare to writing and speaking in explanation and defense of evolution." Conklin guessed that he had delivered "a thousand public lectures on this subject," making it "one of the 'causes' to which I pledged my best effort."[6] There was nothing unique about the opportunity to reply to Bryan in the *New York Times.*

Methodism provided the primary context for Conklin's view of religious opposition to evolution. Most of the faculty under whom he studied at Ohio Wesleyan were ordained Methodist ministers, and during his final year Conklin obtained a Local Preacher's license. When the new college president, a theistic evolutionist, offered him a job upon completing his doctorate in 1891, Conklin accepted, "but on condition that I should be permitted to teach evolution," to which the president "readily agreed," but the situation "naturally created some disquiet" among older faculty. Three years later, Conklin placed the same condition on another school founded by Methodists, Northwestern University, when he became their first zoology professor. He faced no opposition on campus, but his teaching led "certain clerical members of the Rock River Methodist Conference" to attack the university. Soon he moved again, this time to the nonsectarian University of Pennsylvania, where he stayed until Woodrow Wilson recruited him for Princeton. (Schmucker was already doing postdoctoral work in biology at Penn, and he became Conklin's teaching assistant, leading to a lifelong friendship.[7]) Not long after arriving in Philadelphia, Conklin spoke to the Philadelphia Methodist Preachers Meeting on the topic "Science and Religion," only to find "that some of their members were as violently opposed to [evolution] as was William Jennings Bryan some twenty-five years later."[8] Recounting this episode to Fosdick, Conklin said, "I was told that there was no room in the Methodist Church for any such doctrines as I proposed, and that I had better keep out of their church. Since that time I have followed that advice, and have been waiting for the light of reason to dawn on their minds."[9] Reflecting on it again in his spiritual autobiography, he added, "Under these circumstances I did not transfer my church membership from Evanston to Philadelphia, but I still continued to speak occasionally on science and religion in churches and church congresses."[10]

Perhaps the most important such occasion was the address on "Evolution and Revelation" he gave at the first national Methodist Episcopal Church Congress, held in Pittsburgh in November 1897. According to a newspaper account, "the paper came near producing a sensation" among the several hundred attendees. During the address, "the attention of the audience was divided and the general discussion which followed was spicy."[11] Arguing eloquently for a robust theology of divine immanence, at one point Conklin echoed the final paragraph of Darwin's *Origin of Species*: "There is grandeur in this view of the Creator and of his relation to the world. Consider the eternal patience, wisdom, goodness, lawfulness, which has through countless ages wrought out our present world; consider the continual process of creation, the continual presence of the Creator in all natural forces—and then do but contrast with this the idea of an artificer-Creator, a machine-universe with a God who stands outside!" Conklin had not yet completely rejected a traditional understanding of God, for he interpreted Genesis as telling "all peoples and ages that back of the creature stands the Creator, back of the natural the supernatural,"[12] using without hesitation a word (supernatural) that he later rejected emphatically. (When he reused parts of this address in 1921, the phrase following "Creator" was taken out.[13]) He also expressed belief in personal immortality and explicitly endorsed the religious position of Asa Gray, who had unhesitatingly affirmed the Nicene Creed. In short, at that point Conklin was still an orthodox Christian, despite the controversy he provoked by rejecting special creation in favor of evolution. However, the negative reception he got from fellow Methodists could not have encouraged him to maintain such a position.

Over the next two decades he clearly changed his mind, the culmination of a process that had begun several years earlier during his graduate studies at Johns Hopkins, where he was deeply influenced by the naturalistic credo of his mentor, biologist William Keith Brooks. As Conklin put it, Brooks held that "The term supernatural is due to a misconception of nature; nature is everything that is." At the same time, Brooks viewed the natural world in terms of an immanent Aristotelian teleology, such that "the essence of a living thing is not protoplasm, but purpose."[14]

Conklin's own religious position came increasingly to resemble that of Brooks. "With the progress of science," he noted near the end of his life, "the area of the supernatural and miraculous has gradually grown smaller," yet "supernatural agencies or occurrences constitute the very foundations of many religions." Conklin saw his own spiritual journey as one that "orthodox friends" might interpret as "descending steps," leading him further from the traditional Methodist faith of his youth, describing a process identical to that which Bryan had seen as a necessary result of accepting evolution (see figure 1.2). "My gradual loss of faith in many orthodox beliefs," he recalled, "came inevitably with increasing knowledge of nature and growth of a critical sense."[15] When Conklin shared the contents of his faith with Philadelphians in the mid-1920s, just a few years after his pamphlet was published, he called his lecture, "The Religion of Science," a fitting description of a faith defined by a litany of unbelief: "No personal God, No Miracles, No Supernatural Revelation, No Personal Immorality, . . . No objective Efficacy of Prayer."[16] Clearly, he no longer accepted traditional theism, let alone orthodox Christianity, regarding himself a pantheist.[17] Yet, his universe was still purposeful. With Brooks—but also like Mathews, Fosdick, Compton, and many other modernists—Conklin believed that the universe and life, especially human life, cannot be understood in purely mechanistic terms. As he told classicist Theodore A. Miller, author of *The Mind Behind the Universe* (1928), "The modern attitude that there is no God and no purpose in the Universe has led to a philosophy of pleasure and ultimately of despair and sometimes suicide. I believe that the best antidote for such a philosophy is the conviction that there is mind and purpose behind the universe and every individual life."[18] As he wrote in *The Direction of Human Evolution*, "a purely scientific explanation must be mechanistic, but there is no mechanical explanation for the ultimate mechanism of the universe; mechanism cannot explain itself." He did not hesitate to describe evolution as "progressive," since it "must have been guided by something other than chance."[19] Ultimately, like Millikan and Mather he came to accept a God equivalent to the order in nature. He confessed to an attorney from New Zealand, "it seems to me incredible that the order of nature and the general results of evolution could have been the mere result of chance," and "if anything in the universe is to be called God, it seems to me that it is this underlying principle of law and order."[20]

Significantly, Conklin identified John William Draper's *History of the Conflict between Religion and Science* (1874) and Andrew Dickson White's *A History of the Warfare of Science with Theology in Christendom* (1896) as formative influences on his position. Although historians eventually discredited both works as factually unreliable and highly misleading in tone, for much of the last century they were seen as the best authorities on the history of science and religion. Conklin relied on information from White at certain points in his pamphlet (see the editorial notes) and heartily embraced his confrontational attitude. As he saw it, Draper and White "showed the impossibility of harmonizing many traditional doctrines of theology with the demonstrations of modern science."[21] Indeed, by the early 1920s Conklin had come to think that the only solution to "the conflict between science and theology" lay "in the doctrine of the divine immanence in all natural phenomena."[22] When he described Bryan's beliefs as "medieval" in the final paragraph of his pamphlet, he was using that word pejoratively, just as White had done. Therein, Conklin simply took a standard modernist route, as Fosdick also did in his first pamphlet.

---

Annotated Text

## *Evolution and the Bible*

By Edwin Grant Conklin

I.

/3/ The past few years have witnessed a curious recrudescence of the old theological fight of fifty years ago against evolution. This movement is partly due to the increased emotionalism let loose by the war and partly to the fact that uncertainty among scientists as to the causes of evolution has been interpreted by many non-scientific persons as throwing doubt upon its truth. Ten years ago, who would have thought it possible that any fundamental generalization of science would ever again be declared false because it was not supported by certain literal and narrow interpretations of Biblical texts or that attempts would ever again be made to determine by public legislation that any departure from these interpretations should be severely punished? And yet these things have come to pass. A religious organization has been formed and has gained many adherents and wide publicity, the chief purpose of which is to banish "modernism" and particularly the theory of evolution from churches and schools.[23] Bills have /4/ been introduced in certain State Legislatures forbidding the teaching of evolution, or of Darwinism, as applied to man.[24] Textbooks that teach evolution, even as an incidental part of theology or geology, have been condemned and placed on the *index prohibitus* of this new Inquisition. Scientists who have dared to teach this forbidden subject have been under fire and in some instances have lost their positions, and it is said that funds are being raised to endow and perpetuate this fight against evolution.

All this is done, we are told, to save the religious faith of the younger generation. Apparently the leaders of this movement do not realize that they, and not the evolutionists, are making it impossible for young men and women who are intellectually enlightened to remain in their denominations. Those who wield the sword of a militant faith against science should remember that it cuts both ways. It is a dangerous thing for defenders of the faith to affirm that one cannot be a Christian and an evolutionist, for students of nature who find themselves compelled by the evidences to accept the truth of evolution will be apt to conclude that they must therefore count themselves as hostile to the Church. When will we /5/ learn that the worst form of infidelity is not disbelief in certain doctrines, whether theological or scientific, but disbelief in the power and ultimate triumph of truth? If evolution is false, it cannot be saved by science; if it is true, it cannot be destroyed by theology. The advice of Gamaliel to the Sanhedrin is still good advice: "Refrain from these men and let them alone; for if this counsel or this work be of men, it will come to naught; but if it be of God, ye cannot overthrow it, lest haply ye be found to fight against God."[25]

Few opponents of evolution at the present time have either the technical training or even the desire to weigh critically the evidences for or against its truth. Properly to appreciate these evidences requires some first-hand knowledge of morphology, physiology, embryology, ecology, paleontology and genetics. In biology as well as in other sciences it is necessary to see and handle actual materials and to observe for oneself natural processes in order to appreciate their significance; this is the reason why laboratory work plays so large a part at present in the teaching of the sciences. The advice which Huxley gave to the "paper philosophers" of his day is especially applicable to these /6/ opponents of evolution: "Get a little first-hand knowledge of biology."[26] *These anti-evolutionists not only lack such first-hand knowledge but they often have no desire to get it even second-hand*; I once asked a man who was denouncing Darwin if he had ever read his books and he replied, "I wouldn't touch them with a ten-foot pole." Neither facts, evidences, nor sweet reasonableness can penetrate such an armor.

II.

The whole scientific world long since was convinced of the truth of evolution and every year which has passed since the publication of "The Origin of Species" in 1859 has added to the mountain of evidence which has been piled up in its favor. It is fortunately not necessary here to review the evidences of evolution, for these may be found in many elementary textbooks on biology. The evidences are so numerous and come from so many sources that no intelligent man can study them at first hand and not be impressed with their importance. As a consequence *there is probably not a single biological investigator in the world today who is not convinced of the truth of evolution*. The fact that these evidences /7/ accumulate year after year, often coming from fields which Darwin and his contemporaries never dreamed of, is still more convincing. I once heard Lord Kelvin, the great physicist, say that any hypothesis or theory if true should find new support continually as knowledge advances. This is just what happened in the case of evolution.

These new opponents of evolution make much of the idea that evolution is only an hypothesis, or as they prefer to call it, a "guess." But unless they use the word "guess" in the Yankee sense of practical certainty, this is an erroneous and misleading statement. Evolution is a guess only in the same sense as the doctrine of universal gravitation, or any other great generalization of science is a guess. But can one honestly call that doctrine "a guess" which is supported by all the evidence available, which continually receives additional support from new discoveries and which is not contradicted by any scientific evidence?

It is true that we do not know as much as we should like about the causes of evolution (though we know a good deal more than its opponents assume), but the same may be said with regard to the causes of gravitation, light, electricity, chemical affinity, /8/ life or any other natural phenomenon. The problem of cause is never finally solved by science, for no sooner is one cause discovered than it gives rise to questions concerning the cause of this cause. Strange as it may seem, it is only the cause of supernatural phenomena that are supposed to be fully known! But of course this is due merely to the fact that no attempt is made to analyze such phenomena or causes.

Uncertainty among scientists as to the causes of evolution has been interpreted by many non-scientific persons as throwing doubt upon its truth. It is plain that these causes are complex and that they have not yet been fully discovered; it is even probable that some of the proposed explanations are erroneous and will have to be abandoned but it is not fair or honest to quote the doubts of scientists regarding the causes of evolution as if they constituted an abandonment of the theory itself, especially when the same scientists in the same connection affirm that no informed person can doubt the fact of evolution. Thus Professor Bateson of England in his address before the American Association for the Advancement of Science at Toronto expresses his doubts as to the causes and methods /9/ of the origin of species, but goes on to say:

> I have put before you very frankly the considerations which have made us agnostic as to the actual mode and processes of evolution. When such confessions are made, the enemies of science see their chance. If we cannot declare here and now how species arose, they will obligingly offer us the solutions with which obscurantism is satisfied. Let us then proclaim in precise and unmistakable language that our faith in evolution is unshaken. Every available line of argument converges on this inevitable conclusion. The obscurantist has nothing to suggest which is worth a moment's attention. The difficulties which weigh upon the professional biologist need not trouble the layman. *Our doubts are not as to the reality of evolution but as to the origin of species, a technical, almost domestic, problem.* Any day that mystery may be solved.[27]

The minor stages in evolution, known as mutations and elementary species, have been repeatedly observed in plants, animals and man.

DeVries, Morgan and many others have demonstrated that sudden and very great changes or mutations sometimes occur, that these mutations may be combined to form races or elementary species, and it is probable that the characteristics of these elementary species are combined to form Linnaean species.[28] Among our domestic animals /10/ and cultivated plants such changes have been wrought as amount to specific differences. Darwin says that any naturalist, if he should find our races of domestic pigeons wild in nature, would classify them in not less than twenty species and three different genera. A similar statement could be made regarding fowls and dogs as well as many fruits, grains, and vegetables.[29] In short, *evolution has occurred under domestication.*

Those who urge as an objection to evolution that "it is only a theory" neglect to say that their own views can be dignified by no higher title. *As between evolution and special supernatural creation we have to choose between two theories or hypotheses* and it is merely a question of evidence as to which is the more probable. All the evidence available supports the theory of evolution, it continually receives fresh support from new discoveries, it is not contradicted by any scientific evidence. Can the supporters of the theory of special creation say as much?

### III.

Anti-evolutionists apparently are ready to concede the evolution of rocks and plants and possibly animals, but draw the line at the evolution of /11/ man. When one of their representatives says that there are no evidences of the evolution of man, that "neither Darwin nor his supporters have been able to find a fact in the universe to support their hypotheses," it is hard to understand what he means.[30] Darwin's works are filled with facts in support of evolution, they are composed of little except such facts, and multitudes of similar facts have been accumulated since Darwin's day. Apparently the anti-evolutionist demands to see a monkey or an ass transformed into a man, tho he must be familiar enough with the reverse process. The Hotspurs who demand that evolution be reenacted "while they wait" should emulate the example of Josh Billings who said that he had heard that a toad would live 400 years; he was going to catch one and see for himself![31] The evidences for the major transformations in the evolution of man are not personal demonstrations since they do not fall within the lifetime of a single individual, nor indeed within the era of recorded history, but *they are the same sort of evidences as those for mountain-building, stream erosion, glacial action or any other change involving long periods of time.*

A common misunderstanding is /12/ that man is descended from some existing species of anthropoid ape and the latter from some existing species of monkey and so on back to certain existing species of lower animals. Of course this cannot be true for the whole organic world has been evolving together. *Monkeys, apes, and men have descended from some common but at present extinct ancestor.* Existing apes and monkeys are collateral relatives of man but not his ancestors; his cousins but not his parents. Such evolution may be graphically represented by a tree in which the leaves and terminal branches represent existing individuals and species while the larger branches and trunks represent ancestral forms; one leaf is not derived from another nor one terminal branch from another but these are derived from lower-lying branches. In short there *has been evolution in divergent lines.* The human branch diverged from the anthropoid branch not less than two million years ago and since that time man has been evolving in the direction represented by existing human races, while the apes have been evolving in the direction represented by existing anthropoids. During all this time men and apes have been growing more unlike and conversely /13/ the farther back we go, the more we should find them converging until they meet in a common stock which should be, in general, intermediate between these two stocks.

Whenever it is said that man is descended from apes or monkeys, it must be understood that this is only a brief and crude way of saying that he comes from ape-like or monkey-like forms and not from any existing species of monkey or apes. *Present day monkeys and apes cannot become men because they long since passed the parting of the ways which led to these two different types.* It is as absurd to think that any existing species could become any other existing species as it would be to think that any existing branch of a tree could become any other existing branch. On the other hand, *the resemblances between monkeys, apes, and man are due to the persistent inheritance of certain common traits which they have derived from a common ancestor,* just as the resemblances of cousins are due to the inheritance of traits from common grandparents.

It is generally held by scientists that the existing races of men all belong to one species, *Homo sapiens,* because these races are generally fertile *inter se*. But fertility is not a safe and certain /14/ criterion of a species. Some individuals belonging to distinct species are fertile *inter se* and other individuals belonging to the same species are sterile. But if the differences between different races of men are not sufficiently great to warrant placing them in different species, they are at least great enough to constitute sub-species. *How did these differences arise if not by divergent evolution?* If all men are the sons of Adam, there must have been a good deal of divergent evolution to have produced the marked differences in the white, yellow, red, brown, and black races; between the pygmies of Africa and the giants of Patagonia; between the long-limbed African and the short-limbed Japanese; between the long-heads and the short-heads; the flat-nosed and the sharp-nosed races, etc.[32]

IV.

*Everything which speaks for the evolution of plants and animals speaks plainly for the evolution of man.* In the structure of the human body there is scarcely a bone, muscle, nerve, or any other organ that does not have its counterpart in the higher primates and especially in the anthropoid apes. Romanes, who is often mentioned as having lost and regained /15/ his religious faith, though he never lost his faith in evolution, says of these similarities between the body of man and that of the higher primates: "Here we have a fact, or rather a hundred thousand facts, that cannot be attributed to chance, and if we reject the natural explanation of hereditary descent from a common ancestor, we can only suppose that the Deity in creating man took the most scrupulous pains to make him in the image of the beasts."[33]

Not only the structure but the functions of the human body are fundamentally like those of other animals. We are born, nourished and develop, we reproduce, grow old and die, just as do other mammals.

Specific functions of every organ are the same; drugs, diseases, injuries affect man as they do animals, and all the wonderful advances of experimental medicine are founded upon this fact.

Development from a fertilized egg to birth goes through the same stages in man and other mammals even to the repeating of gill slits, kidneys, heart and blood vessels like those of fishes and amphibians. Indeed *development from the egg recapitulates some of the main stages of evolution—in it we see evolution repeated before our eyes.* The fact that certain /16/ embryonic structures do not repeat the evolutionary history does not destroy this general principle of embryonic recapitulation.

It is a curious fact that many persons who are seriously disturbed by scientific teachings as to the evolution or gradual development of the human race accept with equanimity the universal observation as to the development of the human individual—*The animal ancestry of the race should be no more disturbing to philosophical and religious beliefs than the germinal origin of the individual,* and yet the latter is a fact of universal observation which cannot be relegated to the domain of theory and which cannot be successfully denied. *If we admit the fact of the development of the entire individual from an egg, surely it matters little to our religious beliefs to admit the development or evolution of the race from some animal ancestor,* for who will maintain that a germ cell is more complex, more perfect, or more intelligent than an ape?

*If the evolution of a species is an atheistic theory,* as some persons assert, *so is the development of an individual, for natural development involves identically the same principles as does evolution.* If one concedes /17/ the fact of individual development without supernatural interference, one might as well concede the fact of organic evolution without supernatural creation, so far at least as its effects on theology are concerned. It is surprising that the "Fundamentalists" have not denied the fact of individual as well as of species development, and if they are consistent, they will demand that we return to the teachings of the "preformationists" of the eighteenth century, to the idea of endless encasement of one generation within another and hence to the special and supernatural creation of every child of Adam in the creation of Adam himself.

When that comes to pass, there will probably be a demand that the teaching of embryology shall be abolished in all schools and colleges.[34]

The discovery of *fossil remains of man have proved conclusively that other species of men, more brute-like than any existing at the present time, preceded the present species.* The Neanderthal skull, with its low forehead and flat cranium, prominent eye-brow ridges, and large, square orbits was said by many opponents of evolution to be the skull of an idiot. Since then similar skulls and whole skeletons have been found in /18/ various parts of western Europe and anthropologists generally are agreed that this Neanderthal type is so different from existing races that it must be classified as a distinct species, *Homo neanderthalensis.* Portions of the skeletons of still more primitive species of men have been found in Germany, England, and Africa, while the erect ape-man of Java belonged to a distinct genus, *Pithecanthropus.* It is a vastly important fact that the older these species are, the more ape-like they are. Likewise their handiwork, implements, and flints are coarser and cruder the earlier they occur. How can the opponents of evolution explain these facts? They cannot be truthfully denied and they demand some rational explanation.[35]

All the evidences of evolution drawn from morphology, physiology, embryology, paleontology, homology, heredity, variation, etc., speak for the evolution of man as much as for that of any other organism. If evolution is true anywhere, it is true also of man.

V.

Against all this mountain of evidence, what is there in support of the view of special creation? Only this, that evolution denies the Biblical account /19/ of the creation of man. What is that account? Here it is in a sentence: "And the Lord God formed man of the dust of the ground, and breathed into his nostrils the breath of life; and man became a living soul." Observe, ye literalists, that this does not say that God spoke man into existence, as when He said, "Let there be light; and there was light."[36] But a process is described by which man was formed or moulded from the dust, as the Egyptian and Babylonian deities are said to have moulded man from clay on a potter's wheel, and then to have breathed life into his nostrils.[37]

*Since the Scriptures describe a process in the creation of man, the opponents of the theory of evolution ought to be able to conceive of a dignified and divine way in which the Creator fashioned man,* but this they do not do. The idea that the Eternal God took mud or dust and moulded it with hands or tools into the human form is not only irreverent, it is ridiculous. How much more like the usual workings of that Power, by whom and through whom are all things, is the view of evolution that God made the first man as He has made the last, and that His creative power is manifest just as truly and /**20**/ as greatly in the origin of the last child of Adam, as in the origin of Adam himself.[38] *Is it any more degrading to hold that man was made through a long line of animal ancestry than to believe that he was made directly from the dust?* Surely the horse and the dog and the monkey belong to higher orders of existence than do the clod and the stone.

Whether we accept the teachings of evolution or the most literal interpretation of the Biblical account, we are compelled to recognize the fact that our bodily origin has been a humble one; as Sir Charles Lyell once said, "It is mud or monkey."[39] *But this lowly origin does not destroy the dignity of man; his real dignity consists not in his origin but in what he is and in what he may become.*

VI.

If only the theological opponents of evolution could learn anything from past attempts to confute science by the Bible, they would be more cautious. It was once believed universally that the earth was flat and that it was roofed over by a solid "firmament," and when scientific evidence was adduced to show that the earth was a sphere and that the "firmament" was not a solid roof, it was denounced /**21**/ as opposed to the Scriptures. Those who have visited the Columbian Library in the Cathedral of Seville will recall the Bible of Columbus with marginal notes in his own handwriting to prove that the sphericity of the earth was not opposed to the Scriptures, and a treatise written by him while in prison to pacify the Inquisition.[40] Today only Voliva and his followers at Zion City maintain that the earth is flat, and the heavens a solid dome, because this is apparently taught by the Scriptures.[41]

The central position of the earth in the universe with all heavenly bodies revolving around it was held to be as certain as holy writ. All the world knows the story of "Starry Galileo and his woes" at the hands of the Inquisition, but the Copernican theory was opposed not only by the Roman Catholic Church but also by the leaders of the Reformation.[42] Martin Luther denounced it as "the work of a fool."[43] Melanchthon declared that it was neither honest or decent to teach this pernicious doctrine and that it should be repressed by severe measures,[44] and John Wesley declared that it "tended toward infidelity."[45] Even as late as 1724 the Newtonian theory of gravity was assailed by eminent authorities as "atheistic" since "it /22/ drove God out of his universe and put a law in his place."[46]

The conflict between geology and Genesis as to the days of creation and the age of the earth lasted until the middle of the last century, and students of Dana's Geology will recall the reconciliation between the two which that great man devoutly undertook. But, by the ultra-orthodox, he and other Christian geologists were denounced as infidels and as impugners of the sacred record.[47] It took three hundred years to end this conflict, if it may be said to be wholly ended now, but certainly no intelligent person now believes that the earth was made just 4004 years B. C. and in six literal days.[48]

### VII.

And now come men in this twentieth century of enlightenment preaching a new *auto da fe*, attempting to establish an Inquisition for the trial of science at the bar of theology! They propose to prohibit the teaching of evolution in tax-supported institutions by fine and imprisonment, to repeal a law of nature by a law of Kentucky.[49] They propose to gather into the fold of their theology all existing public and private schools, colleges and universities, and to allow /23/ evolutionists and agnostics to found their own schools. In view of the fact that, with the exception of a few strictly sectarian institutions, all our colleges and universities are dedicated to "the increase and diffusion of knowledge among men," that for a generation at least they have turned away from the teaching of dogmatic theology to the cultivation of science, literature and art, that they have during this period received great benefactions for the expressed or implied purpose of carrying on this work in the spirit of freedom to seek, to find and to teach the truth as God gives men to see the truth,— in view of these considerations it may well be asked whether it would not be more fitting for such opponents of evolution to establish their own institutions for teaching their own views of science and theology, as Dowie, for example, did at Zion City, rather than to attempt to convert existing institutions to that purpose.[50]

*Scientific investigators and productive scholars in every field have long since accepted evolution in the broadest sense as an established fact.* Science now deals with the evolution of the elements, of the stars and solar system, of the earth, of life upon the earth, of various types and species of /24/ plants and animals, of the body, mind and society of man, of science, art, government, education and religion. In the light of this great generalization all sciences, and especially those which have to do with living things, have made more progress in the last half century than in all the previous centuries of human history. Even progressive theology has come to regard evolution as an ally rather than as an enemy.

In the face of all these facts, anti-evolutionists turn to their mediaeval theology. It would be amusing if it were not so pathetic and disheartening, to see these "defenders of the faith" beating their gongs and firing their giant crackers against the ramparts of science.

# *Evolution and Mr. Bryan*

Harry Emerson Fosdick (1878–1969)

> This is one world, God's world throughout, whose law-abiding regularities, whose amazing artistries, whose evolution of ever higher structures, whose creation of personality, whose endless possibilities of spiritual growth and social progress indicate that it is a spiritual system. God is here, not an occasional invader of the world but its very soul, the basis of its life, its undergirding purpose, its indwelling friend, its eternal goal. That way of conceiving God saved my faith, after supernaturalism had well nigh ruined it.
>
> —*Harry Emerson Fosdick,* Dear Mr. Brown: Letters to a Person Perplexed about Religion

## Introduction

Harry Emerson Fosdick's first pamphlet, *Evolution and Mr. Bryan*, appeared simultaneously with those by Edwin Grant Conklin and Shailer Mathews in September 1922.[1] The AISL printed 30,000 copies of each title, followed by two similar print runs in the next two months to meet a high initial demand. There were no further printings by the AISL, but the same material appeared elsewhere several times during the turbulent decade of the 1920s.[2]

The pamphlets by Fosdick and Conklin were based on editorials in the *New York Times* the previous winter. Toward the end of February 1922, the Sunday editor, Ralph H. Graves, had invited Fosdick to respond to William Jennings Bryan's op-ed, "God and Evolution." Within three weeks Fosdick produced a typescript that is identical in content to both the newspaper article and the pamphlet.[3] The decision to approach the pastor of the First Presbyterian Church in Manhattan could not have been more appropriate, for the challenge posed by modern science to Christianity lay at the heart of his own spiritual journey from traditional faith into modernism. The Presbyterian and

Methodist churches Fosdick had attended as a youth in Buffalo had confronted him with what he later described as an "incredible Bibliolatry that put religion hopelessly at odds with science." Following his freshman year at Colgate University, he questioned "everything I had been taught, and a consequent spiritual upheaval shook my faith to pieces." A course in metaphysics two years later, however, led him to conclude that "a materialistic philosophy is incredible, . . . and there is a God." Driven by "experiences too deep for irreligion to explain, [and] urgencies too potent for irreligion to express," including a nervous breakdown and "deep depression, four months of it spent in a sanatorium," he "learned more . . . about human nature than any theological seminary can ever teach." When he entered the pastorate not too long afterward, he found "the new dynamic psychology, with its issue in psychiatry, . . . a godsend." Putting it eagerly to work, he discovered that "religion was not simply a set of ideas, but a living force substituting faith for fear, strength for weakness, hope for despair, lifting the burden of unforgiven sins, building foundations under shaken spirits, transforming character, and before one's very eyes winning victories for what the New Testament calls, 'the spirit . . . of power, and of love, and of a sound mind.'" Wholly convinced "that theologies are always psychologically and sociologically conditioned, and that dogmatism in theology, whether 'liberal' or 'orthodox,' is ridiculous," Fosdick made his bed with Walter Rauschenbusch's social gospel and Borden Parker Bowne's personalistic idealism, letting the chips fall where they may.[4]

*Evolution and Mr. Bryan* expresses some core concerns arising out of Fosdick's own story, starting with his confidence in science, including the still somewhat hypothetical science of evolution, and his rock-bottom conviction that the Bible should not be brought into scientific matters. Stark opposition to Bryan's stance is evident in both instances. Nevertheless, he shared Bryan's deeply rooted fear that science posed a danger to Christian students—not as Bryan thought by forcing the unproved theory of evolution on impressionable minds, but by providing opportunities for materialistic scientists to advance "dreary philosophies which reduce everything to predetermined mechanical activity," thereby denying the reality of the very type of religious experience that Fosdick himself had undergone.[5] Fosdick often railed against reductionism and insisted on limiting science to its proper domain. In a very widely read devotional book written five years earlier, he had cautioned psychologists not to "leap to the conclusion, which lies outside their realm, that personality is an illusion, freedom a myth and our mental life the rattling of a causal chain forged and set in motion when the universe began. *All this is not science; it is making hypotheses from a limited field of facts masquerade as a total philosophy of life.*"⁶ Ultimately, his biggest difference with Bryan was theological: where Bryan needed a transcendent God to intervene miraculously in order to create human beings in his own image, Fosdick worshipped the immanent God of Henry Drummond, who had warned Christians not to find God only in occasional miracles rather than in the ongoing, constant work of creation.

I have found no evidence that Fosdick ever preached a sermon based directly on this pamphlet, but in some ways his most famous sermon—"Shall the Fundamentalists Win?," delivered several weeks later on May 21—was a sequel to the editorial, for it proclaims identical concerns. In Fosdick's opinion, fundamentalist views of the Bible were driving educated people, especially younger folk, entirely out of the churches. Science only made matters worse: spoon-feeding packaged answers to honest inquirers was not a winning proposition. If it was to survive in the modern age, Christianity had to find new ways of doing things.

---

## Annotated Text

### *Evolution and Mr. Bryan*

#### By Harry Emerson Fosdick

/3/ The editor of The Times⁷ has asked me to reply to Mr. Bryan's statement on "God and Evolution." I do so, if only to voice the sentiments of a large number of Christian people who in the name of religion are quite as shocked as any scientist could be in the name of science at Mr. Bryan's sincere but appalling obscurantism.

So far as the scientific aspect of the discussion is concerned, scientists may well be left to handle it. Suffice it to say that when Mr. Bryan reduces evolution to a hypothesis and then identifies a hypothesis with a "guess" he is guilty of a sophistry so shallow and palpable that one wonders at his hardihood in risking it. A guess is a haphazard venture of opinion without investigation before or just reason afterward to sustain it; it is a *jeu /4/ d'esprit*. But a hypothesis is a seriously proffered explanation of a difficult problem ventured when careful investigation of facts points to it, retained as long as the discovered facts sustain it, and surrendered as soon as another hypothesis enters the field which better explains the phenomena in question.

## A Hypothesis

Every universally accepted scientific truth which we possess began as a hypothesis, is in a sense a hypothesis still, and has become a hypothesis transformed into a settled conviction as the mass of accumulating evidence left no question as to its substantial validity. To call evolution, therefore, a guess is one thing; to tell the truth about it is another, for to tell the truth involves recognizing the tireless patience with which generations of scientists in every appropriate field of inquiry have been investigating all discoverable facts that bear upon the /5/ problem of mutation of species, with substantial unanimity as to the results so far as belief in the hypothesis of evolution is concerned. When Darwin, after years of patient, unremitting study, ventured his hypothesis in explanation of evolution—a hypothesis which was bound to be corrected and improved—one may say anything else one will about it except to call it a "guess." That is the one thing which it certainly was not. Today, the evolutionary hypothesis, after many years of pitiless attack and searching investigation, is, as a whole, the most adequate explanation of the facts with regard to the origin of species that we have yet attained, and it was never so solidly grounded as it is today. Dr. Osborne is making, surely, a safe statement when he says that no living naturalist, so far as he knows, "differs as to the immutable truth of evolution in the sense of the continuous fitness of plants and animals to their environment and the /6/ ascent of all the extinct and existing forms of life, including man, from an original and single cellular state."[8]

## The Real Situation

When, therefore, Mr. Bryan says, "Neither Darwin nor his supporters have been able to find a fact in the universe to support their hypothesis," it would be difficult to imagine a statement more obviously and demonstrably mistaken. The real situation is that every fact on which investigation has been able to lay its hands helps to confirm the hypothesis of evolution. There is no known fact which stands out against it. Each newly discovered fact fits into an appropriate place in it. So far as the general outlines of it are concerned, the Copernican astronomy itself is hardly established more solidly.

My reply, however, is particularly concerned with the theological aspects of Mr. Bryan's statement. There seems to be no doubt about what his position is. He proposes to take his /7/ science from the Bible. He proposes, certainly, to take no science that is contradicted by the Bible. He says, "Is it not strange that a Christian will accept Darwinism as a substitute for the Bible when the Bible not only does not support Darwin's hypothesis, but directly and expressly contradicts it?" What other interpretation of such a statement is possible except this: that the Bible is for Mr. Bryan an authoritative textbook in biology—and if in biology, why not in astronomy, cosmogony, chemistry, or any other science, art, concern of man whatever? One who is acquainted with the history of theological thought gasps as he reads this. At the close of the sixteenth century a Protestant theologian set down the importance of the book of Genesis as he understood it. He said that the text of Genesis "must be received strictly"; that "it contains all knowledge, human and divine"; that "twenty-eight articles of the Augsburg Confession are to be found in it"; that "it is an /8/ arsenal of arguments against all sects and sorts of atheists, pagans, Jews, Turks, Tartars, Papists, Calvinists, Socinians, and Baptists"; that it is "the source of all science and arts, including law, medicine, philosophy, and rhetoric," "the source and essence of all histories and of all professions, trades, and works," "an exhibition of all virtues and vices," and "the origin of all consolation."[9]

## Luther and Bryan

One has supposed that the days when such wild anachronisms could pass muster as good theology were past, but Mr. Bryan is regalvanizing into life that same outmoded idea of what the Bible is, and proposes in the twentieth century that we shall use Genesis, which reflects the pre-scientific view of the Hebrew people centuries before Christ, as an authoritative textbook in science, beyond whose conclusions we dare not go.

Why, then, should Mr. Bryan complain because his attitude toward evolution /9/ is compared repeatedly, as he says it is, with the attitude of the theological opponents of Copernicus and Galileo? On his own statement, the parallelism is complete. Martin Luther attacked Copernicus with the same appeal which Mr. Bryan uses. He appealed to the Bible. He said: "People gave ear to an upstart astrologer who strove to show that the earth revolves, not the heavens or the firmament, the sun and the moon. Whoever wishes to appear clever must devise some new system, which of all systems is, of course, the very best. This fool wishes to reverse the entire science of astronomy, but sacred Scripture tells us that Joshua commanded the sun to stand still, and not the earth."[10]

Nor was Martin Luther wrong if the Bible is indeed an authoritative textbook in science. The denial of the Copernican astronomy with its moving earth can unquestionably be found in the Bible if one starts out to use the Bible that way—"The world /10/ also is established, that it cannot be moved" (Psalm 93:1); "Who laid the foundations of the earth, that it should not be moved forever" (Psalm 104:5). Moreover, in those bygone days, the people who were then using Mr. Bryan's method of argument did quote these passages as proof, and Father Inchofer felt so confident that he cried, "The opinion of the earth's motion is of all heresies the most abominable, the most pernicious, the most scandalous; the immovability of the earth is thrice sacred; argument against the immortality of the soul, the existence of God, and the incarnation should be tolerated sooner than an argument to prove that the earth moves."[11]

### The Hebrew Universe

Indeed, as everybody knows who has seriously studied the Bible, that book represents in its cosmology and its cosmogony the view of the physical universe which everywhere obtained in the ancient Semitic world. /11/ The earth was flat and was founded on an underlying sea (Psalm 136:6; Psalm 24: 1–2; Genesis 7:11); it was stationary; the heavens, like an up-turned bowl, "strong as a molten mirror" (Job 37:18; Genesis 1:6–8; Isaiah 40:22; Psalm 104:2), rested on the earth beneath (Amos 9:6; Job 26:11); the sun, moon, and stars moved within this firmament of special purpose to illumine man (Genesis 1:14–19); there was a sea above the sky, "the waters which were above the firmament" (Genesis 1:7; Psalm 148:4) and through "the windows of heaven" the rain came down (Genesis 7:11; Psalm 78:23); beneath the earth was mysterious Sheol where dwelt the shadowy dead (Isaiah 14:9–11); and all this had been made in six days, each of which had had a morning and an evening, a short and measurable time before (Genesis 1).

Are we to understand that this is Mr. Bryan's science, that we must teach this science in our schools, that we are stopped by divine revelation /**12**/ from ever going beyond this science? Yet this is exactly what Mr. Bryan would force us to if with intellectual consistency he should carry out the implications of his appeal to the Bible against the scientific hypothesis of evolution in biology.

### The Bible's Precious Truths

One who is a teacher and preacher of religion raises his protest against all this just because it does such gross injustice to the Bible. There is no book to compare with it. The world never needed more its fundamental principles of life, its fully developed views of God and man, its finest faiths and hopes and loves. When one reads an article like Mr. Bryan's one feels, not that the Bible is being defended, but that it is being attacked. Is a 'cello defended when instead of being used for music it is advertised as a good dinner table? Mr. Bryan does a similar disservice to the Bible when, instead of using it for what it is, the most noble, useful, inspiring and inspired /**13**/ book of spiritual life which we have, the record of God's progressive unfolding of his character and will from early primitive beginnings to the high noon in Christ, he sets it up for what it is not and never was meant to be—a procrustean bed to whose infallible measurements all human thought must be forever trimmed.

### Origins and Values

The fundamental interest which leads Mr. Bryan and others of his school to hate evolution is the fear that it will depreciate the dignity of man. Just what do they mean? Even in the Book of Genesis God made man out of the dust of the earth. Surely, that is low enough to start and evolution starts no lower. So long as God is the Creative Power, what difference does it make whether out of the dust by sudden fiat or out of the dust by gradual process God brought man into being. Here man is and what he is he is. Were it decided that God had dropped him from /14/ the sky, he still would be the man he is. If it is decided that God brought him up by slow gradations out of lower forms of life, he still is the man he is.

The fact is that the process by which man came to be upon the planet is a very important scientific problem, but it is not a crucially important religious problem. Origins prove nothing in the realm of values. To all folk of spiritual insight man, no matter by what process he at first arrived, is the child of God, made in his image, destined for his character. If one could appeal directly to Mr. Bryan he would wish to say: let the scientists thrash out the problems of man's biological origin but in the meantime do not teach men that if God did not make us by fiat then we have nothing but a bestial heritage. That is a lie which once believed will have a terrific harvest. It is regrettable business that a prominent Christian should be teaching that. /15/

### Danger of Materialistic Teaching

One writes this with warm sympathy for the cause which gives Mr. Bryan such anxious concern. He is fearful that the youth of the new generation, taught the doctrine of a materialistic science, may lose that religious faith in God and in the realities of the spiritual life on which alone an abiding civilization can be founded. His fear is well grounded, as every one closely associated with the students of our colleges and universities knows. Many of them are sadly confused, mentally in chaos, and, so far as any guiding principles of religious faith are concerned, are often without chart, compass, or anchor.

There are types of teaching in our universities which are hostile to any confidence in the creative reality of the spiritual life—dreary philosophies which reduce everything to predetermined mechanical activity. Some classrooms doubtless are, as Mr. Bryan thinks, antagonistic, in the /16/ effect which they produce, alike to sustained integrity of character, buoyancy, and hopefulness of life and progress in society. But

Mr. Bryan's association of this pessimistic and materialistic teaching with the biological theory of evolution is only drawing a red herring across the red trail. The distinction between inspiring, spiritually minded teachers and deadening, irreligious teachers is not at the point of belief in evolution at all. Our greatest teachers, as well as our poorest, those who are profoundly religious as well as those who are scornfully irreligious, believe in evolution. The new biology has no more to do with the difference between them than the new astronomy or the new chemistry. If the hypothesis of evolution were smashed tomorrow, there would be no more religiously minded scientists and no fewer irreligious ones.

### The Heart of the Problem

The real crux of the problem in university circles is whether we are going to think of creative reality in physical or in spiritual terms, and /17/ that question cannot be met on the lines that Mr. Bryan has laid down. Indeed, the real enemies of the Christian faith, so far as our students are concerned, are not the evolutionary biologists, but folk like Mr. Bryan who insist on setting up artificial adhesion between Christianity and outgrown scientific opinions, and who proclaim that we cannot have one without the other. The pity is that so many students will believe him and, finding it impossible to retain the outgrown scientific opinions, will give up Christianity in accordance with Mr. Bryan's insistence that they must. Quite as amazing as his views of the Bible is Mr. Bryan's view of the effect of evolution upon man's thought of God. If ever a topsy-turvy statement was made about any matter capable of definitive information, Mr. Bryan's statement deserves that description, for it turns the truth upside down. He says: "The theistic evolutionist puts God so far away that he ceases to be a present influence in the life... Why should we /18/ want to imprison God in an impenetrable past? His is a living world. Why not a living God upon the throne? Why not allow him to work now?" But the effect of evolution upon man's thought of God, as every serious student of theology knows, has been directly the opposite of what Mr. Bryan supposes. It was in the eighteenth century that men thought of God as the vague, dim figure over the crest of the first hill who gave this universal toboggan its primeval shove and has been watching it sliding ever since. It was in the eighteenth century that God was thought of as the absentee landlord who had built the house and left it—as the shipwright who had built the ship and then turned it over to the master mariners, his natural laws. Such ideas of God are associated with eighteenth century Deism, but the nineteenth century's most characteristic thought of God was in terms of immanence—God here in this world, the life of all that lives, the sustaining energy of all that lives, as our spirits are in our /19/ bodies, permeating, vitalizing, directing all.

### God Is Not a Carpenter

The idea of evolution was one of the great factors in this most profitable change. In a world nailed together like a box, God, the creator, had been thought of as a carpenter who created the universe long ago; now, in a world growing like a tree, ever more putting out new roots and branches, God has more and more been seen as the indwelling spiritual life. Consider that bright light of nineteenth century Christianity, Henry Drummond, the companion of D. L. Moody in his evangelistic tours. He believed in evolution. What did it do to his thought of God? Just what it has done to the thought of multitudes. Said Drummond: "If God appears periodically he disappears periodically. If he comes upon the scene at special crises, he is absent from the scene in the intervals. Whether is all-God or occasional-God the nobler theory? Positively the /20/ idea of an immanent God, which is the God of evolution, is infinitely grander than the occasional wonder-worker who is the God of an old theology."[12]

Mr. Bryan proposes, then, that instead of entering into this rich heritage where ancient faith, flowering out in new world views, grows richer with the passing centuries, we shall run ourselves into his mold of mediaevalism. He proposes, too, that his special form of mediaevalism shall be made authoritative by the state, promulgated as the only teaching allowed in the schools. Surely, we can promise him a long, long road to travel before he plunges the educational system of this country into such incredible folly, and if he does succeed in arousing a real battle over the issue we can promise him also that just as earnestly as the scientists will fight against him in the name of scientific freedom of investigation, so will multitudes of Christians fight against him in the name of their religion and their God.

# *How Science Helps Our Faith*

Shailer Mathews (1863–1941)

> Laboratory science did something more than lead to research. It undermined habits of thought and substituted the tentativeness of experiment for authoritative [theological] formulas.... It was only when educational processes had ceased to be controlled by the study of classical literature and grew more contemporary, that orthodox theology was felt to be incompatible with intellectual integrity.
>
> —*Shailer Mathews,* New Faith for Old

## Introduction

Chicago theologian Shailer Mathews not only edited the "Science and Religion" pamphlets, he wrote one of the first three, published in September 1922 along with those by Conklin and Fosdick.[1] Thirty thousand copies of each title were printed, followed by similar print runs in each of the next two months. Another 50,000 copies were printed in fiscal year 1926–1927, and a Chinese translation by Shizhang Zhang was published in 1934. Unlike the other two, however, Mathews's essay was written specifically for the AISL and no other English version exists.

Despite significant attention given to the ways in which science has expanded our knowledge of the universe and enhanced our conception of God, Mathews's pamphlet is ultimately about theodicy: "Our religion . . . is the law of life itself. He who does not live in accordance with the laws of his own being, in sympathetic, helpful relationship with the environment upon which he depends, suffers." This viewpoint—that we are responsible for our own suffering, to the extent that we do not accept the regularities of the universe, comply with them, and use them to alleviate suffering—has been described as "natural law theodicy."[2] Theodicy was central to his faith, as it has been central to the faith of many other Christians for two millennia, but his solution was different owing to the emphasis he placed on divine immanence in an infinite universe and the inexorable lawlikeness of nature—two prominent themes in his pamphlet. He fully realized the difficulty of formulating a solution to the problem of evil in such a universe, so he expanded on it in the seventy-three-page section he wrote two years later for a collection of essays he gathered from thirteen scientists, aptly titled *Contributions of Science to Religion* (1924). There he admitted frankly, "whoever thinks of God as immanent (if such a word is permissible) in infinite activity, hesitates to speak easily of His will in the details of life." Nevertheless, the cosmic process had formed personalities like ourselves, and "we work according to the will of God when we use natural forces in accordance with their true nature and to further personality."[3] That was how science helped his faith.

The same hard-nosed attitude was no less evident in Mathews's overall view of Christianity. A theologian and seminary graduate who was never ordained, no other pamphlet author took a more radical approach to Christian beliefs. He had come to theology from a background in social science and history that steered him away from traditional creeds and toward the social gospel. After graduating from Newton Theological Institution in 1887, Mathews returned to his alma mater, Colby College, where he briefly taught rhetoric, history, and economics. Then he went to the University of Berlin, where he studied history with two disciples of Leopold von Ranke, Hans Delbrueck and Ignaz Jastrow, and political economy with the academic socialist and politician, Adolf Wagner. The historians gave him the "scientific" view that historical truth emerges from primary sources, untainted by philosophical or theological bias, while Wagner taught him how to criticize free market capitalism and to value altruism, making fertile soil for the social gospel that Mathews later embraced: if theology could not meet the real needs of the modern world, then it was not worth doing. Returning to Colby, Mathews briefly taught history, studied the new field of sociology, and became interested in the New Testament—all before accepting a faculty position in New Testament at the University of Chicago Divinity School in 1894. Only ten years later did he become professor of systematic theology and then dean of the Divinity School.[4]

Thus, it is hardly surprising that the definition of God from Mathews's most substantive work about theism, *The Growth of the Idea of God* (1931), was heavily tinged with social theory: "*God is our conception, born of social experience, of the personality-evolving and personally responsive elements of our cosmic environment with which we are organically related.*"[5] He was hardly the only pamphlet author with such an abstract God—Mather and Millikan immediately come to mind—nor was he bashful about it. His faculty colleague at Chicago for a quarter century, Millikan remembered someone asking Mathews whether he believed in God. Mathews replied, "That, my friend, is a question which requires an education rather than an answer."[6] In reality, Mathews's concept of God evolved throughout the 1920s, starting from the God of the Boston personalists, evident in his first pamphlet, and becoming more abstract in the next decade.[7] Was the God identified in his later work with the cosmic process that had produced our personalities unambiguously impersonal? It is hard to say. Nor is it entirely clear whether that God amounted to any more than a helpful social construction, although his defense of God's existence in this pamphlet seems sincere. If he was not entirely averse to natural theology, it was a natural theology more devoted to justifying *religion* than proving God: "Science warrants religion because it affords evidence of immanent reason, purpose and personality in the cosmic environment and its discovery of the laws of human life."[8] His approach to Christian doctrine was equally informed by his earlier studies. Creedal statements were simply historically embedded beliefs held for a time by groups of Christians, not permanent statements of absolute truths, leading him to ask, "What should the modern church do with inherited beliefs which did not square with scientific conclusions and social needs?"[9] Nevertheless, despite an almost unchecked tendency to let social psychology and evolution dictate the terms of conversation with theology, Mathews (like Fosdick) flatly rejected the reductionism that he saw at the heart of mechanistic science. As he put it once with typical wit, "Man has always felt himself to be something more than a peripatetic chemical laboratory driven by the sex instinct."[10]

Mathews also participated in the Easter 1930 symposium resulting in another AISL pamphlet, *Life After Death* (1930). The size of the very slim book he devoted to that topic three years later seems consistent with the serious, yet almost carefree, attitude he then held toward the possibility of his own eternal existence.[11] At his funeral, his former pastor Charles Gilkey reminded mourners of Mathews's famous sense of humor by telling stories, "the most characteristic and memorable of all" involving a time when Mathews was ensconced in his office with an unnamed colleague shortly before a chapel service at the Divinity School. Then Mathews's secretary entered, reminding him that he was giving the sermon that day on immortality. "Good Lord," Mathews responded, and a few minutes later he delivered what Gilkey described as "the most memorable address he ever heard" in chapel, just two sentences long. "What gives me most concern about personal immortality, is not so much the question whether I shall have it, as the question what, if it is given me, I shall do with it. Let us pray."[12]

In keeping with that brief second sentence, Mathews understood that immortality is ultimately a matter of faith, incapable of scientific proof. As he said in that pamphlet, "sooner or later one must make up one's mind as to whether or not in the cosmic process of which we are a part, which starts nobody knows where, and goes nobody knows whither, there is any meaning."[13] At the same time, he believed that immortality might actually be part of the fully scientific cosmic process that had produced the human personality in the first place. For Mathews, who certainly did not believe in heaven, hell, or the bodily Resurrection of Jesus, immortality was "no more a matter of religion than is birth or the structure of the universe. If we are immortal, or if we become immortal, it will not be because we want to, but because in a universe like ours, we can no more help it than we can help feeling warm before a fire. So considered personal immortality can be treated as any other alleged fact."[14] The salient fact that the ashes of a lifelong Baptist were interred in the crypt of a Unitarian church is all too fitting to Shailer Mathews's soul.

---

## Annotated Text

## *How Science Helps Our Faith*

### By Shailer Mathews

/3/ Science always has been making its contribution to religious thought. Every religion has had its view of the universe. One can see this in its literature. True, such views do not constitute the most valuable element in religion. For example: the writers of the Old Testament speak of the earth as flat and of a solid firmament resting like an inverted bowl above it. They speak of Sheol, a great pit under the earth where the spirits of the dead went, and of water beneath and around the flat earth. Underneath the firmament were the sun and the moon and the stars. In order to believe the religious teaching of the Bible one does not need to accept such views of the physical universe as these. The really important thing is that *man believed that God was a creator and a preserver of the universe as they knew it*. As far as /4/ their knowledge went, they found God. He was in the uttermost parts of the earth. He was in Sheol.[15]

The early Christians did not feel the need of scientific study in order to believe in Jesus Christ. They accepted the scientific views which were set forth by the teachers of the day. In the course of time these views were, as it were, taken up into theology and received the approval of the church. When a new theory of the universe like that of Copernicus was broached, or when Galileo declared that the earth rotated on its axis, the conflict over such views was not exactly between science and religion, but between those views of science which the church held and other views of science which the church did not hold.[16] It was a feud between scientists as truly as between scientists and ecclesiastics. Yet it was not long before the Copernican theory was accepted as true and built into theology. A new content was thus given to religious thinking. /5/

Today there is the same conflict between views of nature which have been approved by the church and new facts and explanations which science has discovered. There is no doubt as to what the outcome will be. Truth always fits into truth. Theology no longer thinks of the earth as flat or denies that it circles around the sun.[17] Our religious thought has been broadened by the substitution of the one view for the other. But we are only at the beginning of such enlargement of our religious thinking. Science can make contributions to our religious thinking today just as truly as in the past. As our knowledge of the universe extends, so can our faith extend. Unless we can believe in a God coextensive with reality, there will be some place where we must believe God does not exist or work. Such a belief would destroy Christianity.

Few religious teachers have seriously undertaken to utilize the results of recent science. Many of them unfortunately have opposed scientific /6/ conclusions on the basis of theology. Both attitudes are unfortunate, for modern science has much to give us in the support of our religious faith.

Especially is this true as regards our belief in God. It is a familiar fact, that philosophy is very dubious about certain of the arguments used by the defenders of Christianity in the eighteenth century. In late years, new doubt has arisen as men have become more psychological, and the *idea* of God has been said to be the real God.[18] Some have socialized as well as psychologized God out of existence.

But it does not require any particular gift of prophecy to see that unless the word "God" stands for something more than an idea, we soon will not believe in any God whatever. Unless we can find arguments for his existence in the great universe in which we live, we are not likely to feel any desire to pray. *The greatest need of our civilization is an absorbing conviction as to the existence of /7/ God.* Can we be sure that He is? Has civilization lost its reason for faith in his existence, for reliance upon him, for obedience to his laws?

These are fundamental questions, in comparison to which everything else seems trivial. If there is not a real God in existence, we shall never get any help from religion.

But our modern scientific thought is giving us a new basis for our faith in the existence of an infinite God—the very God revealed to us by Jesus Christ. It is giving us new material for our theology.

1. *Chemical and physical sciences have made it certain that there is no such thing as dead matter, but that there is universal activity.* A few years ago men thought of atoms as little pellets of dead matter which in some way or other joined to make molecules. But today the atom is itself analyzed into a nucleus of positive discharge, around which circle one or more electrons of negative discharge.[19] /8/ What looks like solid, dead matter we now know is really active.

This is true of all the universe, so far as the most accurate instruments can search out reality. Everywhere are force and action.

2. *Science is giving us a conception of infinity.* This universe that is pulsating with activity is no abstract infinity reached by speculation. Our minds are of course incapable of defining infinity except negatively, but it is possible to expand our thought so as to bring us to the place where boundaries have practically disappeared. This is what our astronomers are doing. They tell us of single stars so far away that it takes thousands of years for their light, traveling six million million miles per year to reach us; of what look like single stars which are in reality clusters of thousands of stars, billions of millions of miles in diameter. They tell us the number

of stars of what they call our galaxy is probably one billion five hundred million; and then they tell /9/ us of a galaxy of such galaxies.[20]

One's mind grows dizzy as it tries to grasp such facts, and yet that is the universe in which we actually live! And by means of scientific instruments of all but miraculous accuracy we are able to see that the same forces are active, the same elements are present through it all. It is a universe of infinite activity and force.

3. *Science is showing us that this infinite activity in all its variety is reducible to formulas or laws.* No intelligent person would now say that a natural law does anything. A law, in science, is a formula describing how things take place. The activities of all this universe can be expressed by mathematics. Our mathematicians figure out when the comets are to appear, as accurately as they figure the phases of the moon. The same laws describe the movements of electrons as they circle around the nucleus within the atom, which describe how planets circle around the sun. Everywhere /10/ is what we are forced to call *sanity* and *reason*; and so immanent is this rationality within these forces that it is possible to make it an assumption for further investigation. All science rests upon this conviction.

A formula based upon what we already know, used as a means of finding out that which we hope to know, is called an hypothesis. It has been said that an hypothesis is only a theory or guess; that one man's guess is as good as another's.[21] Whoever talks thus is either uninformed, or dishonorably clouding vital issues and weakening faith. Even the tyro in science knows what hypotheses have accomplished. By their means we have found a new planet as well as new elements of matter; we have developed means for utilizing electricity and radio-activity; we understand better the meaning of life, and the cure of various diseases. No scientific man will organize an hypothesis which is not already made probable from the study of facts. A scientific /11/ hypothesis is a reading of cosmic forces.

But if the activity within this infinite universe is thus capable of rational ordering, it is impossible to believe that it does not have within itself that which must be called rational. And thus we gain a new and grander belief in a cosmic Intelligence.

4. *Science is also showing us processes and tendencies within the field of that which can be known.* Ultimate origins and destinies are beyond our knowledge, but within the limited area of experiment it is possible to see tendency and process. Just as the universe is not like the action of an insane man, it is like the action of a man with purpose and design. As one studies this wonderful continuous succession of changes where everything has its cause, called process, or development, or evolution, one supreme conviction is thrust upon us: the only thinkable explanation of this wonderful fact is that of purpose. We have thus added to our /12/ conception of activity, infinity, and rationality, that of purpose. Thus we come inductively to a belief in an Infinite God.

5. This is not all. *Science is also making plain that this process, in so far as we have been able to explore it, culminates in personality; that is to say, we human beings are its outcome.* But if there is that within this rational and purposeful activity which in the course of evolution results in personal life, then *there must be that within the activity itself which is at least as personal as we.* It is logically impossible to think that an impersonal universe should produce personality. It is imperative to think that where personalities like ours actually exist, there must be similar elements within the universe. That though the universe is not God, God is in the universe. For by these facts the great conviction is forced upon us that *just as we men and women have personality within that mass of chemical and physical activity which we call our body, so in this infinite universe* /13/ *is there a Person whose existence and character account for its rationality and its purpose.* And thus one comes to the conception of a personal God. It is the same line of thought followed by men who knew less of the universe than we do, but it culminates in a faith which is co-extensive with a universe we have come to know more actively. Nowhere is there a no-God land.[22] *Wherever there is activity, there is God.*

6. But even this is not all. *Our scientists are making it plain that an organism that has resulted from life in a certain environment must live in accordance with that environment* if it is to continue. This is the great law of the personal life in the universe where personality has been found. We must live in this universe personally, because only thus can we come into

right relations with an environment which has made personality in us possible, and is itself full of personal activity. So to live is to be religious.

Thus we reach a conclusion which is of the utmost value. Our religion is not mere survival of imperfect ancestors, /14/ neither is it mere custom or the maintenance of artificialities. It is the law of life itself. He who does not live in accordance with the laws of his own being, in sympathetic, helpful relationship with the environment upon which he depends, suffers. We are thus not merely using conventional terms when we say that a man can lose his soul. He *can lose his life* in the same way that a plant that does not live in the sunlight loses its color and vigor. He who would save his life must live with the infinite God immanent in this infinite universe of activity. Sin is seen to be more than a mistake. It is a violation of the very laws of being.

Reflection will make it plain that on such a magnificent conception of God suggested by our increased knowledge of our universe, can be raised a *new understanding and faith in Jesus as a revealer of the infinite personality*. Confidence in the gospel is strengthened rather than weakened by our knowledge of the forces in the midst of which we live. /15/ It is idle to be concerned over origins. They can never be changed. Religion is interested in destinies, and these we may control. *The gospel is in harmony with evolution*, for both are revelations of God's will.

He who understands the Bible in accordance with actual facts has no difficulty in realizing the truth of its teaching that God is in the processes which have produced and sustain mankind. It is only those who are ignorant, both of the origin and nature of the Bible and of the facts in our universe, who are terrified lest science should make them lose their faith. To all others, knowledge is light rather than darkness. Just as we live in a universe of light, and make our own night as the earth gets between us and our sun, so do we walk in religious darkness only when our prejudices and our ignorance get between us and reality. *The faith which Jesus evokes is a faith in the Father who made the heavens and the earth*. The Christian's God of Love is the scientist's God of Law.

# A Scientist Confesses His Faith

Robert Andrews Millikan (1856–1953)

> It is a sublime conception of God which is furnished by science, and one wholly consonant with the highest ideals of religion, when it represents Him as revealing Himself through countless ages in the development of the earth as an abode for man and in the age-long inbreathing of life into its constituent matter, culminating in man with his spiritual nature and all his Godlike powers.
>
> —[Robert Andrews Millikan], "Deny Science Wars against Religion," New York Times, May 27, 1923

## Introduction

The fourth pamphlet, by Nobel laureate physicist Robert Andrews Millikan, was written like the first three in direct response to fundamentalist criticisms of evolution.[1] In the summer of 1921, Millikan resigned his professorship at the University of Chicago, where Shailer Mathews had been a longtime friend and colleague, to accept a position as director of the Norman Bridge Laboratory at the California Institute of Technology and de facto president of Caltech. As Bryan's antievolution crusade got off the ground over the next twelve months, Millikan discussed his concerns with his brother-in-law, Robert Elliott Brown, a prominent Congregationalist pastor from Waterbury, Connecticut. When Millikan urged Brown to persuade his clerical colleagues to speak out in favor of evolution, Brown took the suggestion, while asking Millikan to write a statement for scientists to endorse, advancing a positive view of religion. A few months later, in July 1922, the World's Christian Fundamentals Association met in Los Angeles, resolving to "wage a relentless warfare on Evolution and Modernism."[2] The following winter, Millikan drafted the statement Brown had requested and, after circulating it to dozens of his friends, it was published by the New York Times in May 1923. At about the same time he prepared an address on "Science and Religion" for a Los Angeles church, incorporating parts of the statement, that was quickly printed in the Caltech alumni magazine and *The Christian Century*. The AISL printed an unknown number of copies in September 1923, with a second impression of 50,000 in February 1927. They also reprinted Millikan's statement from the *New York Times* no later than September 1926.[3]

Millikan saw a great need for a middle way between fundamentalism and atheism, a point he stressed in a commencement address he gave at his alma mater, Oberlin College, about a year after *Scopes*: "The fundamentalist, denying evolution, holds to the six-day creation; the atheist denies any purpose in a universe he thinks is governed by blind force. Fortunately, a choice of either is unnecessary."[4] The optimistic, open-minded faith of modernism held the solution. Certainly he resonated with the modernist emphasis on practical religion and its lack of emphasis on traditional Christian doctrines. "*The most important thing in the world*," he said at a Washington reception to honor Marie Curie in 1921, "*is a belief in the reality of moral and spiritual values.*"[5] "I have myself belonged to two churches," he said on another occasion, "both of which were unhampered by a creed of any sort." In Chicago, he was active in Hyde Park Congregational Church (now part of the United Church of Hyde Park), where the pastor was Theodore Gerald Soares, a liberal Christian ethicist. In Pasadena, he was a highly influential member of a Congregationalist church that soon merged with a Unitarian church at Millikan's urging. In 1930, Millikan brought Soares to Pasadena to be the next pastor of his church and simultaneously appointed him professor of ethics at Caltech, a private institution that had been founded by a Universalist preacher with a board of trustees that originally included nine Universalists.

The significance of having a world-famous physicist as a spokesperson for the AISL derives mainly from the high public visibility that he helped to create for the general attitude of mutual cooperation of science and religion that he very eagerly endorsed. Science for Millikan was more than merely compatible with religion. What he called "the *spirit* of religion" and "the *spirit* of science (or knowledge)" were "the two supreme elements in human progress."[6] His concept of a fully immanent God, fully compatible with science, must also have appealed to Mathews and other modernist clergy. Like Albert Einstein, however, Millikan identified "The God of Science" as "the spirit of rational order and of orderly development, *the integrating factor in the world of atoms and of ether and of ideas and of duties and of intelligence.*"[7] Unlike Einstein, Millikan gave God two specific roles in his theology of nature. First, God was the benevolent providence working through the long process of evolution, "culminating in man with his spiritual nature and all his Godlike powers."[8] Second, Millikan's research on cosmic rays led him to conclude that atom building in interstellar space balanced the destruction of atoms in stellar interiors. This picture of the universe "as in a steady state now," he said two decades before Fred Hoyle and others advanced a cosmology of that name, would perhaps "allow the creator to be continually on his job."[9] Readers of his pamphlet did not encounter this latter idea, which he only developed a few years later, but the rest of these ideas were clearly expressed there.

---

Annotated Text

## *A Scientist Confesses His Faith*

By Robert A. Millikan

/3/ In speaking upon this theme I am clearly somewhat out of my normal orbit. Most of my life has been spent in experimental work in the physical laboratory, devoted to the study of pure science, and in all such work the first aim is to eliminate all unnecessary complexities, to get rid of all secondary disturbing causes, to reduce the study of a particular phenomenon to its simplest possible terms in order to get at fundamental underlying principles so that when conclusions are drawn they are obvious and inevitable. The result of such a method has been to build up a certain body of knowledge in physics which is assented to by all intelligent men who take the trouble to study it. I do not mean by this that there are no controversies in physics, but rather that there has been produced a very considerable body of non-controversial material. At the risk of being uninteresting because of the fact that I deal only with the obvious, I shall attempt to keep in my accustomed orbit sufficiently /4/ to use the same method in discussing the relations between science and religion, for I think at least nine-tenths of the controversy which rages in this field is due to a confusion which arises from the failure to eliminate purely extraneous and incidental matters, or to simple misunderstanding of facts which have been quite definitely established, or are at least practically universally recognized by the well informed.

## An Ancient Controversy

There seems to be at the present time a strange recrudescence of a point of view which is completely out of keeping with the developments of the age in which we live, a point of view which thoughtful leaders of both sciences and religion have in all ages realized never had any basis for existence. In the time of Galileo, it is perhaps understandable, in view of the crudity of the sixteenth century, that certain misguided religious leaders should have imagined that the discovery of the earth's motions might tend to undermine in some way the basis of religion and who therefore attempted to suppress Galileo's teachings.[10] Yet it is to me not a little surprising that men even of such opportunities as Galileo's persecutors could have got religion upon /5/ such an entirely false basis in their thinking as to make its fundamental verities, its very existence, dependent in any way upon any scientific discovery. It is not a question of whether Galileo was right or wrong. That is a scientific matter with which religion as such has nothing whatever to do, and which should not have given it the slightest concern. Science could be counted upon to take care of that. It is its business to doubt, and it always does so as long as there is any room left for uncertainty. That even those inquisitors were far behind their own times in supposing that there could be any real contradiction between religion, properly understood, and the findings of astronomers cannot perhaps be better demonstrated than by the following quotation from St. Augustine, who lived 1,200 years earlier, about 400 A.D., and is probably recognized as the most influential authority, next to Jesus and St. Paul, of the early Christian church.

In commenting upon the entire distinctness from his point of view of the two great lines of thought, the natural and spiritual, Augustine says: "There is some question as to the earth or the sky, or the other elements of this world . . . respecting which one /6/ who is not a Christian has knowledge derived from most certain reasoning or observation: and it is very disgraceful and mischievous, and of all things to be carefully avoided, that a Christian, speaking of such matters as being according to the Christian scriptures, should be heard by an unbeliever talking such nonsense that the unbeliever, perceiving him to be as wide from the mark as east and west, can hardly restrain himself from laughing."[11]

### Old Disputes Renewed

That this same controversy that Augustine thus saw nearly 1,600 years ago had no basis for existence, because it is outside the proper field of religion, but which nevertheless flared up so violently in Galileo's time and then died out as men grew in intelligence, should have appeared again in as enlightened a country as America, in the year 1923, is one of the most amazing phenomena of our times.[12] But it is not less amazing than it is deplorable, for the damage which well-meaning but small-visioned men can do to the cause of organized religion, as represented in the Christian church, through the introduction inside the organization of such a disintegrating influence, is incalculably greater than any which could possibly be done by attacks from /7/ outside. Indeed, should the movement succeed, the church would inevitably soon lose all its most vital elements, and society would be obliged to develop some other agency to do the work which the church was organized to do, and which to a very large extent it now does, namely, the work of serving as the great dynamo for injecting into human society the sense of social responsibility, the spirit of altruism, of service, of brotherly love, of Christ-likeness, and of eliminating, as far as possible, the spirit of greed and self-seeking.

### Error of some Scientists

But I am not going to place the whole blame for the existence of this controversy upon misguided leaders of religion. The responsibility is a divided one, for science is just as often misrepresented as is religion by men of little vision, of no appreciation of its limitations, and of imperfect comprehension of the real role which it plays in human life—by men who lose sight of all spiritual values and therefore exert an influence upon youth which is unsettling, irreligious, and essentially immoral. I am ready to admit that it is quite as much because of the existence of scientists of that type as of their counterparts in the field of religion, /8/

that the "fundamental" controversy has flared up today, and it is high time that scientists recognize their share of the responsibility and take such steps as they can to remove their share of the cause.[13]

I do not suppose that anything which I may say will exert much influence upon the groups whose prejudices have already been aroused, and who are therefore not interested in an objective analysis of the situation, but I may perhaps hope that some of the youth whose minds have been confused by the controversy may profit somewhat from a restatement of what seem to me the perfectly obvious and indisputable facts.

## No Real Conflict

The first fact which seems to me altogether obvious and undisputed by thoughtful men is that there is actually no conflict whatever between science and religion when each is correctly understood. The simplest and probably the most convincing proof of the truth of that statement is found in the testimony of the greatest minds who have been leaders in the field of science, upon the one hand, and in the field of religion, upon the other. Suppose, for example, that we select the greatest names in the last two centuries of the /9/ history of British sciences, or, for that matter, of world science. Every one would agree that the stars that shine brightest in that history, as one's glance sweeps down from 1650 to 1920, are found in the names of Newton, whose life centered about 1680; Faraday, living about 1830; Maxwell, 1870; Kelvin, 1890, and Raleigh, who died three years ago. No more earnest seekers after truth, no intellects of more penetrating vision, can be found anywhere, at any time, than these, and yet every one of them has been a devout and professed follower of religion.[14]

It was Kelvin who first estimated the age of the earth at something like a hundred million years without seeing the least incompatibility, in spite of the first chapters of Genesis, between that scientific conclusion and his adherence to the church, of which he was a lifelong member and a constant attendant. Indeed, in 1887, when he was at the very height of his powers, he wrote: "I believe that the more thoroughly science is studied the further does it take us from anything comparable to atheism." Again in 1903, toward the end of his life, he wrote: "If you think strongly enough you will be forced by science to the belief in God, which is the foundation of all religion. /10/ You will find it not antagonistic, but helpful, to religion." His biographer, Silvanus P. Thompson, says: "His faith was always of a very simple and child-like nature, undogmatic and untainted by sectarian bitterness. It pained him to hear crudely atheistic views expressed by young men who had never known the deeper side of existence."[15] Just as strong a case of the same sort can be made by turning to the biographies of any of the other men mentioned, and these were chosen, let it be remembered, not because they were religious men, but because they are universally recognized as the foremost of scientists. Indeed, I doubt if the world has ever produced in any field of endeavor men of more commanding intellects than two of them, Sir Isaac Newton and James Clerk Maxwell.

### Testimony of Pasteur

If some one says that I am calling only on the testimony of physicists and of Englishmen, then listen to the man whom the French nation has repeatedly voted the foremost of all Frenchmen, and who is also easily the peer of any biologist who has ever lived anywhere, Louis Pasteur, of whom his biographer says, "Finally, let it be remembered that Pasteur was a deeply /11/ religious man."[16] Over his tomb in the Institute Pasteur are inscribed these words of his: "Happy is he who carries a God within him, an ideal of beauty to which he is obedient—an ideal of art, an ideal of science, an ideal of the fatherland, an ideal of the virtues of the gospel."[17]

Or, again, if I am accused of calling merely on the testimony of the past, on the thinking which preceded the advent of this new twentieth century in which we live, I can bring the evidence strictly up to date by asking you to name the dozen most outstanding scientists in America today and then showing you that the great majority of them will bear testimony, not only to the complete lack of antagonism between the fields of science and religion, but to their own fundamental religious convictions. One naturally begins with the man who occupies the most conspicuous scientific position in the United States, namely, the president of the National Academy of Sciences, who is at present both the head of the Smithsonian Institution of Washington and the president of the American Association for the [A]dvancement of Science, Dr. Charles D. Walcott, one of the foremost of American students of the evolution of life in the /12/ early geologic ages. He is personally known to me to be a man of deep religious conviction and has recently written me asking that he be described for the purposes of this address, which he has seen, as "an active church worker."[18]

## A Cloud of Witnesses

The same is true of Henry Fairfield Osborn, the director of the American Museum of Natural History of New York, and one of the foremost exponents of evolution in the country. Another rival for eminence in this field is Edwin G. Conklin of Princeton, who in recently published articles has definitely shown himself a proponent of the religious interpretation of life. In the same category I know, also from direct correspondence, that I may place John C. Merriam, president of the Carnegie Institution of Washington and America's foremost paleontologist; Michael Pupin, the very first of our electrical experts who has "approved every word" of this address and recently delivered a better one at Columbia University on this same subject; John M. Coulter, dean of American botanists; A. A. and W. A. Noyes, foremost among our chemists; James R. Angell, president of Yale University, an eminent psychologist, with whom I have had an exchange of letters on this subject; /13/ James H. Breasted, our most eminent archeologist, who served with me for years on the board of trustees of a Chicago church, upon which also T. C. Chamberlin, dean of American geologists, was a constant attendant; Dr. C. G. Abbott, Home Secretary of the National Academy of Sciences, eminent astronomer and active churchman; and so on through the list of most of the scientists of special eminence in this country.[19]

Turn now to the other side of the picture and ask what have been the views of the most outstanding and most inspired religious leaders upon the relations of science to religion, and you obtain altogether similar testimony. Was it not Jesus himself who said, "You shall know the truth and the truth shall make you free"?[20] There is not one syllable in all that he taught nor one idea which he introduced into human life which would justify one in arraying him on the side of those who would see antagonism between any scientific truth and the deepest spiritual values. There were no creeds in Jesus' teaching, no verbal inspirations of any sort. Religion was to him a life of love and duty, the simple expression of the golden rule. /14/

### Augustine and John Wesley

Turning to the next great religious personalities since Jesus' day, I have quoted Augustine to show how he warned against religious leaders of such narrow insight as to make religion a laughing stock by the presentation of an antagonism which did not exist. John Wesley, the founder of the Methodist church, in the chapter of his Compendium of Natural Philosophy on "A General View of the Gradual Progression of Beings," after speaking of "the ostrich with the feet of a goat which unites birds to quadrupeds" says, "By what degrees does Nature raise herself to man? . . . How will she rectify this head that is always inclined toward earth? How change these paws into flexible arms? What method will she make use of to transform these crooked feet into skillful and supple hands? Or will she widen and extend this contracted stomach? In what manner will she place the breasts and give them a roundness suitable to them? The ape is the rough draft of man, this rude sketch, an imperfect representation which nevertheless bears a resemblance to him, and is the last creature that serves to display the admirable progression of the works of God. . . . But mankind have their /15/ gradations as well as other productions of our globe.[21] *There is a prodigious number of continued links between the most perfect man and the ape.*" (Italics mine.) I am not here asserting that Wesley was right. For our present purposes that is quite immaterial. But he was a supreme leader and the quotation shows that he saw too clearly to allow his scientific thinking to be trammeled by any man-made religious dogmas.[22]

Again, in our own time, there has been no more spiritual religious leader than Henry Drummond, whose most inspiring work was in showing the contribution of science to religion, and I think I might name practically all of the outstanding religious leaders now living and say that there is not one in ten of them who would not take his place beside Jesus and Augustine and Drummond and Beecher and Lyman Abbott and Fosdick and Soares and King and Brown and Burton and Mathews and a host of other men of broad vision and deep experience who have seen science and religion as twin sisters which are effectively cooperating in leading the world on to better things.[23]

My argument thus far has been merely this, that there can be no conflict /16/ between science and religion if the greatest minds in the two fields, the minds to which we look for our definitions of what both science and religion are, have not only not seen such a conflict but have clearly seen and clearly stated that there is none.

### Separate Tasks of Science and Religion

But now let us go to my second obvious fact and show why in the nature of things there can be no conflict. This appears at once as soon as one attempts to define for himself what is the place of science and what the place of religion in human life. The purpose of science is to develop without prejudice or preconception of any kind a knowledge of the facts, the laws, and the processes of nature. The even more important task of religion, on the other hand, is to develop the consciences, the ideals, and the aspirations of mankind.[24]

The definition of science, I think all will agree with. The definition of religion is in essence that embodied in the teachings of Jesus, who unlike many of his followers of narrower vision, did not concern himself at all with creeds, but centered his whole teaching about a life of service and the spread of the spirit of love. It is of course true that /17/ the scientific and the religious sides of life often come into contact and mutually support each other. Science without religion obviously may become a curse, rather than a blessing to mankind, but science dominated by the spirit of religion is the key to progress and the hope of the future. On the other hand, history has shown that religion without science breeds dogmatism, bigotry, persecution, religious wars and all the other disasters which in the past have been heaped upon mankind in the name of religion: disasters which have been so fatal to organized religion itself, that at certain times and in certain countries the finest characters and the most essentially religious men have been found outside the church. In some countries that is the situation today, and whenever this is true it is because the essence of religion has been lost sight of, buried under theologies and other external trappings which correspond exactly to the "mint, the anise, and the cummin" of Jesus' day.[25] If anyone wishes to see what disaster these excrescences can bring upon the cause of real religion let him read the history of the church in Asia Minor for the first six centuries, and see for himself what sects and schisms and senseless quarrels over the /18/ nature of the person of Jesus can do, in the way of sucking the life-blood out of the spirit of his teachings, and out of the effectiveness of the organization which was started for the sole purpose of spreading that spirit.

### Vital Christianity Untouched

Yet in America, at least, it is not primarily those inside the church who thus misinterpret and misunderstand it, though we must sorrowfully admit that such a group does exist there. It is, however, for the most part the outsiders, the critics who have never seen the inside of church walls, and many of whom know so little about the church in America as to actually believe that Christianity is to be identified with medieval theology, when the fact is so obvious that he who runs may read, that all that is vital in Christianity has remained altogether untouched by the most complete revolutions in theology, such as have gone on, for example, during the past hundred years.[26] Many of us were brought up under creeds and theologies which have now completely passed on, as such things will continue to do as the world progresses, and yet, as we look back, we see that the essential thing which the churches of our childhood were doing for us and for our neighbors then, is precisely what they /19/ are doing now, namely, stimulating us to right conduct, as each of us sees it, inspiring us to do as we know we ought to do, developing our ideals and our aspirations. There is a very simple and a very scientific way of finding out for yourself what is the heart and center of the Christian religion, the fundamental and vital thing which it stands for in human society, and that is to get far enough back so that details are lost sight of and then to observe what is the element which is common to all Christian churches in the United States. He who does that will see at once that it is the life and the teachings of Jesus which constitute all that is essential to Christianity; that the spread of his spirit of unselfishness, of his idealism, and of his belief in the brotherhood of man and the fatherhood of God is the great purpose of the Christian religion. In other words, that religion exists, as stated above, for the sake of developing the consciences, the ideals, and the aspirations of mankind.

### Development in the Old Testament

My third obvious fact is that both science and religion have reached their present status through a process of development from the crudest beginnings. This will be universally recognized in the case of science, and in the /20/ case of religion the most superficial study of history shows that this is true. The religious ideals and practices of the American Indians and of all other primitive tribes, with their totem-poles and incantations, have obviously been of the most primitive type. The ideas of duty, responsibility, have always been involved in these religions, but the motives of right conduct, as primitive man conceived it, have been, from our present point of view, of the most unenlightened and even unworthy sort.

But is it not altogether obvious that religion cannot possibly rise higher than the stage of development of the people of whose ideals it is the expression? Nothing could show that process of development better than the Bible itself, for the early books of the Old Testament reveal the conception of God, characteristic indeed of the age, but not at all satisfying to us, for it was a God who was indeed benevolent and just toward his own chosen people, but vindictive and cruel and utterly regardless of the welfare of those outside this chosen group. This imperfect conception is developed and refined through the history of the Jews as portrayed in the Bible until it culminates in the all-embracing love and fatherhood preached by Jesus. He who /21/ would deny this development process going on in both science and religion and clearly revealed in all the records of the past which we have, must shut his eyes to the indisputable facts as they are presented in all history, including sacred history.

To me it has always been of the utmost interest and profit, especially when I was disposed to judge severely great religious leaders of the past, like Paul or Moses, to try to conceive myself living in their surroundings, with their lack of scientific knowledge, interpreting life from the limited point of view which they had, formulating rules of conduct relating, for example, to matters of hygiene, such as those dealt with in Deuteronomy, trying to interpret mysterious phenomena of nature like eclipses, the possession of evil spirits, etc. And when I do this my wonder always is that these men saw as clearly as they did, and succeeded as well as they did in separating the fundamental from the incidental. Difficult as it is to judge the great leaders of the past by their standards rather than by ours, it is imperative that we do so if we are to form any just appreciation of them and of their contributions to the development of the race. Indeed, this is the essence of the whole /22/ problem. Once get this point of view and you will never think of asking whether Genesis is to be taken as a modern text-book of science. It was written long before there was any such thing as science. It is of the utmost importance from every point of view to realize that the Bible itself makes no claims whatever of scientific correctness or for that matter of verbal inspiration. It is rather the record of the religious experiences and development of a race.

### All Thinking Men Believers

My fourth obvious fact is that every one who reflects at all believes in one way or another in God. From my point of view, the word atheism is generally used most carelessly, unscientifically, and unintelligently, for it is to me unthinkable that a real atheist should exist at all. I may not, indeed, believe in the conception of deity possessed by the Congo negro who pounds the tom-tom to drive away the god whose presence and influence he fears; and it is certain also that no modern religious leader believes in the god who has the attributes which Moses, Joshua and the Judges ascribe to their Deity. But it seems to me as obvious as breathing that every man who is sufficiently in his senses to recognize his own inability /23/ to comprehend the problem of existence, to understand whence he himself came and whither he is going, must in the very admission of that ignorance and finiteness recognize the existence of a Something, a Power, a Being in whom and because of whom he himself "lives and moves and has his being."[27] That Power, that Something, that Existence, we call God. Primitive man, of course, had anthropomorphic conceptions of that being. He was not able to think of a god who was very different from himself. His God became angered and had to be appeased, he was jealous and vindictive and moody; but man's conceptions have widened with the process of the suns, and as he has grown up he has slowly been putting away childish things.[28]

### Agreement Not Necessary

I am not much concerned as to whether I agree precisely with you in my conception or not, for "can men with thinking find out God?"[29] Both your conception and mine must in the nature of the case be vague and indefinite. Least of all am I disposed to quarrel with the man who spiritualizes nature and says that God is to him the Soul of the universe, for spirit, personality, and all these abstract conceptions /24/ which go with it, like love, duty, and beauty, exist for you and for me just as much as do iron, wood and water. They are in every way as real for us as are the physical things which we handle. No man, therefore, can picture nature as devoid of these attributes which are a part of your experience and mine, and which you and I know are in nature. If you, then, in your conception identify God with nature, you must perforce attribute to him consciousness and personality, or better, superconsciousness and superpersonality. You cannot possibly synthesize nature and leave out its most outstanding attributes. Nor can you get these potentialities out of nature, no matter how far back you go in time. In other words, materialism, as commonly understood, is an altogether absurd and an utterly irrational philosophy, and is indeed so regarded by most thoughtful men.

### Two Great Influences in History

Without attempting, then, to go farther in defining what in the nature of the case is undefinable, let me reassert my conviction that although you may not believe in some particular conception of God which I may try to give expression to, and although it is unquestionably true that many of our /25/ conceptions are sometimes childishly anthropomorphic, every one who is sufficiently in possession of his faculties to recognize his own inability to comprehend the problem of existence, bows his head in the presence of the Nature, if you will, the God, I prefer to say, who is behind it all and whose attributes are partially revealed to us in it all, so that it pains me as much as it did Kelvin "to hear crudely atheistic views expressed by men who have never known the deeper side of existence."[30] Let me then henceforth use the word God to describe that which is behind the mystery of existence and that which gives meaning to it. I think you will not misunderstand me, then, when I say that I have never known a thinking man who did not believe in God.

My fifth obvious fact is that there have been just two great influences in the history of the world which have made goodness the outstanding characteristic in the conception of God. The first influence was Jesus of Nazareth; the second influence has been the growth of modern science, and particularly the growth of the theory of evolution. All religions, including Christianity, have impersonated the spirit of evil and the spirit of good, and /26/ in many instances the former has been given the controlling influence. All of us see much in life which tends to make us pessimists. The good does not always prevail. Righteousness does not always triumph. What is the meaning of existence? Is it worth while? Are we going anywhere? Jesus and modern science have both answered that question in the affirmative—Jesus took it as his mission in life to preach the need of the goodness of God. He came in an age which was profoundly ignorant of modern science. He used the terms, in dealing with disease and evil, which were appropriate to his day, the only terms which his audiences could have understood, but he saw a God who was caring for every sparrow and who was working out through love a world planned for the happiness and well being of all creatures.

### God and Evolution

Similarly science in the formulation of the theory of evolution has the world developing through countless ages higher and higher qualities, moving on to better and better things. It pictures God, however you may conceive him, as essentially good, as providing a reason for existence and a motive for making the most of existence, /27/ in that we may be a part of the great plan of world progress. No more sublime conception of

God has ever been presented to the mind of man than that which is furnished by science when it represents him as revealing himself through countless ages in the development of the earth as an abode for man and in the age-long inbreathing of life into its constituent matter, culminating in man with his spiritual nature and all his god-like powers.[31]

But let me go a step farther. Science in bringing to light the now generally admitted, though not as yet obvious and undisputed fact, that this is not a world in which things happen by caprice, but a world governed throughout by law, has presented the most powerful motive to man for goodness which has ever been urged upon him, more powerful even than any which Jesus found. That "whatsoever a man soweth that shall he also reap" is no longer merely a biblical text; it is a truth which has been burned into the consciousness of mankind by the last hundred years of the study of physics, chemistry and biology.[32] Science, then, not only teaches that God is good, but it furnishes man with the most powerful of motives to fit in with the scheme of goodness which God has provided /28/ in nature. It teaches him not only that disease breeds disease, but also, by inference at least, that hate breeds hate, that dishonesty breeds dishonesty that the wages of sin is death, and on the other hand that love begets love.[33] It teaches him that the moral laws and the physical laws alike are all laws of nature, and that violation of either of them leads to misery.

Men Believe in a "World Scheme"

In closing this brief statement of the faith of the scientist, let me present a situation and a question. In the spring of 1912 the great ship Titanic had collided with an iceberg and was doomed. She was about to sink. The lifeboats were insufficient. The cry went up, "The women first!" The men stepped back. The boats were loaded and the men sank with the ship. You call it an heroic act. Why did they do it? Perhaps you answer, because it was the law of the sea and the men preferred to die rather than to live after having broken that law. Then take a simpler case, for I want a more fundamental answer. Two men were clinging after the wreck to a floating piece of timber. It would not support them both. One of them voluntarily let go and sank. Heroisms of /29/ just this sort happened thousands of times during the war. Men threw away their lives for a cause. Such events happen every day in times of peace. Why do they happen? Because men and women prefer to die rather than to live in the consciousness of having played the coward, of having failed to play their part worthily in the great scheme of things. It is true that not all men are like that, but I am optimist enough to think that most men are. But now come back to the question, why are most men like that? Simply because most men believe that there is such a world scheme; that they are a part of it, that their deaths are going to contribute to its development, in short, because most men believe in God. This is the obvious inference from the fact that men are willing to die for a cause. They may not know whether there is personal immortality for them or not, but they do know with absolute certainty that they live on in memory and in influence; many of them, too, have faith to believe that they live on in consciousness, but in either case they are a part of a plan of development which gives meaning to life. In other words, men who have the stuff in them which makes heroes all believe in God, in "a power in the /30/ world which makes for righteousness."³⁴ Without that belief there is no motive for heroism or for self-sacrifices of any sort, nor any such thing as "the development of the consciences, the ideals, and the aspirations of mankind," which I said above, was the task of religion, for there is then no basis for ideals or aspirations. This is why Kelvin said that "belief in God is the foundation of all religion."³⁵

If there be a man who does not believe, either through the promptings of his religious faith or through the objective evidence which the evolutionary history of the world offers, in a progressive revelation of God to man, if there be a man who in neither of these two ways has come to feel that there is a meaning to and a purpose for existence, if there be such thorough-going pessimism in this world, then may I and mine be kept as far as possible from contact with it. If the beauty, the meaning and the purpose of this life as revealed by both science and religion are all a dream, then let me dream on forever.

# *The Heavens are Telling*

Edwin Brant Frost (1866–1935)

> Everything that we learn from the observational point of view in the study of astronomy seems to me to point precisely and always towards a purposeful operation of nature. When you accept this, it seems to be inconsistent with physical sciences not to believe in a mind behind the universe.... If the universe is purposeful, then it is plain to me that man, who is the highest form of development on the earth, must himself be distinctly a result of purpose rather than of accident. His evolution, whether it is by procedures which are clear to us or not, must be consistent with purpose and not with chance.
>
> —*Edwin Brant Frost,* An Astronomer's Life

## Introduction

Edwin Brant Frost was the ultimate contradiction in terms: a blind astronomer.[1] In December 1915, seven months before his fiftieth birthday, he suffered a detached retina in his right eye while photographing the spectrum of a star at Yerkes Observatory in Williams Bay, Wisconsin, a condition that quickly led to a full loss of vision in that eye. Over the next several years a cataract developed in his left eye, leaving him almost completely blind for the rest of his life.[2] Since his home was located several hundred feet from the observatory grounds, a wire was strung along a path through the woods as a guide.[3] Despite this handicap, he continued his duties as Director of Yerkes and editor of the *Astrophysical Journal* until 1932. Perhaps even more surprisingly, according to longtime friend and former associate Philip Fox, in this same period "he lectured widely, instructively, and entertainingly. The solid science presented with dignity but often enlivened with whimsical humor and ready wit–quaint, homely, informal address soaring at times to lofty expression—carried a unique impressiveness. Blindness passed from his eyes in making his audience see and understand."[4] Specific evidence of such activity is limited, but Frost was already speaking to popular audiences early in the century and he probably did so more often in the 1920s. Always assisted by his wife and often using lantern slides, he spoke at museums, churches, colleges, seminaries, and meetings of civic groups and professional organizations.

The topic of Frost's public talks was usually purely scientific and not at all religious. An important exception occurred in 1923, when he spoke on "Science as an Aid to Religion" at McCormick Theological Seminary in Chicago, but the text of this address is unknown.[5] The previous year Shailer Mathews had invited Frost to write a chapter for the anthology, *Contributions of Science to Religion* (1924). In fifteen thousand words Frost said nothing at all about religion, leaving us entirely in the dark about his remarks at McCormick.[6] Frost had a further opportunity to speak at McCormick in 1929, when his friend John Timothy Stone was inaugurated as president. Stone asked Frost to speak on "Science and Religion," but again the text is lost. This event led the Southern Baptist Seminary in Louisville, Kentucky, to invite Frost to give three lectures sponsored by the Norton Foundation in 1931. Titled "On Some Complementary Aspects of Science and Religion," they were never published and their content is unknown.[7]

Likewise, Frost's writings on science and religion were not very extensive. The AISL pamphlet was his longest published statement, and the larger part has no direct bearing on religion. The only other remarks of any importance are found at the end of his William Vaughn Moody Lecture, "Fragments of Cosmic Philosophy," which Frost delivered at the University of Chicago on December 3, 1930. It contains only a few hundred words directly pertinent to the title and was appended to his autobiography only at the suggestion of others.[8]

Overall, there is not much to go on. Frost believed in a cosmic intelligence of a spiritual nature, similar to Einstein's conception, which he mentioned favorably in "Fragments of Cosmic Philosophy." He believed "that we may gain a finer conception of the Creator by the study of His works," but the word "God" is notably absent from the "Fragments" and appears just three times in the pamphlet. His universe was a cosmos, not a chaos; evolution had not happened simply by chance, but he found "evidence of the perpetuity, future and past, of the universe in a way of cycles such as are suggested in other realms of organic nature" and the "divine power," whatever it may be, "does not need to have in it any supernatural elements." Indeed, "For Omniscience and Omnipotence there are no miracles."[9]

Despite his apparent hesitation to mention God, Frost was a lifelong Congregationalist. The younger son of a Dartmouth College professor of medicine, he attended church "in the somewhat chilly Calvinistic atmosphere of a New England college town." His wife's father, Marshall C. Hazard, was editor of Pilgrim Press (the Congregationalist publishing house) for a quarter century, and Frost was a trustee in the Congregational Church (now United Church of Christ) of Williams Bay.[10]

Mathews invited Frost to write a pamphlet in April 1924, three months after Frost had sent him the typescript of his chapter for *Contributions of Science to Religion*. Enclosing copies of the first four pamphlets, Mathews spelled out what he wanted: "We now feel that we should have a pamphlet which gives some of the great facts, the marvels one might say, of modern astronomy and which is at the same time prepared with the purpose of helping people to see that scientific discovery increases rather than diminishes knowledge of God and the religious impulse." Mathews added that he hoped to distribute 100,000 copies in the next year.[11] However, the first print run in September 1924 probably amounted to 50,000 copies.[12] Apparently there was a second impression at some point, but I have never seen a copy, nor have I located information about it. A third impression of unknown quantity was carried out in September 1930.

---

## Annotated Text

### *The Heavens are Telling*

By Edwin B. Frost

/3/ The spectacle of the starry heavens is the privilege of everyone who has eyes with which to see.[13] It is only necessary to escape from the glare of artificial light which modern civilization finds needful for its streets and public places. In the quiet of the country, or on the upper deck of a steamer, or, best of all, from the vantage of a mountain peak, we can obtain without instruments and without guidebook, if the night be clear and if there be no moon, a view of the richness of the stellar heavens which well repays the slight effort required. If it is a transparent night, the eye beholds the stars of many degrees of brightness and of varied shades of color, shining steadily from above and twinkling near the horizon, while some brilliant planet with serene radiance may be a dominant factor in the scene, or as Milton says:

. . . now glowed the firmament
With living sapphires; Hesperus that led
The starry host rode brightest.

/4/ Athwart the arch of the heavens will appear the soft outline of the Milky Way,

Which nightly as a circling zone thou seest
Powder'd with stars.[14]

This scene is one of beauty not soon to be forgotten. Such a view can hardly fail to inspire in the observer a feeling of wonder and of worshipful admiration of this partial exhibit of the external universe.

If the watch of the glittering sky is maintained for some time, or repeated at a later hour, the stately motion of the stars becomes apparent as our planet turns beneath them. Such views, as we have said, are the heritage of all who will take them, and have been shared alike by the untutored savages and by those of cultured mind since human eyes first began to look upward. The pleasure is the greater for those who have often watched the stars and have come to know the constellations, so that each hour of the night brings new but familiar stellar friends into view. It cannot be questioned that every increase of our appreciation of the creation must enhance /5/ our respect for its Creator. It is equally clear that the expansion of our knowledge, due to deeper study with the use of powerful instruments and a better interpretation of the significance of what we see, can only add reverence to our respect. As was said by the philosopher Kant, one of the first to formulate a theory of the development of the solar system,

> Two things there are that inspire wonder and constantly increasing reverence the oftener and the more they are considered—the starry heavens above me and the moral law within me.[15]

The marvelous photographs which are now being obtained reveal a wealth of celestial splendor that is almost appalling. Interpretation is necessary to understanding, and we shall try to give an idea of the enormous span of the material universe, of the immense number of luminous objects it contains, and of the vast reaches of time which are required for their development.

The brightness of the stars gives us no certain indication of their distance. This must be determined by the most precise of measurements made with the most powerful of instruments. The stars /6/ differ in their glory because some are enormous giants while others are dwarfs like our sun. Some are white stars with intensely brilliant surfaces, while others are ruddy, with fading radiance. Some are known to be non-luminous—relatively if not absolutely dark.

Our ideas of size may be quite incorrect. The sun and the moon, for example, appear to be of the same size in the sky, but the proper measurements show that the sun has a diameter 400 times that of the moon. Because it happens to be also 400 times as far away, it presents the same angular diameter to the eye. We, of course, recognize that the sun is very much brighter than the moon; we should hardly guess what the measurements show, that it is more than 500,000 times as bright. Our interpretation of what we see must therefore always be dependent upon most patient and minute measurements extending over long periods of time.

As our appreciation of the sizes and distances in the universe has increased, new units of length, mass, and time have to be employed to express the vast quantities in question. It was once the /7/ custom, and still is in some parts of the world, to state the distance between two points on the earth by the time it would require to walk that distance. Thus, a Swiss native will tell you that it is two hours to the next village. In astronomy we have to adopt similar expedients for giving the idea of the immense distances involved. It is hardly possible for us to grasp the significance of such a moderate distance as that from the earth to the sun, often taken as the unit of length in measurements of the solar system. That it is 93,000,000 miles means little, for few of us can appreciate the quantity known as 1,000,000. Let us try to realize how enormous this system is by the use of a modern illustration:

The speed of 100 miles an hour is not uncommon for an airplane. Suppose, then, that some tireless aviator in a machine carrying an inexhaustible supply of fuel could fly on without stop at this rate. Hurrying on by day and night, he would pass around the earth at the Equator in 10 days, traversing 2,400 miles per day. If we now suppose that his machine could fly across the vast airless space beyond the earth's atmosphere /8/ at the same speed, he would reach the moon, the earth's little brother, in 100 days, but to arrive at the sun, our aviator would need also the gift of perpetual youth, for at this constant speed of 100 miles per hour, 106 years would be required to reach the sun. The planet Saturn would be passed in 1,000 years, and Neptune, which is at the frontier of the solar system so far as is known at present, would be attained in 3,000 years of ceaseless flight. Leaving the solar system, our imaginary aviator would now begin his real passage through the desert waste of space, and it would take him 27,000,000 years to reach our nearest stellar neighbor, the system of Alpha Centauri. To cross this stretch of void, the swift rays of light require 4 1/3 years.

For stellar distances the unit known as the light-year has been found to be most convenient. It is the distance that light would cover in a year traveling in empty space at the uniform speed of 186,000 miles per second, or 11,000,000 miles per minute. This unit is 63,000 times the astronomical unit, or the distance from the earth to the sun; the /9/ light-year is, in fact, only a little short of six million million miles. Only 500 seconds are required for light to pass from the sun to the earth; it would take light to reach us from Neptune slightly more than 4 hours; from the nearest known star, 4 1/3 years, as mentioned above; from the average star visible in a small telescope, about 800 years, or, in other words, the distance would be 800 light-years. Our whole conception of the scale of the sidereal universe has increased very greatly since the beginning of the present century. Previous to that it had been possible to measure with accuracy the distance of hardly more than a dozen stars; now we have fairly reliable determinations of the distances of several thousand.

Careful investigations have recently been made which show that, counting our own sun as one and at the center, there are only 105 stars situated in an imaginary sphere in space having a diameter of 65 light-years. This illustrates the enormous vacant spaces around each star.

Although the stars may seem innumerable to one who is looking upon /10/ them on a perfect night from a well-chosen station, yet, as a matter of fact, hardly more than 2,000 are thus visible to one person at any one time. The fainter stars are vastly more numerous than those seen by the naked eye. In fact, only one-twentieth of the starlight which may guide the wayfarer at night is due to the stars which he can see; nineteen-twentieths of the illumination is due to stars too faint to be seen as such. Estimates of the total number of stars must necessarily be very uncertain, and allowance has to be made for those which are beyond the reach of our present telescopes, but the evidence is that *they are not infinite in number*. A recent estimate has placed the number of luminous stars at about fifteen hundred million; several million have already been catalogued on the photographic charts of the heavens.[16]

Let us now consider the dimensions and qualities of a star, taking our sun as an example, although it is smaller than the average star. It contains over 300,000 times more matter or mass than the earth, and its volume or bulk exceeds that of the earth more than 1,000,000 /11/ times; in other words, over 1,000,000 planets like our earth could be put inside our sun. It emits heat as would a perfectly radiating surface at a temperature of about 11,000° F., but in the interior its temperature may reach several hundred thousand degrees. The continuous radiation from a square yard of the sun's surface expressed as power is about 1,000,000 horse-power. This is vastly more than could be produced if the sun were of the best coal burning in pure oxygen. In fact, the temperature of the sun is so high that there could be no burning or oxidization. The source of the sun's heat and its steady maintenance are not yet fully understood. If a gaseous ball like our sun contracts, as it must under the enormous pressure of its own gravitation, which is twenty-seven fold that at the earth's surface, then the present supply of heat resultant from this continuous compression would last perhaps for

50,000,000 years, but the geological evidence indicates that one thousand million years or more are required for the development of our earth to its present condition. This is from twenty to fifty times the time available /12/ on the contraction theory. New discoveries of radio-activity suggest new sources of heat from the transformation of the elements in the sun, which are probably adequate for its maintenance for as many billions of years as may be required.[17] The sun's material is chiefly or wholly gaseous with an average density not twice that of water; it contains vast amounts of vapors of iron and of metals familiar on the earth, together with enormous amounts of permanent incandescent gases like hydrogen and helium. In fact, about one-half of the elements found on the earth have already been demonstrated by the spectroscope to be present in the sun; the others are doubtless there, but are not brought out under the conditions of the sun's radiation.

We have thus tried to give an idea of the immense scale of our sun and of its characteristics; but we must remember that it is less than an average star. It has been found, however, that the mass in the stars is not so much greater than that of the sun, but the size and hence the volume may be vastly greater. Thus, recent measurements of the diameters /13/ of a few of the stars have indicated that such stars as Betelgeuse and Antares have a bulk of from thirty to forty million times that of the sun! Other stars vastly exceed our sun in brilliance; thus Rigel, one of the glories of the constellation of Orion in the winter sky, would be at least ten thousand times as bright as our sun if we could come as near to it as we are to the sun.[18]

The spectroscope, one of the magic instruments of modern research, spreads out the light of a star into a band of color with red at one end of the band, violet at the other, and myriad hues between. There are gaps in the band, some very narrow and some broad, which are called the lines of the spectrum and which teach us many important things about the physics and chemistry of the stars. When photographs of these stellar spectra are compared with each other, it is found that they may be arranged in a long series merging from one into the next, and representing successive stages of the evolution of the stars. These changes must occur in general very slowly, requiring periods of /14/ time far beyond any possibility of record in the lifetime of a race, but we may be as confident of the progress of the stars' evolution as a child at an early age is certain that human beings pass through various stages of life from youth to age; the child realizes that these changes must occur although he has not lived long enough to see them. There are many points in stellar evolution which require for their explanation much more than our present knowledge, and, as a matter of fact, we are just now decidedly unsettled in our views as to the relation of the stars to the nebulae, but all the evidence points toward an orderly evolution of celestial bodies.

Much attention has been given in recent years to the globular clusters of stars. These are immense systems roughly spherical in their shape, containing vast numbers of brilliant suns. They appear in the telescope so crowded that they overlap each other and become a hazy mass in the central portion. In reality they must be billions of miles apart, although clearly members of one relatively compact system in which the movements of each is under the gravitational /15/ control of all. There may well be a million stars in the great cluster in Hercules, which is faintly visible to the naked eye as a little patch of hazy light. Fairly accordant estimates of the distance from us of these clusters made by different methods place the nearer of these remarkable objects at a distance of some 20,000 light-years, and the more remote at some 200,000 light-years. In other words, if a star in one of the more distant clusters should suddenly explode, 200,000 years would be required for the rays of light to bring us the visible news of that catastrophe. The fact that there is no great difference in the conditions in the nearer of these clusters as compared to those most remote shows that their development proceeds very slowly, requiring periods of time almost beyond our imagination. We can sweep over vast periods of time as well as depths of space.

There is another class of celestial objects which are of the greatest interest and which have been studied in the last few years with success by photography. These are the spiral and spheroidal nebulae. Perhaps a million of them are /16/ within reach of our present photographic telescopes. Estimates of their distance are very difficult to make, and vary greatly according to the method used, but it cannot be doubted that some of them may be a million light-years from us. One of these is visible to the naked eye as a fuzzy little spot in the constellation of Andromeda. It may be one of the nearest of the spiral nebulae. The spectroscope demonstrates that it is approaching us with a speed in the line of sight of one one-thousandth that of light-rays–that is, about 186 miles per second. If this speed and direction have been maintained for the past billion years, then this spiral has come nearer by a million light-years in that time. It is agreed among astronomers that there may be visible objects at a distance of a million light-years. This, then, may give us some little idea of the modern findings as to the scale of God's universe.[19]

Can anyone doubt that the recognition of the immense size and detail of the celestial universe must give to all who consider it a new and larger idea of the Cause behind it? There is no adequate evidence known to the writer that the /17/ universe is automatic, that it has within itself the power to make the laws which govern it. Mere matter cannot be imagined to be endowed with such capacity. The universe is a cosmos, as has been indicated in what we have already said, and from evidence which we shall cite farther on. It is not a haphazard aggregation of fortuitous and accidental bodies moving without system or order. It is perfectly true that we cannot comprehend in spite of all the efforts of science the whence and the whither or the why of all this. Nor is it to be supposed that these problems will ever be solved by the human mind. Each generation of students may contribute its little part, and sometimes the questions are pushed a little farther back toward the cause, but Omniscience would doubtless be required to understand the works of Omnipotence.

The thought may occur to some that in a universe on such a tremendous scale there would be no place for small objects, that the minute might be regarded as of no importance. The answer is that the laws of the universe apply equally to the almost infinitesimal parts, the molecule, /18/ and the atoms of which it is composed, as they do to the gigantic bodies which it includes. The marvelous experimental researches in physics of recent years have taught us that the atom is a very complicated structure and that the electrons move within the atom somewhat as do the planets around the sun, and indeed with no greater crowding than occurs in the case of the planets; yet these inconceivably minute electrons follow the laws of the universe quite as truly as do the major structures.[20] The Omniscience which we predicate can take within its ken the infinitesimal as well as the infinite. Does not this give us a more spiritual view of the Author of the Universe? It is so evidently "a house not made with hands"; it transcends so far anything that could be produced by any infinitely magnified model of the human form that we must recognize that only a spiritual power can lie behind it.[21] It is not surprising that, in the dawn of civilization, men, in their very limited knowledge, imagined their Creator as merely a superman, to whom they attributed many of the faults of human nature, whose domain embraced only /19/ that of their own tribe. Yet some of the writers caught at times the idea of the grandeur of the universe, as the psalmist when he said:

When I consider thy heavens, the work of thy fingers, the moon and the stars, which thou hast ordained;
What is man, that thou art mindful of him? and the son of man, that thou visitest him?[22]

Now, although the earth is so insignificant and man and his works are so small, even in relation to the earth, it is interesting to consider what astronomy teaches as to the unity of the universe. Might we not expect that among the millions of suns we should find a great many kinds of matter representing thousands of chemical elements? We do not, however, find such variety, and the testimony of the spectroscope is sufficient. The atom of hydrogen is shown to be the same throughout the universe, and it is found in every self-luminous celestial object yet observed. Similarly, the atoms of the other elements do not vary from star to star. With only one or two exceptions, all the chemical elements are found on earth that are /20/ known to exist in any of the celestial bodies. One exception is the gaseous element nebulium, which is a permanent constituent of one class of nebulae. It is still possible that this will be found on the earth, although there is no place for it in the present series of elements of the chemist. Perhaps it is not an element, but a combination of elements, a molecule.[23] But aside from such exception, we learn that our sun is like the most distant star, and that our earth is chemically quite the same as the sun, and finally that our own bodies are composed of the more common elements of the earth. Thus, we may truly regard ourselves as samples of the whole universe. This may well give us the sense of a new dignity as citizens of the cosmos. We may even go so far as to think that the combination of spirit in material body such as we possess may not be of great variety, and that we may be not vastly dissimilar from beings which may inhabit other planets circling around their appointed suns. Science at present cannot either affirm or deny the existence of such planets; they are beyond the reach of modern telescopes. The largest planet /21/ of our own system, Jupiter, would be wholly beyond detection with our telescopes from the distance of the nearest star. Our sun occupies no central or pre-eminent position among the stars—merely one of thousands of suns which seem to have passed somewhat beyond the middle point of their evolution. There is no logical reason to suppose that our sun is any better fitted to have planets about it than thousands of others, or that the planet earth should be highly exceptional. It is granted that mathematicians find great difficulty in understanding how a system of planets can develop around a star, but we know that such a system exists in the case of our sun, and it is difficult to believe that a similar development has not occurred for vast numbers of other suns. In the words of Pope:

He, who through vast immensity can pierce,
See worlds on worlds compose one universe,
Observe how system into system runs,
What other planets circle other suns,
What varied being peoples every star,
May tell why Heaven has made us as we are.[24]

If it may be permitted to an astronomer to consider the social development of /22/ the earth from a planetary standpoint, that is, as merely one of many presumptive planets with possible inhabitants, then we shall have to admit that this social evolution has been very slow, almost discouraging, in spite of all that Christianity has done for it. Paleontologists teach us that man has lived upon the earth for some 75,000 years. Nevertheless, one of man's chief occupations during much of this period has been to take from his neighbor, by force if necessary, even to the point of murder, that which he desires for himself. He has been suspicious of his fellow-men living in different parts of the earth with ways and habits unlike his own. He has had the strongest of prejudices against those who may differ from him in race and color. This is not a reproach to Christianity, for Christianity has been the principal cause of improvement in these matters. Nevertheless, collective murder is still current in the world, and it seems difficult enough to get rid of it. We can imagine planets that have developed otherwise, with brotherly love as the main motive in life, planets where the altruistic /23/ principles of Christianity have reached a far higher development than here. But this is the only planet with which we can have any contact, so far as science foresees at present, and although we may be heartily ashamed that there is so much evil on the earth, we must make the best of the situation and try to improve it. Let not the reader misunderstand us to be advocating the policy of pure pacifism. The destructive tendencies of individuals and nations must be controlled, and controlled by force if necessary. The good Samaritan of the parable was commended; but would he not have been a better Samaritan if he had done his duty as a citizen and had taken up responsibilities which the priest and the Levite had shunned, and had seen to it that the road to Jericho was so policed and so protected that robbery would have been impossible on that highway?[25] From the planetary point of view, the most thorough application of the principles of Christianity seems to be the only way to bring our planet up to the moral standard to be expected of it.

The principle of faith is not at all foreign to science and to its workers, and /24/ this faith is quite akin to that of religion. The labor of scientists is chiefly directed by their faith in great principles, and by their firm belief that the universe is one of law. Otherwise why should men and women give up their lives in patient research in the endeavor to discover the principles which hold true and are part of the laws of nature? There would be no encouragement to such research if the operations of nature were purely capricious and without law. The process of scientific discovery is that of following some incomplete working hypothesis or theory based upon previous experiment or observation, and then of testing this hypothesis by further experiment and observation, holding to that part of it which proves to be good, and improving it as the work proceeds. Such faith in the eternal principles of the Creator, as we have said, differs not much from Christian faith in the eternal moral principles of God.

The idea of immortality of spirit is also not far different from some fundamental beliefs of the physicists. The name is different, and conservation of energy is the principle. This affirms that the /25/ sum total of energy in the universe is constant, changeable from one form to another, but essentially immortal. Now this principle, to our thinking, cannot be rigorously demonstrated any more than can the immortality of spirit. Those who have crossed into the realm of spirit do not communicate with mortals; at least, there seems to be no thoroughly scientific evidence of such communication. The conservation of energy can be tested by experiment only within a limited field.

Another illustration is the theory of an all-pervading aether, a theory which has been tenaciously held by physicists for a century past. It is difficult, if not impossible, to demonstrate the objective reality of aether, but the mind demands the existence of a medium by which the waves of light may transfer their energy across the almost limitless void of space from distant suns or nebulae. Although the recent theory of Einstein, which postulates the aether in its initial stages as a theory, abandons it before it is through, it still remains to be seen whether or not a substitute for the aether has been found which /26/ will be permanently acceptable to science.[26]

We are contending for the view that the scientific study of the material world tends to separate the spiritual realm from that world—in other words, that it tends to make spiritual conceptions less material. In physics the fundamental notions or qualities which are not purely numerical are expressed in terms of mass, of length, and of time, that is, of quantity of matter, of extent in space, and of duration of time. These are technically known as "dimensions" of physical units. Thus, velocity is length divided by time; energy is mass multiplied by the square of length and divided by the square of time. By definition or common understanding of spirit, shall we not regard it as wholly divorced from mass or quantity of matter in the first of these dimensions? Secondly, can we associate length of extent with spirit?[27] Does it not seem illogical, if not absurd, to think of a large spacial extent of spirit? Finally, may we not question whether spirit has any relation to time or the succession of material events? In other words, a spirit is wholly distinct /27/ from matter; must it not also be beyond the restraint of the material "dimensions" of matter? If the possibility of this speculation is admitted, there would be no question *where* to answer in regard to spirit, and presumably no question *when*.

With the advancement of science, our recognition of the supremacy of law in the material world has greatly increased. The motions of the celestial bodies were among the first to be established by the immortal discoveries of Copernicus, Kepler, and Newton.[28] Cause and effect are much more difficult to discover in biology, but the penetrating researches of Pasteur and many others have revealed the micro-organisms which cause diseases in men and animals.[29] Many of the factors controlling men's actions and relations have been analyzed; much that was formerly attributed to a capricious deity has now been seen in a far more dignified light as the operation of natural law. Certain phenomena were long considered as beyond law, such as the lightning stroke and the tornado. These things were regarded as a divine retribution for human sin. Electric /28/ discharges of a million volts are now produced in the laboratory and the weather conditions which develop tornadoes are gradually being understood so that they may be predicted with some degree of certainty. Human destiny is subject to these physical laws. Human lives are wiped out by storms, by floods, by disasters, but all as a consequence of laws as yet only partially understood. Science reveals critical stages in the changes of matter; for instance, water suddenly turns to steam as a certain definite temperature is reached. After a regular progress, an abrupt change may occur. In astronomy we have seen with our own eyes the sudden climax in the history of a star, when within thirty-six hours its light has increased many thousand fold, rising to be the rival of the brightest star in the heavens, and then fading away until it returns to its previous condition of relative inconsequence. Tycho saw such a star, as did Kepler, and two have attained the brilliance of the first magnitude within our own experience in the past quarter of a century.[30] These cataclysms seemed at first purely fortuitous, /29/ sporadic, supernatural. But as minute study has been given to the changes in the spectra of these stars, we begin to see law and regularity in what was previously accident. In fact, it may be the destiny of any star to have its day of glory in the course of its evolution. Or such outbursts may occur periodically, repeating themselves after a few thousand years or longer. The record of science is still far too short to give a history of such cosmic phenomena. Thus, as science progresses, order and law become more and more recognizable, in the test tube or in the stars.

It seems to be a sacred and reverent duty to thoughtful men to go on with the study of these laws and to find man's relation to them. When they bear most heavily against him, he will seek to adapt himself to them, and he will always have as a supreme illustration these words of Christ, "Thy will, not mine, be done."[31]

The quest of science is the Truth within the material universe; that of religion can be no less limited in its range, but refers more particularly to moral truths and the relation of man to the Creator and to his fellow-men. It is hard to see /30/ why there should be an apparently deliberate effort to develop or to maintain an estrangement between these two forms of effort to gain useful knowledge.

We cannot be expected to believe that the inspiration of the seekers for truth has been confined only to those of certain limited periods, or of a certain race, or that the inspiration was ever complete, or that it ever has ceased to be given in some degree to honest workers in the quest for truth. For us, the two friends, Milton and Galileo, were glorious illustrations of inspired men of their time.[32]

We have no worry about miracles. The marvels of life and growth that are daily seen about us in a natural way seem miraculous almost beyond belief, yet they are observed facts. Who can find the mathematical formulae under which that little bulb in my garden, subject only to the environment of soil, of water, and of air, can send up its shoots in April of each year, and develop in its exquisite detail the trumpet narcissus, always essentially the same and gathering up from its surroundings that delicate odor characteristic of it? Beside it may /**31**/ grow the dandelion, subject to the same external conditions, but so different in the result.

Again, we are still very far from understanding how that frail, bright-hued bird can find his way across some hundreds of miles of sea and thousands of miles of land, back again to that maple tree where last year the nest was built. Yet, when at last such mysteries are solved after patient research, it will, of course, be found that they are operations of natural law in God's world. For Omniscience and Omnipotence there are no miracles. The development of a human being is doubtless as complicated as that of a star, but from the atom to the star and from the microbe to the man, we can believe that the same divine power holds sway.

# *Through Science to God, The Humming Bird's Story, An Evolutionary Interpretation*

Samuel Christian Schmucker (1860–1943)

> Somehow the infinite power, acting with infinite resource, having infinite matter through which to operate and infinite time to accomplish its unendingly increasing results is behind all of this activity. I like the happy phrase "creation by evolution." It is not a Godless phrase. It does not imagine a Godless process. It is our impotent way of reaching one step farther towards a glimmer of an understanding in the workings of an Infinite Power whom we lovingly name God.
> —*Samuel Christian Schmucker,* Heredity and Parenthood

## Introduction

The author of the sixth pamphlet, naturalist Samuel Christian Schmucker of West Chester (Pennsylvania) State Teachers College, is the only one whose name is all but unknown today.[1] During the 1920s, however, few scientists were more widely visible to the American public than Schmucker. A very popular speaker at the Chautauqua Institution and its regional affiliates for more than three decades, Schmucker wrote five books about evolution, eugenics, and the environment for major publishing houses. *The Meaning of Evolution* (1913) was printed eight times in a dozen years by Macmillan, and the first edition was published simultaneously by the Chautauqua Literary and Scientific Circle as a national course text in 1913–1914. *Man's Life on Earth* (1925) was a Chautauqua text in 1925–1926.[2] Shailer Mathews served as a trustee of the Chautauqua Institution and directed its religious work for much of this period. He spent many summers there and knew Schmucker, perhaps quite well—they were less than three years apart in age, and Mathews would have had numerous opportunities to attend his lectures, which drew very large crowds and sometimes received top billing in advertisements for the annual program.³ Schmucker also spoke frequently at other prestigious venues, including the Brooklyn Institute of the Arts and Sciences, the American Museum of Natural History, The Cooper Union, and the Wagner Free Institute of Science in North Philadelphia. In academic circles, Schmucker was nationally known for his commitment to nature study, an early form of environmental education with a large spiritual component that was well represented on the Chautauqua circuit. He was president of the American Nature Study Society in 1917 and 1918.

Schmucker's theological pedigree was even more impressive: his grandfather, the controversial theologian and abolitionist Samuel Simon Schmucker, founded the first Lutheran seminary and college in the United States at Gettysburg. His father, Beale Melanchthon Schmucker, was instrumental in starting an alternative Lutheran seminary in Philadelphia. Although Samuel Christian Schmucker was an Episcopalian, not a Lutheran, and he never pursued a clerical career, his pamphlet contains some of the most theologically sophisticated material in the "Science and Religion" series, almost presaging process theism for the way in which it denies the divine creation of the laws of nature. The AISL printed 50,000 copies in September 1926; no further printings are known.

A Christian pietist with a profound sense of ecology, Schmucker was a post-Darwinian natural theologian, arguing from the order and beauty found in nature to the wholly immanent God he saw behind it. This is especially evident in his pamphlet, which was probably based on a lecture about birds that he had been giving for many years. Commenting on the interplay of natural and sexual selection in the evolutionary process, Schmucker noted how goatsuckers are camouflaged to hide from predators, while hummingbirds are brightly colored to attract mates. This exemplifies "two overpowering emotions, fear and love," in which "one is at war with the other," leading Schmucker to conclude, "Here at last, in the bird-world as in the human world at its highest, perfect love has cast out fear." Beauty seems to increase throughout evolutionary history, and the whole trend of evolution is "steadily upward, through long, succeeding ages," leading to the immanent God, who is almost reduced to the evolutionary process itself: "The laws of nature are not the decisions of any man or group of men; not even—I say it reverently- of God. They are co-eternal with God, "the manifestation in nature of the presence of the indwelling God."⁴

Consistent with his view of God and nature, Schmucker placed ultimate hope for humanity in evolutionary progress. Through the struggle for existence "God is ever raising his great creation to higher and nobler levels."[5] The process of evolution would eventually achieve moral perfection, as we slowly cast off our animal nature. "Slowly the brute shall sink away, slowly the divine in him shall advance, until such heights are attained as we to-day can scarcely imagine." The process of evolution, driven for Schmucker "by a Spirit that groans and travails through all creation," took all living things "to higher and higher levels, without leaving unoccupied the lower ranks." The ultimate result was "a creature capable of recognizing the Power which has made and is making him what he is, and filled with a striving to work towards His likeness."[6]

The neo-Lamarckian conception of evolution underlying Schmucker's view of moral progress was still common among scientists of his generation. So was enthusiasm for what he called "the new, and as yet very imperfectly developed practical science of Eugenics." As he saw it, the "high hope" of eugenics was to "increase the proportion of fine strong beautiful upright human families and diminish the ratio of shiftless, weak, defaced, unmoral people," and thereby "the world will be bettered for ages." Since "there is no limit to human perfectability," the key was simply to persuade enough people "of the truth of what the scientist knows and to act on it."[7] Although his pamphlet did not introduce readers to this evolutionary eschatology, it did show them the God who lay entirely within the processes that would ultimately perfect them.

---

Annotated Text

## *Through Science to God, The Humming Bird's Story, An Evolutionary Interpretation*

By Samuel Christian Schmucker

### The Facts

/3/ To anyone who really cares for birds, our common nighthawk is full of interest. Whether on the wing, capturing his prey, resting between flights, or nesting, he has an individuality that marks him as very different from the robins, the bluebirds, the catbirds, and the orioles.

His method of flight is quite his own. He seems to fly easily and to keep it up as long as he wishes, but his course is so erratic, so constantly shifting in its direction, that in the South he has gained the name of bull bat. The bat part of this name is due to the vagrant flight, so different from that of most birds that spend much time on the wing, like the swallows or the hawks. The bull part of the name has been given as the result of a strange sound he makes at intervals in his flight. In the late afternoon one may see him careening about the sky at a height of several hundred yards and uttering a /4/ very frequent "peet! peet! peet." Suddenly all is changed. Almost vertically downward he swoops, on a clear curve far different from his previous dodging course. Then his progress is suddenly arrested, while a deep, booming "gawk" sweeps down through the air, and he is away on his old, careening flight, with its steady "peet! peet" only to be interrupted again with the quick drop and its concomitant "gawk." The reason for this curious procedure is that his food consists almost entirely of insects, and he catches them always in flight. As he goes wildly about he doubtless sees beneath him a droning beetle, a pollen-laden bee, a buzzing fly, or even such small prey as the flitting mosquito. His wanderings are interrupted. He starts with precision, drops with accurate aim and, with wide open mouth, captures his food. A quick outthrust of wings, and of tail rigidly set, breaks his fall. The "peet" was his voice, but it is the quivering, firm-set quills of his wings, with their resistance to the drop, that probably give off the "gawk" as the air rushes through them.

In his resting this bird has a trick of his own, imitated in our neighborhood by but /5/ one other bird, strangely like him in appearance and habit, the whippoorwill, which is a member of the same bird-family as the nighthawk. Most birds, when they rest on a tree, catch their toes about the limb and sit crosswise of their support. The nighthawk always sits lengthwise of the branch, usually with his tail toward the trunk, his body closely pressed to the bark, and his head well sunken into the feathers of the neck. His plumage is very soft and fluffy, dark brown in color, much specked and mottled with small flecks of gray and buff. So entirely does he look like an old knot on the branch that it is almost always only by accident that one discovers him. Usually this occurs when one happens to strike the limb with stick or stone, or to climb into the tree, when the supposed knot suddenly is up and away.

Through much of the day the nighthawk sits thus quietly, often day by day in exactly the same spot. In the late afternoon many kinds of insects which, for safety, have been keeping themselves out of sight of their bird enemies, now come forth to search for food, or still more likely, for a mate. The nighthawk /6/ has gained almost a monopoly of these insects by himself taking to night flight. The later afternoon sees him started, but all night through one may hear the frequent "peet" and the much rarer "gawk" even over the heart of a great city.

On the top of the Grand Central Station in New York, on the pebble-dashed roof of a factory in Cleveland, but of course much more frequently on stony ridges all over the country, this bird may be found nesting, if such name can be applied when she makes no nest but seems to lay her two dark, mottled eggs on the ground and to trust to her remarkably groundlike color to hide her and them.

After the discovery of America, and the bringing home to Europe of "curiosities" by thousands from the new world and the more recently discovered parts of the old, the museums of the world became cluttered with specimens. Carl von Linne, botanist of the University of Upsala, devised a system of arranging specimens, based on their similarities.[8] While he spoke of animals and plants as belonging to the same "family," he meant no more than if a furniture manufacturer /7/ were to put together in his catalogue chairs, stools, benches, and sofas and call them all the "Chair Family." Linnaeus put into the same group animals or plants that seemed to him much alike. His successors have carried his idea very far and any discoverer of a plant or animal previously undescribed, not only gives it a double Latin name, but assigns it to the family it seems to him most closely to resemble. The bird which, in Europe, is most like our nighthawk is called the goatsucker, from the absurd notion that it deserves the name. This name has been applied to the entire group which is now known as the "Goatsucker Family." These birds range all over the world, except the cold north where its food is too scarce, and a few detached Pacific islands to which they seem never to have spread. When in any great collection of birds one comes to this family with its nearly one hundred species, the observer is struck with the marvelous similarity, in form and in color, though great diversity of size, among the goatsuckers.

In such grouping of birds in collections and in catalogues, students are now fairly /**8**/ well agreed that the next family to the goatsucker, because most like them in actual structure, is a group represented in our neighborhood by a bird much better known than the nighthawk, namely the chimney swift. His flight, in some respects like that of the swallows, has led many people to call him the chimney swallow, but this is a mistake. He is in many essential characteristics unlike the swallow and he should be called a swift. The bones of his wings are very long, much longer than those of most birds in proportion to the body length. In this, and many other respects, he is much like the goatsuckers. Like them he is an insect eater, and catches his food on the wing. Like them, too, he is dull of color. But his flight is more accurate by far, he dodges better and flies faster. He is very small and hence may content himself with smaller insects. These he gathers in his unending flight, carried out most eagerly in the early morning and the late afternoon, but lasting all day through. At night he drops into a crevice of the rock, a hollow tree, or best of all in closely settled neighborhoods, into a chimney, disused in summer. Here the swifts build /**9**/ their nests made of small twigs, often broken from the parent tree without arresting the flight of the swift. These sticks are glued to each other and to the wall with a sticky, varnish-like substance which is really saliva. In Japan and China some of these swifts have discarded the twigs in their nest-building, and use only the saliva which hardens into a gummy substance which can be dissolved in boiling water, thus making the much-prized "bird's-nest soup" of those countries.

Like the goatsuckers the swifts are found all over the world. Again there are nearly a hundred kinds of them, but they are much less uniform in color and in shape than are the goatsuckers.

Still more closely like the swifts than are the goatsuckers, are those marvelous gems of the bird world, the humming birds. These would seem to the untrained bird student to be almost in a class by themselves. Their colors are wonderfully brilliant and the feathers seem to have over their colors a metallic luster, while the wings quiver so rapidly as to make them seem almost only a blur in the air. They are so exquisitely small that they /10/ stand in marked contrast to all the rest of the American bird world. True, we find in southeastern Asia and on the adjacent islands a group of sunbirds that to the ordinary observer would certainly seem to be in the same group as the hummers. Their habits are the same, their coloring is as marvelous, their size almost as exquisite. However, the careful student, who goes beneath appearances and studies the form of bone and muscle, knows these sunbirds are widely different from the hummers, which latter are found only in America. The humming birds have long wing bones, deep keels on their breasts, and a peculiarly shaped wishbone and palate found most like them in the swifts. So next to the swifts, and very near to them, they must be placed in our collections.

This exquisite flier has carried to perfection the quick flutter of the wings which allows the creature to hold its body poised in the air. Most birds can do this momentarily and infrequently. The hummer does it as his constant practice. His bill too is very long and his tongue is a tube. He hovers in front of a honeysuckle, an azalea or, best of all, a trumpet /11/ flower, probes the depths of the corolla with his long, slender tongue and sucks up the nectar. When one knows his habit of eating this, the daintiest of foods, it comes with quite a shock to find the stomach of the hummer stuffed with insects. Sometimes these are the bees and flies that visit the flowers, but this is not always the case. I have seen our ruby-throat gleaning the eggs and larvae of insects from the trunk of an oak tree, in front of which it poised and up and down which it moved, darting repeatedly at the bark. The nest of our hummer is as dainty as the bird itself. Perched on a slender limb, constructed of fernhair and thistledown, and covered concealingly with lichens, it is a fitting home for its jeweled inmate.

The variety of these birds equals their brilliancy. There are about five hundred kinds of them. While a few of them are found far down the continent, most of them inhabit the tropical forests of the northwestern corner of South America. They are abundant also in Central America, grow fewer in kinds in Mexico, and only a dozen or more species extend up into California. But one, the ruby-throat, /12/ is found in eastern North America. Over all the rest of the world there is not one hummer, nor any fossil evidence that there ever was any.

### The Interpretation

Is there any significance discernible beneath these facts? Must the scientist stop here? Yes, if he is only to teach what he actually knows to be the "fact." But no student ever stops with the facts. Biography would be a long and terribly dull "Who's Who" if we stopped with the facts. History would be a stupid chronicle if we stopped with the facts. Astronomy would be a barren catalogue of stars and chemistry a list of reactions if we stopped with the facts. There would be no laws, no principles, no impulse to further study, no interest in things acquired if we stopped with the facts. We would have no literature, no poetry, no fiction, no fables, no parables if we stopped with the facts. Of course we must realize the difference between the facts and our inferences, but inferences may be, and often are, quite as true and much more interesting than the facts on which they are based.

Suppose we let the evolutionist's fancy /13/ (if you so wish to call it—he would call it constructive imagination) play for a while on the facts given above.

The story must begin very far back; quite before there were even the most primitive people on the earth. At this time (probably the late Pliocene) a group of birds was very widespread and already had the characters of our goatsuckers. Their great powers of flight and their abundant food made life possible almost anywhere. But already the competition was heavy for insect food and those who learned to come out at night for their prey found it in greatest abundance. These grew strongest and lived longest and left behind them most progeny. Finally those with these habits were the only survivors. Flying as they did, by night, they must rest in the daytime. Yet it is the resting bird that is most easily caught by its enemies, particularly the hawks. Hence it came about that all that were in anywise conspicuous in their resting habits were first killed off. As a natural result of this tendency those that were colored most like their background were the ones regularly surviving. However any of them chanced on the habit, /14/ unusual in birds, of resting lengthwise rather than crosswise of the limb, such as did this escaped attention most thoroughly. When these two qualities blended, lengthwise position on the limb and coloration like the bark, there was a double bid for persistence and a bid that won security for them over a very wide portion of the earth's surface. So the goatsuckers were about as widespread as their food. The external family likeness remained so uniform because it was so simple and so efficient.

It is the regular bird habit to fly by day. There must have long persisted in the goatsuckers a liability to revert to the general and, doubtless, ancestral habit. This was more possible for such of them as were very small. The likelihood was greater if they were also especially active on the wing. Some of these smaller goatsuckers began to have a power of continued flight, like the swallows, when in pursuit of their food. They had, moreover, a quality in their goatsucker flight which the swallows had never gained. This was the power of erratic flight. It is this which has made it safe for butterflies to flit about in the sunshine /15/ in spite of their gaudy color. Any enemy striking at a moving prey, must aim ahead of its position. If then the pursued dodges, the pursuer misses his aim. The swifts dodge with wonderful skill. Many observers believe the swift sometimes uses only one wing at a time. I have never been sure I detected this, but I have keen-sighted friends who are quite certain they have done so. I know of no other birds suspected of this power. So the swifts came back into the light of day. This made their evolution more rapid, because it separated them in mating, from the goatsuckers, and hence sooner overcame the swamping effect of the presence of new and old forms in the same area.

The smaller the body and the quicker the flight, the greater was their power of scattering widely. This lead to the formation in various areas of various great groups of swifts. So it comes about that, while there is great uniformity among goatsuckers, there are three quite distinct groups of swifts: a general form, a tropical form, and the tree swifts of the East Indian region.

All of these developments in the bird-world /16/ must have happened long ago, not later than the Miocene or Pliocene, the periods in which the majority of the great groups of mammals and of birds became fixed. This was perhaps no longer ago than three million years nor less than one. Only such time will account for their wide distribution, over the earth, and for their diversity. There has been but one period since then, the Pleistocene, and that was the period of the glaciers. Then the northern part of both continents was invaded repeatedly by the polar icecap. This formed an impassable barrier to migration between the two great land masses, the eastern and the western. Only animals that were accustomed to the icy north could now pass. A few, living close to the ice, might have gone over in the interglacial periods, but such migration must have been very limited. Even so very adaptable a creature as man seems to have been held back. Human habitation on the Western Hemisphere seems to have been confined to the period since the last glacier, when an offshoot of the great division of men with yellow skin, slant eyes, and straight black hair came across the Bearing, on the /17/ Aleutian bridge, and developed into the American aboriginal.

Meanwhile, in the tropical portions of the Western World, the curtain was lifting on the last act of our drama. There seems to have been on the Andean highlands a race of swifts uncommonly small and unusually quick of flight. The rapid beats of their tiny wings evinced the utmost energy, and they became the darters of the bird-world. A new discovery on their part gave the modern and characteristic turn to the race. They found that one of the best places to obtain their prey was in the throats of the flowers the insects visited for their nectar. Many of the insects these quick-witted swifts captured had been gathering nectar, and the birds who ate them often tasted its delicious flavor. What more natural than that they should add this dainty to their heavier food? What started their bills and tongues to lengthening I cannot even surmise. A bird cannot by thinking add an inch to his bill any more than can a man add a cubit to his stature by the same process. However it started, the innovation was a fortunate one, for it enabled the visitor to the flower more readily /18/ to draw up the nectar. Those swifts that grew this novelty won out most completely over their shorter billed cousins, and the new form impressed itself on the race. That it is a very recent addition is quite evident. New changes do not quickly work back into the young. The human race has acquired the upright position and the power of speech too lately for it to have appeared in the young. When the baby hummer comes out of the tiny egg in which it took form, it is billed like a swift. In the short time it remains on the nest the bill stretches out and with it grows its tubular tongue.

Visiting the flowers in search first of insects and then of nectar, these birds naturally became uncommonly sensitive to color. What more natural than that when a brighter sheen appeared on any of the birds themselves, a process probably aided by their intense activity, it should render them peculiarly attractive to their mates? This principle, whenever it occurs, is its own hastener. Preferential mating tells quickly on heredity. Differences in matters of taste are notorious. Hence variations due to taste will be most abundant. Our humming birds have divided, /19/ within the short space of their evolution, into more than two hundred species.

They are essentially the children of the tropics. The old home in the Andes still remains their chief haunt. South and north they slowly spread, the few forms that became adjusted to greater cold moving down to Terra del Fuego and up to Alaska. These forms are very few and are strikingly less brilliant. Their abundant cousins haunt the tropical flowers, their diademmed bodies quivering in the brilliant sunlight.

It at first sight seems strange that these dainty little creatures should be pugnacious. They are, uncommonly so. This is especially true when their homes are invaded. Then they dart fiercely at creatures as big even as man himself.

There are two great instincts in the animal world, the instinct of self-preservation and the instinct for the continuance of the race. There are two great color-schemes corresponding to these instincts, the protective and the attractive. The former allows the possessor to fade into his background for his own safety, but it also hides him from the glance of /20/ his possible mate. The attractive color draws the mate, but it makes its owner conspicuous to his enemies. Correspondingly there are two overpowering emotions, fear and love. In each of these pairs, one is at war with the other. In most animals the instinct of self-preservation, the protective color, the wary life predominates. In many others both intermingle or alternate. In a few the latter member of each pair has won the victory. Here the race claims marked precedence over the individual.

In the goatsuckers the protective coloration almost alone prevails. In the humming birds attractive coloration has completely won out. Here at last, in the bird-world as in the human world at its highest, perfect love has cast out fear.

### Going Deeper

We have examined the facts as the bird student sees them. We have considered the interpretation the biologist puts on them. Is there not still a lack? What makes all this happen? Why is there this tendency in nature to fit animals to their surroundings? What is there that makes animals spread as widely as they can? /21/ Why did these higher creatures get warm blood and quickened emotions? Why do they also care for beauty and why does beauty increase as time goes on? Why is the whole trend of the history of the animal and plant-world, in spite of occasional lapses, steadily upward, through long, succeeding ages? The plants could not have planned it. The animals could not have determined it. Even man, in the mass, seems unable to work for the betterment of his own race—though his leaders apparently have glimpses of the direction in which such advance could easily be made.

We sometimes speak of the "law of their being," or of the "laws of nature" as bringing this about. But here law does not have the same meaning as it has in human legislation. In our political affairs a law is the expressed determination of a man or of a group of men in authority. Thereafter such laws are obeyed—or disobeyed.

The laws of nature are not the decisions of any man or group of men; not even—I say it reverently—of God. The laws of nature are eternal even as God is eternal. The tendency of two bodies to draw together /22/ is not only as old as Newton, but as old as the bodies. It is inherent in the nature of the bodies. It was not "put there" by a higher power.[9]

The laws of nature are not the fiat of almighty God, they are the manifestation in nature of the presence of the indwelling God. They are the principles of his being as they shine out, declaring his presence behind and within and through the whirling electrons. These eternally restless particles are not God: but in them he is manifest. Science, in studying them, is studying him. Science is man's earnest and sincere, though often bungling, attempt to interpret God as he is revealing himself in nature.

# Creative Co-ordination

Michael Idvorsky Pupin (1854–1935)

> We can either believe that cosmos, the beautiful law and order, is simply the result of haphazard happenings, or that it is the result of a definite Intelligence. Now, which are you, as an intelligent being, going to choose? Personally, I choose to believe in the coördinating principle, the Divine Intelligence. Why? Because it is simpler. It is more intelligible. It harmonizes with my whole experience.
> 
> —*Michael Idvorsky Pupin, quoted by Albert E. Wiggam,*
> *"Science Is Leading Us Closer to God,"*
> The American Magazine *(September 1927)*

## Introduction

At the time of the Scopes trial in 1925, few American scientists were better known than the President of the American Association for the Advancement of Science, Columbia University physicist Michael Idvorsky Pupin.[1] Of the seven scientists who wrote pamphlets, only Nobel laureate Robert Millikan had as high a public profile.[2] Articles about Pupin's life and work had been appearing in popular magazines since the turn of the century. His autobiography, *From Immigrant to Inventor* (1923), was written with evident pride in the many accomplishments of an immigrant whose success story made him a household name in America. This book won the Pulitzer Prize in 1924 and was followed shortly after by two more books, *The New Reformation* (1927) and *Romance of the Machine* (1930).[3] All three were serialized in *Scribner's Magazine*, arguably the leading literary magazine of its era. Consequently, Pupin published more articles in *Scribner's* during the 1920s than almost any other author, including Ernest Hemingway; only Henry van Dyke and Edith Wharton had more. In addition, *The New Reformation* was listed by the Religious Book Club as a selection in the category, "On Religion and Science," alongside works by (among others) Kirtley Mather and the great Quaker astrophysicist Arthur Eddington, which only helped to promote it among liberal Protestants.[4]

Not surprisingly, Pupin's pamphlet was also based on ideas he had already published in *Scribner's Magazine*, his autobiography, and elsewhere.[5] The AISL version was based on a commencement address at Lafayette College that was subsequently published in James McKeen Cattell's magazine, *School and Society*.[6] Presumably, that is where Georgia L. Chamberlin noticed it and brought it to the attention of Shailer Mathews.[7] The pamphlet is identical to that version, except that the opening and closing paragraphs of the article are omitted. The initial run of 25,000 copies in May 1928 was followed by a further 25,000 copies in the next fiscal year (1928–1929). Less than nine pages long, *Creative Co-ordination* was the shortest pamphlet, yet one of the most original.

The idea of "creative co-ordination" was the principal leitmotif in Pupin's many popular writings about science and religion, linking the activities of living things with the incarnate Son of God. As he explained it in a funeral oration for a son of the famous poet Julia Ward Howe, "modern science shows that all terrestrial organisms are endowed with a power of co-ordination which transforms the chaotic, the non-co-ordinated, forms of energy radiated to us by the sun into co-ordinated forms, and thus supplies a sustenance for all organic life on the earth." Isaac Newton had "co-ordinated the scientific achievements of his predecessors ... and created the cosmos, the beautiful harmony of law and order of modern dynamics." Moses had "co-ordinated the moral and ethical experience of many, many centuries and formulated the Ten Commandments." Likewise, Christ's command to "Love thy neighbor as thyself" led Christians to recognize "a transcendental power of co-ordination, a heavenly harmony, which cannot originate in the souls [*sic*] of a mere mortal man, and hence our firm Christian belief that its possessor, our Lord Jesus Christ, is the Son of God."[8]

This passage shows that Pupin was no religious modernist, unlike the other pamphlet authors. Nor was he a Protestant or even a native-born American. The son of an illiterate farmer from Idvor in the Banat region of the Austro-Hungarian empire (later part of Yugoslavia, now Serbia), Pupin was a devout Serbian Orthodox believer whose many writings about science and religion presented his mainly Protestant American audiences with a vision that was both theologically orthodox and subtly Orthodox. Beauty and order in the universe manifested the transcendent divine Word (the Λόγος of John's gospel) through which all things were made, the same divine Word revealed most clearly in Christ. Furthermore, Orthodox doctrine teaches that the divine "energies" express divine action and bring creatures into unity with God. It was probably no accident that Pupin gave special attention to wave motion and other forms of physical energy, which for him were the principal means by which the immanent God creates and maintains order. Finally, his belief that science enhanced the believer's ability to participate in the mystery and beauty of heaven reflects Orthodox worship and contemplation.[9]

Although he differed markedly with the modernists on some central theological questions, Pupin shared their attitude toward Bryan's assault on evolution. Just two years before the Scopes trial, he wrote, "I feel the heavy hand of the fundamentalist pulling me down, and the icy chill of his disapproving voice reminds me that his theology will not permit an interpretation of Genesis which cannot be understood by people whose knowledge of science is about the same as that of the Assyrians and Chaldeans of several thousand years ago."[10] His pamphlet is not about evolution, but the trial was over and Mathews was taking the "Science and Religion" series in other directions. The version of "creative co-ordination" that it presents is less explicitly incarnational than some other versions and therefore less likely to offend modernist sensibilities.

## Annotated Text

## *Creative Co-ordination*

By Michael I. Pupin

### The Universe as a Cosmos

/1/ The prophets of ancient Greece believed that in the beginning this universe was a chaos and that the Olympian gods transformed it into a cosmos, a universe of simple law and beautiful order.[11] The earliest idea of the cosmos was undoubtedly suggested to the human mind by the starry vault of heaven. "The firmament," says the psalmist, "sheweth His handiwork."[12] The stars in the firmament were to the psalmist the visible parts of an ideally perfect structure, a picture of the unchangeable, the eternal. This picture is undoubtedly the origin of the ancient belief among the Indo-Europeans that after death the immortal soul of man rises along the Milky Way to its abode /2/ above the eternally unchangeable stars, where the divinity resides.

It was the sight of the starry vault of heaven which inspired Copernicus, Kepler, Galileo, and Newton. This inspiration was the cradle of the beautiful science of dynamics, which, two hundred years ago, revealed to the human mind the physical reality of matter in motion obeying laws of childlike simplicity. Newton's dynamics gave us the first scientific representation of the planetary cosmos, when it demonstrated that the members of our solar system move with a precision never approached by any mechanism constructed by the hand of man.

Since Newton's time other physical realities were revealed which modified our views of the orderly universe suggested by Newtonian dynamics. Electricity in motion, chemical reactions, electrical radiation, electrical structure of matter and the granular structure of living organisms are the physical realities which have been revealed since Newton's time. These realities have taught us that the universe has not yet become a cosmos, but that there is still an all-pervading invisible chaos. /3/

### The Universe as a Microcosmic Chaos

A glimpse of this chaos was caught by the ancient philosophers like Democritus and Lucretius, when by a strange vision they prophesied that the ultimate components of matter were tiny atoms; that matter, therefore, is granular in structure.[13] Modern science has transformed this dream into a reality, which is much more far-reaching than all the dreams of the ancient philosophers, poets, and prophets. It tells us that not only is matter granular in structure, consisting of atoms and molecules; but that electricity is also granular, consisting of electrons and protons; and that the structure of living organisms is granular, consisting of tiny cells and their microscopic components, call them the molecules of organic life. In other words, science teaches us that the visible universe is a macrocosm, consisting of invisible granular microcosms, and that the cosmic granules, the atoms and molecules, the electrons and protons, the organic cells and their microscopic components of life are all endowed with the power of practically autonomous action. A countless swarm of autonomous granules is obviously chaotic. /4/ Heat is chaotic, because it is the action of the chaotic motion of the molecules of the hot body; light is chaotic, because it is the action of the chaotic motion of the electron, in the atoms of the luminous body; chemical reactions are chaotic, because they are the actions of the chaotic atoms and molecules. Science of the last hundred and fifty years reveals, therefore, that everything that exists derives its breath of existence from the all-embracing microcosmic chaos of heat, light, and chemical energies. And yet, in the midst of this invisible all-embracing chaos, there rise here, there, and everywhere the visible beauties of orderly creation just like so many blessed islands of beautiful order rising from the turbulence of a storm-tossed sea. There is, therefore, in every nook and corner of the universe a never ceasing transformation of the primordial chaos into a cosmos of simple law and beautiful order.

Creative Co-ordination in the Inorganic Universe

This is a new view of the universe never dreamed of by the ancient philosophers and prophets. Let me illustrate: The cylinder of your automobile /5/ engine will tell you that every one of its explosive puffs unchains a chaos of molecular motions, but that the piston averages up their chaotic pulses and transforms them into a steady pressure which drives your automobile smoothly and speedily over the country road, where you can feast your eyes upon God's beautiful creation. This transformation is a creative co-ordination, because it creates a beautiful cosmos of visible motion out of an invisible molecular chaos. This is the process which is at work in every domain of cosmic phenomena. Every snowflake tells us that it is the offspring of a creative coordination which is going on in the upper atmosphere where the chaotic vapor molecules give up their erratic motions to the chilliness of the surrounding space, and provide an opportunity to the internal molecular forces to assign to each molecule its proper place in the beautiful snowflake crystals. The chilling process and the internal molecular forces are the coordinators which transform the chaos of the vapor molecules into the beautiful cosmos of the snowflake crystals. The chaos of solar radiation is transformed into the beautiful cosmos of /6/ heavenly clouds which carry the blessed summer showers to the mountain sides and to the thirsty fields, meadows, and pasturelands. Each blazing star in the heavens is an atomic and electronic chaos which radiates its inexhaustible store of energy into the chilliness of the interstellar space. This is the chilling process which guides the destiny of the star until some day it will coalesce like a celestial snowflake and, perhaps, become like our mother earth a congenial home for the world of organic life. These illustrations suffice to exhibit the operations of creative co-ordination in the inorganic world.

## Creative Co-ordination in the Organic World

The whole organic world is the product of creative co-ordination. Each tiny organic cell and each one of its microscopic units of life feeds upon the chaotic energy supply of heat, light, and chemical activities and, like a skilled workman in an industrial plant, it transforms them into orderly structures performing orderly functions. This is indeed a creative co-ordination which never ceases until the cells have finished their creative mission and the /7/ organic body rises from the cosmic chaos just as beautiful Aphrodite rose from the shapeless foam of a turbulent sea.

## Creative Co-ordination in the World of Consciousness

Pass now from the external physical world to the internal world of our consciousness. Here again you will find creative co-ordination at work transforming the chaotic events of the external world into a cosmos of our consciousness. The beauty of the sunset, for instance, is a creation of our consciousness. The origin of that beauty is the chaos of solar radiation reflected, refracted, and scattered by the tiny corpuscles floating in our atmosphere. But its image is not a chaos; it is a thing of beauty and joy created by a co-ordinator which resides and rules supreme in the world of our consciousness. We call it our soul, and although we know nothing about its ultimate nature we know that it exists and that its creative power constructs the cosmos in the world of our consciousness.

### Creative Co-ordination in the Spiritual World

As soon as human reason, guided by experience, began to recognize that the /8/ creative power of our soul is only a tiny part of an immeasurably superior creative power then the embryo of a new world began to form in the world of our consciousness. It is the spiritual world, revealed to man by his belief in God and by the worship which this belief nourishes in the soul of man: The spiritual world with its spiritual realities and spiritual forces is just as real as the physical world with its physical realities and physical forces. They are both governed by similar laws which are the precious extracts of human experience. Just as the activities of physical forces have been summed up in Newton's dynamics, Maxwell's electrodynamics, and Carnot's thermodynamics, so the activities of the spiritual forces have been summed up by Christ in his spiritual dynamics.[14] He formulated its two fundamental laws when he said:

> Thou shalt love the Lord thy God with all thy heart, and with all thy soul, and with all thy mind.
> Thou shalt love thy neighbor as thyself.[15]

Employing the language of science we can say that these two commandments are the message from Christ's spiritual dynamics informing us that /9/ creative co-ordination rules supreme in the spiritual just as it does in the physical world. Love, according to Christ, is the greatest of all spiritual forces. It is the force, which like the gravitational force in the physical world, guides our power of spiritual co-ordination. The spiritual chaos vanishes from the soul of man when it is guided by the love of God. The life of humanity will be rescued from chaos and transformed into a cosmos of beautiful law and order, a kingdom of God on earth, if the love for our neighbor is the creative co-ordinator which guides the life of our Christian nations.

## Religion's Debt to Science

Harry Emerson Fosdick (1878-1969)

> Religious modernism is the endeavor to rationalize religion; it starts with science and with this new world-view which science brings and, taking that as test and standard, says, We will have nothing in our religion contrary to that.
> —*Harry Emerson Fosdick, "Beyond Reason,"*
> *Religious Education 1 (May 1928)*

### Introduction

Fosdick's second pamphlet is based on a sermon delivered at Park Avenue Baptist Church (now Central Presbyterian Church) in February 1927 that appeared in *Good Housekeeping* thirteen months later. The typescript of the sermon has been compared to the pamphlet. Changes in wording are summarized in footnotes, which also provide biblical and other citations from the sermon that were omitted in all published versions. The AISL printed 25,000 copies of the pamphlet in May 1928, followed by a second printing of the same size at some point in the next fiscal year (1928–1929).[1]

The opening line speaks volumes: "America today is filled with echoes of conflict between science and religion."[2] This is more than just a segue into the subsequent reference to Bryan's anti-evolution campaign. Fosdick's lifelong interest in science and religion was profoundly shaped by his encounter with the "Conflict" thesis. Four years before his birth, John William Draper had published a historically challenged book of nearly 400 pages, *History of the Conflict between Religion and Science* (1874), while Andrew Dickson White's two-volume screed, *A History of the Warfare of Science with Theology in Christendom* (1896), came out one year after Fosdick matriculated at Colgate University.[3] We must not overlook this salient fact: as much as anything else, White's book directly precipitated Fosdick's famous crisis of faith (see the introduction to his first pamphlet). As he recalled sixty years later, it "finally smashed the whole idea of Biblical inerrancy for me." White "seemed to me unanswerable. Here were the facts, shocking facts about the way the assumed infallibility of the Scriptures had impeded research, deepened and prolonged obscurantism, fed the mania of persecution, and held up the progress of mankind. I no longer believed the old stuff I had been taught. Moreover, I no longer merely doubted it. I rose in indignant revolt against it."[4] Nothing better encapsulates Fosdick's rejection of orthodox theology than this confession.

Like many other modernists, Fosdick lifted the conflict thesis uncritically from his plate and dutifully swallowed it whole, taking its alleged "facts" as gospel truths to illuminate a new path to righteousness. Both of his pamphlets took certain information from White, as Conklin's pamphlet also did (see the footnotes in all three for details), but the overarching scheme of intense conflict between obscurantist theology and progressive science shaped Fosdick's second pamphlet even more than the first. Toward the end of his first pastorate in Montclair, New Jersey, Fosdick wrote, "Science affects religion tremendously. Science lays violent hold on old traditions, . . . and scatters them in scorn to the four winds." It makes "old arguments, long used in defense of the faith, . . . as obsolete as bows and arrows." With "pitiless disregard of anything but the sheer truth, [science] gives old cosmologies the lie, although the church weeps for her dead like Rachel for her children and will not be comforted."[5]

To avoid such catastrophes, Fosdick believed that Christians had to distance themselves from all efforts to tie their beliefs about nature closely to the Bible—the overall theme of his first pamphlet. The second pamphlet showed Christians how to transcend this type of conflict entirely. Harmonious coexistence was not enough, for "religion is deeply in debt to science."[6] Just as Mathews had done in his pamphlet, Fosdick called attention to the vastness of the universe and its lawlike character, using that information in support of different conclusions: God is great, and just as free to act as we ourselves are free—although Fosdick was not about to give God the freedom to work genuine miracles. Even more, science could inspire selfless Christian service, especially in medicine, while giving us new tools to carry it out more effectively than ever before. A dozen years later, as war waged in Europe, Fosdick took this further in a sardonic sermon about "The Real Point of Conflict of Science & Religion." Science has replaced religion as the source of every good and perfect gift. "Science is my shepherd; I shall not want," so that for many, in words he borrowed from an unidentified college chapel address, "God becomes progressively less essential."[7]

This very aspect of science later inspired Fosdick, who had become a pacifist, to expand again his thinking about its relationship to religion. Two months after Hiroshima, he preached on "Science Demands Religion," identifying "four relationships between the two. First, science was in bondage to religion; no scientist dared contradict an established dogma of the church but, imprisoned within the confines of ecclesiastical authority, faced fearful penalties if he should transgress." Next, "Science broke free from dogmatic bondage," breaking down "old world views on which as on a trellis the faith of millions had been entwined," resulting in an "antagonism between science and religion [that] was tense and bitter." In the third phase, "science was religion's competitor," but since they are both "ways of meeting human need," religion has no chance in the competition. Science has taken over one field after another, "supplying what men want and for which in vain they had prayed to the gods for ages." Finally, we have reached the fourth phase. Science has given us "power that chills to the marrowbone thoughtful folk around the world, power utterly to destroy ourselves and our civilization." No longer bound by "dogma," its days as "the enemy of religion" are gone "for intelligent Christians." It is "a new era," where we find "science preaching like an evangelist, with hell and heaven on earth to choose between, saying to mankind, Seek wisdom and character that can control these powers I am giving you for mankind's good!" We must "repent of war, and of all that leads to it," lest we perish, seeking instead "world-wide good will, and world gov[ernmen]t," believing "by faith, that men are not merely brutes who must obliterate in utter ruin all that man has gained." Science says to religion, in the name of God, "if you still believe in God take him seriously, and somehow get control of what I am giving you, or else—or else ye shall all perish!"[8]

As these passages show, ultimately not even the pacifist Fosdick had become was able to jettison his reliance on the conflict view, which still substantially shaped his narratives—it had been the central narrative of his own spiritual and intellectual story. To be sure, Fosdick could deny that modernist Christianity conflicted with science, regardless of what the fundamentalists were saying, but he just could not get past making explicit or implicit references to White. Even in one of his last books, *Dear Mr. Brown* (1961), his reliance on White to decry "supernaturalism" continued unabated.[9]

Annotated Text

## *Religion's Debt to Science*

By Harry Emerson Fosdick, D.D.

/1/ America today is filled with echoes of conflict between science and religion.[10] Continually state legislatures make themselves conspicuous either by passing or refusing to pass laws dealing with the teaching of evolution, so that even if one were not naturally interested in the subject one would find it difficult to avoid thinking of it.[11] Moreover, even when the sense of antagonism between science and religion is not publicly exhibited, multitudes of individuals are acutely conscious of it in themselves. Typically religious people often resent the characteristic activities of science. No sooner are we comfortably settled in a view of the universe to which our religion has succeeded in adjusting itself than along comes some newer science still and, like a disturbing policeman, tells us to move on. That has been happening ever since modern science started. Today some /2/ people are trying to secure laws to prevent professors from teaching evolution; in the seventeenth century our forefathers in Europe made astronomers sign a pledge to teach that comets did not return in regular orbits, but came only by direct and special volition of God.[12]

On the other side, many scientists are as vexed at religion as religionists can possibly be at science. Religious attitudes often seem to a scientist utterly absurd. Religion associates itself with sacred books whose abiding principles and truths are necessarily expressed in the forms of thought of the generations in which the books were written; then religion insists that those forms of thought, representing the science of their time, must be eternally believed. In consequence, Copernicus and his moving earth, or Newton and his gravitation, or Darwin and his evolution, or Einstein and his relativity must be accredited to the acceptance of the religious by discovering them in the book of Genesis. Of course, to a scientific mind that is necessarily anathema. Many a scientist, therefore, says that religion holds up progress, that it blinds people's eyes /3/

to facts, that it fathers credulity and superstition, and in general blockades the forward march of man's mind.

So the unhappy controversy goes on, and printing presses groan with books discussing the marital difficulties of science and religion. The matter becomes a scandal. Divorce seems at times imminent. Religion cries that science is going to make materialists of us all and lifts resounding slogans of attack, like "God or Gorilla," while science claims that religion is keeping the human race in intellectual swaddling bands when it might be growing up.[13]

Out of this situation comes the attitude of the harmonizers. They are trying to keep the two together. They do not want them to break up house-keeping. They suggest all the soothing things that might mollify the mutual suspicion with which science and religion regard each other. They insist that true science cannot be anti-religious and that true religion cannot be anti-scientific.

With the attitude of the harmonizers, so far as it goes, I should suppose we would agree. But it does not go far enough. Let us try a more courageous attitude. So often in this regard one /4/ hears a timorous and quavering note as though perhaps, sometime, somehow, religion and science might manage to get on together. But can a man content himself with such a neutral tone when the plain fact is that religion and science are unpayably indebted to each other? In particular this article is devoted to the thesis that religion is deeply in debt to science.[14]

### We Have a Greater Universe

In the first place, religion is unpayably indebted to science for the new apprehension of the universe in which we live. I well recall the comet of 1884, which, as a very little boy, I was waked in the middle of the night to see. It seemed close at hand. It hung its huge tail over our roof so low that it looked like a shining prize for man to catch. I eagerly urged my father to go out upon the roof and get it for me.[15] So in the childhood of the race this universe must have looked to men. It was small and compact. They even thought they could build a tower to reach the sky and imagined God worried lest they should successfully invade his habitation.[16] But now consider the vast immensity through which our imagination /5/

wings its way, where the light that falls upon our eyes tonight, even from the nearest stars in the Milky Way, left them when Abraham was feeding flocks on Syrian hills.

One easily feels the greatening of the universe when he considers the changed measuring rods that men have used. First they were content with the length of a man's foot; we call it a foot yet. Then they used his stride; we call it a yard. Then, as the range of action grew greater, they took the length that a man could walk in a half-day or a day, and there are tribes who still use that as their longest measurement. But we are using measuring rods that the imaginations of our forefathers could not have dreamed. For the smaller spaces we employ the distance from earth to sun, 92,830,000 miles; for the larger spaces we use the distance that light, traveling 186,000 miles a second, traverses in a year.[17] From an earth that could be measured in feet to a cosmos measured in light-years—how amazing is the greatening of the universe! Alpha Centauri is our nearest stellar neighbor, 275,000 times farther away than the sun. It is not exactly cozy, but it is magnificent. /6/

With this outer enlargement of the world we are familiar, but it is not more marvelous than the world's extension infinitesimally. Who remembers the old Dutch janitor, Leeuwenhoek, who in the seventeenth century, first of all mankind, put together crude lenses in a microscope and viewed the populations that throng in a drop of water?[18] There was a new world unveiled as amazing as any Galileo viewed. So today we live in a cosmos infinite and infinitesimal, and as unified as it is vast. Once polytheism was possible. No more! Whatever else this cosmos is, it is one. All its power comes from one source. All its laws are built on one pattern. A man who believes in God, who is sure that matter is not the last word in this universe, and that the power at the heart of all things is akin to that which rises in us as intelligence, purposefulness, and good will, stands before this vast system feeling with Wordsworth:

A presence that disturbs me with the joy
Of elevated thoughts; a sense sublime
Of something far more deeply interfused,

Whose dwelling is the light of setting suns,
And the round ocean and the living air,
And the blue sky, and in the mind of man;
A motion and a spirit, that impels /7/
All thinking things, all objects of all thought,
And rolls through all things.[19]

To be sure, it is possible for a man's faith to be wrecked by the vastness of the cosmos. If a man carry out into this great universe too small a conception of God, it will not stand the strain. Many have done that and, finding it inadequate, have surrendered God and had nothing left except a vast machine that came from nowhere, meant nothing, and was going nowhither. More numerous are the others, who, not desiring thus to land in blank atheism, nevertheless feel a timorous uncertainty about this great universe:

I remember, I remember
The fir trees dark and high,
I used to think their slender tops
Were close against the sky:
It was a childish ignorance,
But now 'tis little joy
To know I'm farther off from Heaven
Than when I was a boy.[20]

There are defeatists everywhere but there is no reason why a man should be a defeatist in this realm. Long centuries ago upon the walls of Zion a Psalmist on a clear night, when with unaided eye he might have seen 2,500 stars, praised God for the glory of the /8/ firmament. A few months ago I walked those same walls of Zion, when the sun went down in the Mediterranean and the moon came up over Moab, and there, where the Psalms first had their birth, I meditated on this new universe, vast, majestic, evolving, infinite, infinitesimal, and, for all the suffering here which we cannot understand and all the mysteries too deep for our minds to plumb, the old words took flame with a fresh meaning,

The heavens declare the glory of God;
And the firmament showeth his handiwork.²¹

Here is the testimony of a man who does not regret that his fir trees no longer touch heaven:

As wider skies broke on his view,
    God greatened in his growing mind;
Each year he dreamed his God anew,
    And left his older God behind.
He saw the boundless scheme dilate,
    In star and blossom, sky and clod;
And as the universe grew great,
    He dreamed for it a greater God.²²

### We Know the Reign of Law

In the second place, religion is unpayably indebted to science for knowledge of the reign of law.²³ I know well the deadly fear that formerly congealed /9/ the souls of Christians when first they heard about the reign of law. They had lived in a world of miracle where nothing was regular, nothing law-abiding. Then science came upon the scene with its reign of law, involving irrefragable regularity where the same thing happens under the same circumstances always and everywhere. It was dreadful. Christians fought against Newton's idea of gravitation with the same desperation with which many Christians now fight against evolution.²⁴ They saw the shadows of the prison-house closing in, an iron regularity where, as a chemist said, if you ask nature the same question in the same way, she will always give you the same answer.²⁵

Nevertheless, despite all opposition, the laws grew more clear and more certain. They covered realm after realm until nothing was left except a few whimsical and capricious things like comets which God might handle as he pleased. To these the old science and the old religion clung. They were not law-abiding. They were miraculous. In old Massachusetts Bay Colony, Cotton Mather put the fear of the Lord into the people when a comet hung /10/ over the town, thundering, as he did, that it was a special warning of the Almighty against their sin. Then came the crucial test. The astronomers prophesied the exact time of the return of Halley's Comet in 1759. It was a fearful crisis. Many thought that all religion hung in the balance. If Halley's Comet came back on time then there was nothing left but a huge machine and God was gone. And Halley's Comet did come back on schedule time. Alas, cried many Christians, man and God, if there be a God imprisoned in law![26]

As one looks back now upon that old fear, how utterly incredible it seems to an intelligent Christian! Imprisoned by law? Look around you! Who is imprisoned by law? Every time we discover a new law we are set at liberty to do a new thing. Knowledge of natural law has proved to be one of the most liberating experiences that mortals ever achieved. If one doubts it, let him live for a while in Arabia, where the old world still persists, where there is no knowledge of natural law and no control over law-abiding forces—only fate and fatalism with an occasional miraculous event to interrupt the /11/ tedium of submission to the will of Allah. That is a prison-house, from which one returns to America as to liberty. Here we are free to do things, to change our lives, to master our circumstances, and the reason for that liberty is the very thing our forefathers feared: knowledge of law. Every time we get our hands on a new law we have got our hands on a new law-abiding force. Then we are free to fly on the wings of the wind, to speak from New York to London, to bridge rivers, to cure diseases, to create better educational systems. Imprisoned by law! And if man is not, why should God be?

### And Law Is Everywhere

This holds just as truly of law within the personality as it does of law without. Many people today have got themselves fairly well adjusted to the idea of natural law in the outer world but they are desperately disturbed about the reign of law in the inner world. Psychology upsets their religion. They fear that they may be nothing but psychological machines and their thoughts as automatic as the sounds of a striking clock. The same kind of /12/ dread that our forefathers had about law without, they have about law within.

Why should they? Who is imprisoned by psychological law? My children are receiving a better education than I had. The reason is plain.

We know more about psychological law. Every time we discover something new about the regular processes of the human mind we can improve our education. This last week a young man came to report "All's well!" We had talked together two years ago. Two years ago, having tried to commit suicide and suffering desperate mental anguish, he came to see me. After listening to his story for a few minutes, I stopped him and narrated to him something that had befallen him when he was young. "My heavens," he said, "how did you know that?" Then we got to the nub of the matter, and two hours afterward he went out on a new basis of thought and life. How could that be done? Only because we know more about psychological law. Imprisoned by law? There never was a place where the Master's words had more transparent application: "Ye /13/ shall know the truth, and the truth shall make you free."[27]

This leads us naturally to our third statement: we as Christians are unpayably indebted to science for the new tools of service that it puts into our hands. It is true that all tools can be misused. They are as capable of disservice as of advantage, but when we cease emphasizing the perils and begin thinking of the possibilities, when we count up not alone the liabilities but the assets, how wonderful they are! When, for example, people talk about conflict between science and religion as though that were the gist of the matter, we would do well to think of the triumphs of scientific medicine. We ought to speak of that oftener in places where the heroes of the race are remembered and the servants of humanity are put into our inward halls of fame. Our missionary enterprise has produced great souls who valorously pioneered the world for Christ, and we know the roster of their names: Paton of the New Hebrides, Morrison of China, Judson of Burma, Livingstone of Africa.[28] But we do not know so well that scientific medicine also has its /14/ roster of heroes who well deserve to stand beside them.

### Science Inspires Service

Do you recognize the names of James Carroll, Walter Reed, Jesse Lazear, Private Kissinger? Yet they were saviors, too.[29] They saved mankind from yellow fever. For many a long century it had taken its toll of millions of lives. It never will again. They and their successors stopped it.

Moreover, the principle of the Cross held true in their case; you cannot achieve such salvation without sacrifice.

To allow yourself deliberately to be bitten by mosquitoes in Cuba in order to see if, perhaps, that is what causes yellow fever, to do it when you have a wife and children and love life as well as any man—is not that worthy of the Master's encomium, "Greater love hath no man than this"?[30] James Carroll did that and went down into the dark valley of a dreadful death and barely came back again. Then Jesse Lazear, wanting to be scientifically sure that they were on the right track, did that. He went down into the valley of the shadow of the dreadful death and never came back again. Then Walter /15/ Reed wanted to try it to make assurance doubly sure, but he was too old and they would not let him. So he posted a notice in the American camp that he wanted volunteers to face death in the fight against yellow fever. Before the ink was dry, Kissinger, who was a private, and Moran, who was a civilian clerk, had volunteered.[31] Well, they won their fight. They are all dead now except those two, and Kissinger is paralyzed from the effects. They never had glory or reward. Their widows today are living on government pensions of $1,500. But they won their fight. They stopped one of the most devastating scourges that ever cursed mankind.

Conflict between science and Christianity? Think of Lister's work in antiseptic surgery, Morton's work in anesthetics, Jenner's work in vaccination, or the new antitoxins that are stealing terror from old scourges like diphtheria.[32] Would not Jesus rejoice, who cared so much for the bodies of men and spent so much energy upon their health? I say it reverently: he healed a few people after the manner of his day, but how grateful would he be if, coming back, he should see science now /16/ fulfilling his own words, "Greater things than these shall ye do."[33]

This last year I visited ancient Corinth in Greece. Thither, late in the last war, an astonishing American woman had brought two thousand children from the carnage and chaos of Asia Minor. There in Corinth in old Greek army barracks she installed the children. Within a few weeks twelve hundred of them had malaria. Malaria had been there for centuries. Historians say that it caused the disintegration of ancient Greece.

One thinks of all the prayers that have been offered from old pagan shrines, from Christian churches, from Moslem mosques against this insidious foe. But now we have new tools to work with. That American woman sent to Athens for a trained nurse from Johns Hopkins, who was there. She threw her scientific knowledge into the problem, would not leave a single pond uninvestigated, became a nuisance to the government until it joined forces with her. She cleaned up the entire countryside until not a single case of malaria was left.[34] Now there need never be any malaria in Corinth again. Conflict between science and religion? As one thinks of /17/ these new tools in the hands of the spirit of service, such vistas open as never hitherto have opened before the hopes of man.

## We Are Honest with Our Facts

Finally, religion is unpayably indebted to science for the new note of straightforwardness and honesty in dealing with facts. One can be sure that Marcus Dods, a Scotch scholar of the last generation, had been under scientific influence when he said, "The man who refuses to face facts doesn't believe in God."[35] That is the tone of mind with which true science works, and it has been an incalculable benediction to religion. Recall those crisp and noble phrases in which Thomas Huxley once summed up his ideal as a scientist: "To smite all humbugs, however big; to give a nobler tone to science; to set an example of abstinence from petty personal controversies, and of toleration for everything but lying; to be indifferent as to whether the work is recognized as mine or not, so long as it is done."[36] Because of such an attitude, what superstitions that dogged our fathers' footsteps have vanished, what fears of ghosts and demons have /18/ been done away, what ancient ignorance has been lighted up that once was filled with sinister shadows, what honesty, fearlessness, and candor in dealing with facts have come!

For a long while I have wished to strike this positive note about the relationship between science and religion. There are other notes. This is not the whole story, but this is a neglected and important part of it. Christians are unpayably indebted to science. Modern science permeated with religious faith and spirit—there never was so magnificent an outlook on the world. Modern science, however, bereft of religious faith and spirit—there never was a deeper abyss of pessimism. Listen to this scientist, typically modern, who has lost the last shred of his religious faith and has nothing but the science left:

"Brief and powerless is Man's life; on him and all his race the slow, sure doom falls pitiless and dark. Blind to good and evil, reckless of destruction, omnipotent matter rolls on its relentless way; for Man, condemned today to lose his dearest, tomorrow himself to pass through the gate of darkness, it remains only to cherish, ere yet the /19/ blow falls, the lofty thoughts that ennoble his little day."[37]

So old men write, and young men commit suicide.

This great new world in the hands of the Christian spirit—there never was such a chance. This great new world in lesser hands—there never was such a peril.

# *Life After Death*

Arthur Holly Compton (1892–1962), Shailer Mathews (1863–1941), and Charles Whitney Gilkey (1882–1968)

> We could, in fact, see the whole great drama of evolution moving toward the making of persons with free intelligence capable of glimpsing God's purpose in nature and of sharing that purpose. In such a case we should not look upon consciousness as the mere servant of the biological organism, but as an end in itself. An intelligent mind would be its own reason for existence.
> —*Arthur Holly Compton,* The Freedom of Man

## Introduction

The longest pamphlet, *Life After Death* (1930), is the only one with multiple authors and the only one not explicitly identified on the title page as being part of the series on "Science and Religion."[1] However, the back of the pamphlet carries an advertisement for most of the other titles in the series, and there is further evidence that they treated it as part of that series. In a report to the Rockefeller Foundation for the fiscal year ending in June 1929, AISL staff stated, "we still hold to our proposal of last year which has not yet been carried out, of a symposium of several scientists: My Feeling about Immortality," and they said this under the heading, "Science and Religion Series."[2] In any event, the content and authorship justify including *Life After Death* in this collection.

The pamphlet did indeed originate in a symposium, but it did not take place until the spring of 1930, when several University of Chicago faculty offered diverse views on immortality at the university chapel (later named in memory of the donor, John David Rockefeller Sr.). The dean of the chapel at that point was Charles Whitney Gilkey, father of Langdon Gilkey and former pastor of the Hyde Park Baptist Church (now Hyde Park Union Church), a very liberal congregation located just one block east of campus in an upscale neighborhood. Gilkey did not speak in the symposium himself, but his Easter sermon constitutes the final portion of the pamphlet. Four others spoke. Physicist Arthur Holly Compton and theologian Shailer Mathews, both of whom were members of Hyde Park Baptist, argued in favor of immortality, while ethicist Thomas Vernor Smith and physiologist Anton Julius Carlson argued against it. The talks by Compton and Mathews precede Gilkey's sermon in the pamphlet, 25,000 copies of which were printed in September 1930. No other editions of the pamphlet are known. However, all four chapel talks were published in November 1930 by *The University of Chicago Magazine*, and the contributions by Compton and Gilkey were published in various religious periodicals the following year.[3] Mathews elaborated on his position in *Immortality and the Cosmic Process* (1933), and Compton did so in *The Freedom of Man* (1935), based on his Terry Lectures at Yale University in 1931.

Mathews's views on science and religion, including immortality, are surveyed in the introduction to his other pamphlet, and Gilkey did not write on this larger topic, so we focus here on Compton. Although he was a very active Christian, mainly as a Presbyterian, Compton did not believe in the deity of Jesus—in that respect, he was a thoroughgoing modernist. Jesus for him was the supreme example of brotherly love and faithfulness to God, but not the literal Son of God. Owing to his lack of sympathy for traditional doctrines, he declined to recite the Apostles' Creed in church.[4] Nevertheless, he believed in God as the source of order in the universe and of human freedom and dignity. He began speaking and writing extensively about science and religion in the late 1920s, shortly after he received the Nobel Prize for Physics in November 1927. Like Millikan, he used his newly acquired fame as a bully pulpit to address religious questions; unlike Millikan, he developed a nuanced, carefully articulated theology of nature that in some ways anticipated the views of important later voices in the conversation between science and religion.

Compton was convinced that science strongly supported the existence of an intelligence behind nature. At times his arguments sound like early forms of what would later be called the anthropic principle. "The chance of a world such as ours occurring without intelligent design," he said at a Unitarian church in 1940, "becomes more and more remote as we learn of its wonders." In his view, "the study of natural science is the primary source of the raw material for building our idea of God."[5] On another occasion he told a reporter that studying physics had changed his conception of God, but it had also "strengthened my confidence in the reality of God. I feel surer of a directive intelligence than I did at [age] 20." Atoms, molecules, and cells were "all built up out of simple units: electrons and protons. It seems to the $n^{th}$ degree improbable that such an intricate and interesting world could have ordered itself out of particles with random character." Thus, the world revealed by modern physics "can only be the result of an intelligence working through nature."[6] Furthermore, "the intelligent Creator, whose existence as we have seen is by far the most reasonable basis for accounting for our world, [takes] an active interest in the welfare of the uniquely intelligent beings he has created," even sharing responsibility with us for "carrying through the final stages of making this a suitable world and ourselves a suitable race for what is perhaps the supreme position of intelligent life in His world!"[7]

Physics also revealed the world to be "a vast machine" characterized by "immutable" natural laws, and "the world plays [no] favorites by showing partiality toward man."[8] Thus, Compton was confronted by what he regarded as the most significant problem posed by Newtonian physics: How can humans be free, responsible moral agents in a mechanistic universe? The dilemma evaporated, in his view, with the quantum mechanics of Werner Heisenberg. Compton hosted the famous series of lectures by Heisenberg at the University of Chicago in 1929. Echoing Heisenberg's own view of the situation, he said, "I myself should consider it more likely that the principle of the conservation of energy or the second law of thermodynamics would be found faulty than that we should return to a system of strict causality."[9] Holding that human action might be subject to quantum uncertainty, Compton believed that he had found an adequate basis to deny that we are merely automatons with no minds of our own. Although he never explicitly advanced a similar argument about God's freedom, he hinted at that possibility in the last decade of his life.

Compton's views on immortality were closely linked with his views on evolution and free will. The reality of free will convinced him that "there must be at least some thinking possible independent of any corresponding physical change in the brain," so that "consciousness may persist after the brain is destroyed." He also believed "that the evolutionary process is working toward the development of conscious persons rather than toward the development of a physical organism." If so, then "the whole great drama of evolution" terminates in "the making of persons with free intelligence," and "an intelligent mind would be its own reason for existence." Therefore, if "the thoughts of man . . . are conceivably to the Lord of Creation among the most important things in the world," then "we might expect nature to preserve at all costs the living souls which it has evolved at such labor. This would mean the immortality of the individual consciousness."[10] Although I just quoted Compton's Terry Lectures, there are identical passages in the pamphlet. In the book, however, there is a further section about the Resurrection, which he did not mention either in the pamphlet or the chapel address on which it was based—a loud silence for an Easter symposium, reflecting the central fact that he did not believe in the actual reality of the empty tomb. His modernist faith had gone as far as it could go in defending the traditional Christian belief in immortality.

---

## Annotated Text

## *Editor's Note*

Not long ago some students at the University of Chicago requested a group of members of the Faculty from different departments to speak to them on their views of immortality.[11] This pamphlet contains the addresses of two men of this group; Dr. Arthur H. Compton, Professor of Physics, and Dr. Shailer Mathews, Professor of Historical Theology and Dean of the Divinity School. The third contribution comes from Dr. Charles W. Gilkey, Dean of the Chapel of the University of Chicago, and is a sermon preached on Easter Sunday.

To the thousands of young men and women who are raising the question of immortality in their own minds, as well as to those older Christians to whom it has long been an essential element in their faith, the reading of this pamphlet is commended, as representing the attitudes of religious men from three different fields of work—the scientist in his laboratory, the theologian in his creative thinking, the minister in his pastoral relation to human souls.

## *From the Point of View of a Scientist*

### By Arthur H. Compton

/1/ The seeker after religious truth asks earnestly whether science has an answer to its vital problems: Is there a God? Is man morally responsible for his actions? What about a man's soul? Does death end all? The answers of science to questions of this kind are usually hesitating and tentative. Some things point one way and some the other.

So, in discussing the problem of immortality from the standpoint of science, it is not my purpose to draw any conclusions but rather to present as fairly as possible the meager evidence which science offers. Science does not supply a definite answer to this question. If one is to have either a positive faith in a future life or a conviction that death ends all, such beliefs must, it seems to me, be based upon religious, moral, or philosophical grounds rather than upon scientific reasoning. /2/

Mortality and Immortality of Living Organisms

Is it not obvious to one who views without bias the course of life about him that life is invariably followed by death? If, then, science is a description of the way in which things happen, how can science state any other conclusion than that death is the inevitable terminus of life?

But what is it that dies? Each person, or, to be more general, each organism dies; but the race or species lives on unless some world-wide accident occurs which makes the species extinct. Sir James Jeans in his just-published book, *The Universe About Us*, assigns a million million years as the reasonable live-expectancy of the human race on earth.[12] This million million years may not be life-eternal, but it is probably as long a life as most of us are interested in.

The biological center of life is the germ cell, and this, with divisions and subdivisions, grows and lives forever. What the fruit of the apple is to the seed, the body of man is to his germ cell. The apple may decay, but the /3/ seed grows into a new tree, which flowers and begets new seeds. The fruit and the tree will pass away, but there is eternal continuity of life in the cells which develop from seed to tree to flower to seed, over and over again. It is thus because we concentrate our attention upon the tree or the fruit that we say the end of life is death. These are merely the outer wrappings, the hull which surrounds the living germ. Biologically speaking, life, whether it be of an apple seed or of the germ cells of man, is essentially continuous and eternal.

### What Becomes of the Soul When the Body Dies?

"But," you say, "that is not the kind of eternal life in which I am primarily interested. My body may be merely the hull that surrounds the living germ; but what will happen to me when the hull decays?"

To this question science has no straightforward answer to give. For when you ask, "What will happen to me," you are concerned not with your body but with your consciousness, mind, or soul, however you may choose /4/ to name it, which is not material, and regarding which physical science does not directly concern itself. If we are to tell what is the fate of consciousness when the body dies, we must know what the relation is between body and mind.

Certain psychologists use the hypothesis that thought is a function of the brain, in the sense that every idea that we have and every decision we make is a consequence of some action occurring in the brain. On this view it is obvious that destruction of the brain would carry with it the destruction of consciousness.

This hypothesis has been adopted primarily in order to simplify the problem of behavior by reducing it to a set of mechanical laws. If a thought is a by-product of some molecular change in the brain, and if these molecular changes follow the usual definite physical laws, there will be a straightforward sequence of molecular changes starting with the initial stimulus and ending with the final action of the organism. Thoughts may be associated with these various changes, but they cannot alter the end result, for this is /5/ determined by the physical laws which govern the molecular actions. The problem of a man's behavior is thus simplified by reducing him to an automaton.

To the large majority of thinking people it seems that this simplified behavior fails as a complete description of our actions. In some reflex actions and habitual acts we may behave as automata; but where deliberation occurs we feel that we choose our own course. In fact, a certain freedom of choice may, it seems to me, be considered as an experimental fact with which we must reconcile our theories. Because the mechanist's basic hypothesis leaves no room for such freedom, I see no alternative other than to reject the hypothesis as inadequate.

On the other hand, if freedom of choice is admitted, it follows by the same line of reasoning that one's thoughts are not the result of molecular reactions obeying fixed physical laws. For if they were, one's thoughts would be fixed by the physical conditions, and his choice would be made for him. Thus, if there is freedom, there must be at least some thinking possible /6/ quite independently of any corresponding cerebral process. On such a view it is no longer impossible that consciousness may persist after the brain is destroyed.

That there is some correlation between the brain's activity and mental processes is, however, evident. This is frequently assumed to imply that thought is produced by cerebral activity. If this is the case, destruction of the brain would result in the cessation of thought and consciousness. William James has, however, called attention to the fact that the observed correlation is equally consistent with the view that the function of the brain is to transmit the thoughts from a non-physical thinker to the body of the organism.[13] On this view the brain would correspond to the detecting tube of a radio receiver, without which the outfit will not operate. Stopping the sound by destroying the tube would not imply the destruction of the ether waves which carry the music.

An examination of the evidence seems to show that the correspondence between brain activity and consciousness is not very close. Professor Lashley /7/ has pointed out that in certain animals a large portion of the brain may be damaged, or even removed, without destroying consciousness or seriously disturbing the mental processes.[14] On the other hand, such a relatively minor disturbance as a tap on the skull may, so far as we can tell, completely destroy consciousness for a considerable period of time. I understand that it is impossible to distinguish the physical condition of the brain of one who is awake from that of one who is asleep, though the difference between the two states of consciousness is very great. The detailed proof by Professor Bergson that "there is infinitely more in a human consciousness than in the corresponding brain," and that "the mind overflows the brain on all sides, and cerebral activity corresponds only to a very small part of mental activity" (*Mind-Energy*, pp. 41 and 57), seems convincing.[15]

That consciousness must die with the body is thus logically required only if we adopt the mechanistic viewpoint that a definite thought is the result of an equally definite physical change in the brain. The seeming fact of free will /8/ makes this viewpoint appear to me highly improbable. It seems rather that our thinking is partially divorced from our brain, a conclusion which suggests, though of course does not prove, the possibility of consciousness after death.

### Experience of Revived Persons

What might appear to be first-hand evidence regarding the persistence of consciousness after death comes from the experience of those who have been revived after some accident. In the *Atlantic Monthly* some years ago appeared an article by one who claimed to have died nine times. He had been drowned; had fallen down an elevator shaft and stunned into unconsciousness; had died a lingering death on the battlefield, later to be revived; had been knocked out by a blow on the head; had been anaesthetized; and so on. This emulator of the cat described the experience of death as "the mere cessation of consciousness—nothing more."[16]

Yet even such evidence is of doubtful value. I recall one evening when my brother came home from football practice /9/ and sat down to dinner.[17] Soon he began asking us, "What's the matter?" He had received a blow on the head, resulting in a lapse of memory. Five minutes after each explanation would come back the question, "What's the matter?" He recognized everyone in the room, talked rationally, and could play familiar tunes on his mandolin. By every test he was perfectly conscious; only his memory was very short. At about ten o'clock the following morning his thoughts returned to their normal channels. Now he remembered diving into the interference of the opposing team to break up a play, but the subsequent events remained blank. How he got to his home he did not know. He had only the testimony of his friends that he had remained conscious except while he slept at night. Had he been asked he would have described the experience of diving into a massed interference as "the mere cessation of consciousness—nothing more." In a similar way evidence based on the memory of a revived person must be doubted.

Evidence of perhaps equal weight but pointing in the other direction is /10/ given by statements such as the following, made by former President Little of the University of Michigan, himself a biologist of no mean standing:

> The death of my own parents within a day of one another completely wiped out pre-existing logical bases for immortality and replaced them with an utterly indescribable but completely convincing and satisfying realization that personal immortality exists. Such experiences are not transferable, but are probably the most comforting and sacred realizations that can come to any of us.[18]

### Uselessness of Consciousness to Dead Organisms

An argument against immortality which carries considerable weight is based upon the value of consciousness to the organism. From the biological point of view consciousness appears in animals to enable them to compete more successfully in the struggle of life. That is, consciousness is the servant of the biological organism. In the evolutionary process we should on this view expect consciousness to appear only where it can be of some value to the organism with which it is associated. For a babe at birth consciousness is of little if any value, and it seems to be /11/ only feebly developed. In youth and maturity, however, it is of vital importance that the organism be aware of what is going on, and consciousness is accordingly most highly developed. Clearly, consciousness can be of no value to a dead organism. From the biological point of view, therefore, we should expect an efficient evolutionary process to bring about the cessation of consciousness with death.

There is, however, an alternative point of view which is equally tenable and which points toward the opposite conclusion. This is that the evolutionary process is working toward the development of conscious persons rather than toward the development of a physical organism. The old-fashioned evolutionary attitude was that the world as we know it developed as a result of chance, variations of all kinds occurring, some of which would be more suited to the conditions than others, and therefore surviving. More recent thought has found this viewpoint increasingly difficult to defend. To the physicist it has become clear that the chances are infinitesimal that a universe filled with atoms having random /12/ properties would develop into a world with the infinite variety that we find about us. Slight alterations in the properties of the electrons and protons of which the world is made must have resulted in a very dead world indeed. To the chemist, it becomes apparent that the development of protoplasm, whose chemical properties are of the most complex and unstable sort, could have occurred only under narrowly defined conditions—just the right chemical elements, associated in just the right way, at the proper temperature, and probably with suitable illumination by ultra-violet light. If left to chance, such a combination appears highly improbable, even through the immense time of geological history. The paleontologist finds that at least in certain well-authenticated cases the evolutionary procedure has not been the development of many branches with the final survival of only the more favorable variations, but rather the straightforward development from a primitive form through gradual steps to a more highly developed form, without wasting time experimenting with unfavorable variations. This is the phenomenon /13/ known to evolutionists as "orthogenesis."[19] The biologist calls attention to the fact that, as the evolutionary process goes on, phenomena appear which could not have been predicted from our knowledge of earlier stages. The three following are generally recognized to be phenomena of this type: (1) the appearance of life itself in the form of Protozoa; (2) the appearance of multicellular organisms, differentiating into animals and plants; (3) the emergence of animal consciousness. These facts of world-history have been described by the term "emergent evolution."[20]

This situation strongly suggests that the evolutionary process is not a chance one but is directed toward some definite end. If we suppose that evolution is directed, we imply that there is an intelligence directing it. It thus becomes reasonable to suppose that intelligent minds may be the end toward which such an intelligent evolution is proceeding. In such a case we should not look upon consciousness as the mere servant of the biological organism but as an end in itself. An intelligent mind would be its own reason for existence. /14/

A survey of the physical universe, however, indicates that mankind is very possibly Nature's best achievement in this direction. Though astronomers tell us that there are millions of millions of stars in the sky, a planet is a very rare occurrence, and a planet on which life can exist is even more rare. Thus in his recent book, *The Nature of the Physical World*, Professor Eddington, the noted British astronomer, concludes, "I feel inclined to claim that at the present time our race is supreme; and not one of the profusion of stars in their myriad clusters looks down on scenes comparable to those which are passing beneath the rays of the sun."[21]

If in the world-scheme conscious life is the thing of primary importance, what is happening on our earth is thus of great cosmic significance; and the thoughts of man, which have come to control to so great an extent the development of life upon this planet, are perhaps the most important things in the world. On this view we might expect Nature to preserve at all costs the living souls which it has evolved at such labor, which would mean the immortality of intelligent minds. /15/

### Summary of What Science Says

This is about all that present-day natural science can tell us about immortality. There are many if's and but's. While according to the mechanistic view the mind could not survive the brain, the evidence seems against this view, and no cogent reason remains for supposing that the soul dies with the body. The evidence of revived persons brought back from Hades is unreliable. If consciousness is merely the servant of the living organism, we should expect the two to die together; but if, as seems perhaps more plausible, intelligent consciousness is the objective of the evolutionary process, we might expect it to be preserved.

### Analogy of the Light and the Candle Flame

Permit me now to come away from our scientific reasoning and to present a scientific analogy which, though of no value as an argument, may yet be suggestive of possibilities.[22] Where does the light go when you put out the flame? /16/

Let us take the flame to represent the body, and the light which comes from it the consciousness or soul. In a candle flame vapor comes from the wick and air comes from the side, forming a steady stream of burning gases passing continually through the flame. There is an intake of "food" and oxygen at one end, and an outpouring of waste products at the other. It is a kind of metabolism. The material of the flame is continually changing, just as the cells of our bodies change; yet the form of the flame remains the same. It is the same flame. But puff, and the flame is out! Is this the end? The flame is dead. What then?

What is happening to the light? The flame was material made up of atoms and molecules; but the light is a different kind of thing—electromagnetic radiation, flying away at tremendous speed. We know that if the candle was out under the open sky, its light was streaming into interstellar space, where it will keep on going forever. The flame was mortal, but the light which it gave was immortal. More than that, the escaping light carries with it the story of the candle's life. If, on some far distant /17/ planet the light is caught in a spectroscope, it can tell that it was born of the burning carbon in oxygen, and that the temperature of the flame where it lived was some fifteen hundred degrees. By a study of the light an amazing number of things could be found out about the flame from which it came.

Suppose, now, we can observe molecules but are blind to the light. Would we not have said the flame died, and that was the end? Is not this precisely our position regarding the life of man? His body we can observe, his mind we can only infer from the actions of his body. The body dies—is blown out. The light from the candle flame lived on through eternity, though the blind man could not see it. We know we are blind to the soul. How can we know that it does not go on living forever with a fulness of life corresponding to that of the light?

### The Conservation of Character

Let me close with an observation suggested by our discussion of the way in which Nature has been evolving intelligent life. We found strong reasons /18/ for believing that, in spite of his physical insignificance, man as an intelligent person is of extraordinary importance in the cosmic scheme. If we were to use our own best judgment, what would we say is the most important thing about a noble man? Would it be the strength of his body, or the brilliance of his intellect? Would we not place first the beauty of his character? A man's body is at its prime before middle life, and his intellect probably somewhat after middle life. But it takes a whole lifetime to build the character of a noble man. The exercise and discipline of youth, the struggles and failures and successes, the pains and pleasures of maturity, the loneliness and tranquillity [sic] of age—these make up the fire through which he must pass to bring out the pure gold of his soul. Having been thus perfected, what shall Nature do with him? Annihilate him? What infinite waste!

Speaking, now, not as a scientist but as man to man, how can a father who loves his children choose to have them die? As long as there is in heaven a God of Love, there must be for God's children everlasting life. This is not the /19/ cold logic of science but the warm faith of a father who has seen his child on the brink of death.

And so at last, it may be you and I
In some far azure Infinity
Shall find together some enchanted shore
Where Life and Death and Time shall be no more
Leaving Love only and Eternity.[23]

---

## From the Point of View of a Theologian
### By Shailer Mathews

/21/ I do not think that I am particularly worried as to what is going to happen to me after I am dead. I shall find out in a few years. Even those of us who think that a belief in the continuance of personality after death is forced upon us logically by an understanding of the total world in which we live, have times when about all that we can confidently say is that we should be surprised if we woke up and found ourselves annihilated!

We all have only a minimum of interest in the future in so far as it has a direct bearing on any particular act. It would seem that those of us who believe in the continuance of the personal life after death are not trying to frighten people into decency by talking about hell, or endeavoring to make heaven a sort of compensation for defeat on the earth. We await the outcome as we await any event in the order of nature. /**22**/

I think most of us are interested in immortality because of those who are dear to us. We are rather ready to take a chance ourselves, but we hate to think that those whom we have loved, who have actually been of significance in the world, who have risen above the backward pull of their animal inheritance, and have gone on to something noble and pure, are annihilated.

Let us first consider negative conclusions. A professor of theology must scrutinize his religious faith if anybody must. My belief in personal survival after death, if I can estimate it, is not due to any particular wish but to my view of the total situation in which we are. There is no use to talk about what a particular form of matter will do or will not do before you know what matter is. There is no use to talk about pre-conditions of mental life until you know what life is. There is no use to be dogmatic about anything where you do not know. But you must have working hypotheses in life; you must organize sooner or later an answer to the question as to whether there are forces in the cosmos which can develop characteristics superior to those of our immediate /**23**/ predecessors, the animals. If one thinks seriously, sooner or later one must make up one's mind as to whether or not in the cosmic process of which we are a part, which starts nobody knows where, and goes nobody knows whither, there is any meaning.

What then, avoiding both negative and positive dogmatism, seems on the whole the best working hypothesis? The issue cannot be avoided by a study of origins. At best, origins are far from being origins. After you have the atom, you have the proton and electron; and after you have them, what have you? When you look at those tremendous stars a million light years distant, you find yourself face to face with space. But nobody knows what space is; nobody knows what time is. Who dares be negatively dogmatic in an area where true origins are hidden in the obscurity of an activity you do not understand? Who can learn all of the possibilities of the matter of the brain by studying its corpuscles and cells? They are not ultimate.

You are thrust into a choice of working hypotheses, drawn either from the pattern of the machine or from the /24/ pattern of what seems to be the finest product of the cosmic process—yourself—and yourself in relation to other persons. Of course, you will never get any sense of immortality so long as you think of things in terms of machines—mere mechanisms. But neither will you dare organize your ethics on a mechanistic philosophy. You dare not organize your relations with one another as if you were what a strictly mechanistic conception of life says you are. We must treat one another as more than machines; we must treat one another as though others were like ourselves, self-directing, capable of choosing values, acknowledging things like beauty and honor and faith, and the very act of will. But there is no time element in those concepts.

Religion is not an entity any more than science is an entity. What we have is people, men and women, who have a certain behavior, who organize their lives with a certain sense of values. And you have hard work getting beyond folks. No scientist ever showed me an electron; nobody ever showed me anything, except their impressions of something. What we mean /25/ by science is really the experience of scientists. When we talk about religion, we are talking still about a group of experiences, of experiences of people who have tried in some way or other to live in the frame of their personal selves, to utilize in their behavior the power which they possess, not as machines, but as persons.

I know perfectly well how idle and crude have been some of those attempts at living with the universe at the level of personality. Yet the whole scale of life was the same. Sometimes primitive religion seems pathetic, but on the whole I think it was glorious. Somewhere, sometime, some life emerged, by some strange metamorphosis, from the animal; someone found himself consciously facing the universe in which he had to live. He had no scientific training back of him; he had to live as best he could. And appropriating his experience in dealing with his fellows, he tried not only to live with the trees, rocks, and all Nature's mechanistic operations, but as if there were also in the universe the capacity for a personal adjustment. He had chosen his working hypothesis! /26/

Religion did not invent belief in the continuance of life after death. That came out of the urge of life itself, and this is practically universal. Millions, and hundreds of millions, of people, in all grades of culture, have believed themselves immortal. This belief was due to the deeper belief that what had produced personal values was still at work producing personal value. This belief came out in strange ways, for primitive men had no psychological laboratories. They thought in terms of soul and spirit, but still they dared think of personality as worth something in nature and as not ending with death. The development of this belief in personal values gave new thrusts to all life.

Let us not confuse belief in the persistence of personality with pictures of the future. As a child I used to believe that I should play a harp forever with a crown on my head. Later I feared I should feel a good deal of ennui before eternity was over. President William Rainey Harper, when he faced death, as we all have to, tried to reconcile himself to stopping in the midst of an extraordinary career by saying, "There /27/ must be some work for me to do over there."[24] His idea of heaven was an opportunity to be of service of some sort.

It really comes back to this: The creative process and the evolutionary process did not stop back in the first chapter of Genesis. If the human soul, or whatever you want to call it, is the outcome of the evolutionary process, there must be in the universe forces capable of producing what they have produced; and they must still be operating in our environment. The process is still on. Personal development through relation with these forces is still going on. I know the difficulty from the point of view of physiology, but I do not think that physiology knows all there is in the universe. I do not know what matter is—of course, I am only a theologian—but I am not at all afraid of what matter may be. And physiology is about the only science that stands between a conception of matter and this world-view that of necessity includes a belief in the survival of the individual after death.

If it did not seem so hopelessly naïve metaphysically, I should endeavor to set forth a monism which is not ready to /28/ say that matter is dead, even when it is found in dead bodies; but rather as a form of the cosmic energy it is always active.

Now, of course, it is perfectly true nobody can prove that, but you can take your choice as to whether you live in a universe where personality-evolving forces are at work or whether you live in a universe where all you have is chemistry, mechanics, and mechanism. Religion says that immortality as a personal experience is an element in this working hypothesis. All men's crude pictures of what comes after death are efforts to set forth their belief that personal values are supreme and that personality is not at the mercy of impersonal forces. Our faith in the continuance of this process, at least in the case of those who are at one with the personality-producing elements of the total activity, is to the effect that life at our present stage can be better lived on the supposition that the animal and the mechanical aspects of our life are really lower phases of what we really are becoming. We expect the continuance of those elements of our life which are in adjustment to the /29/ personality-producing forces of the universe not subject to time or space.[25] We do not allow imperfect conceptions of matter to deprive us of this one tenable world-view.

As religious persons, we are not trying to frighten folks; we are not trying to deceive folks. We are trying to induce folks to live as if the ultimate values in life were outside of time; that life is not futility or frustration; that the qualities which we should further in ourselves are not those which we share with the beast that dies, or the animal that survives in our own body, but those of the timeless personality-evolving forces of the universe. "This mortal," says Royce, "must put on individuality."[26] Really, that's what religion looks forward to—a life which will be richer and more self-directed because not limited by those particular forms of matter in the midst of which we now move and which we now express, but which will be conditioned by some other form of eternal force as superior to present forms of matter in human brains as these are superior to the crystal and the atom. *We believe that personal adjustment with the personality-producing* /30/ *forces of the universe will carry the* evolution of persons beyond the changes in the form of existence which we call "death." /31/

---

## From the Point of View of a Christian Minister
### By Charles W. Gilkey

One Easter afternoon some years ago a university professor said to me: "I'm very much of a liberal in matters religious. But once a year at least, on Easter morning, I do go to church. And I always come away unsatisfied if what we have heard is just another balancing of the old arguments, when what I'm wanting and needing is some fresh sense of newness of life."

He had put into frank and simple terms the characteristic approach of the religious man, and the distinctive contribution of religion, to the oldest question in the world: "If a man die, shall he live again?"[27] It is a question on which it is not easy to get beyond mere balancing of arguments: as an issue for debate this always has been, and doubtless always will be, an open question. All the appearances are certainly on one side: but on the other side is /32/ the cautioning reminder of modern science, from its early days under Copernicus and Galileo to these latest days of Eddington and the new physics, that appearances are often deceitful and never final. There are plenty of arguments on both sides, and no conclusive demonstration either way, for faith in immortality, like faith in the existence and still more in the goodness of God, moves in a realm where decisive proof and disproof are alike impossible.

So, too, seem at first sight the traditional contributions of religion, and especially of the Christian religion, to this old debate. What shall we moderns make of the New Testament accounts of the first Easter—the familiar story of the empty grave, and of the appearances of the risen Christ to his disciples? Here begins another long debate, for the same evidence that seems entirely conclusive to some among us is unconvincing to others of us. What the former regard as a demonstration has become for the latter a part of the problem itself, adding thus new areas and complications to the old debate. /33/

At this point, therefore, it becomes necessary to clear our minds first of all as to what it is that we may expect from religion in all these ultimate questions where debate continues down the generations, and conclusive demonstration seems to our generation more than ever impossible. Thoreau said once, with profound insight: "It is morning when it is dawn in your soul."[28] Religion brings the dawn of a new day to the soul of man—the rising within him of a new kind and quality of life that he experiences as coming from beyond and above himself, quickening within and around him a faith and hope and love that lighten both his burdens and his perplexities, and enable him to pass even through the valley of the shadow of death, unfearing and expectant of another and brighter dawn. In the classic phrases of the Bible, the rising of the sun of righteousness brings life and immortality to light, and shineth more and more unto the perfect day.[29] The contribution of religion to faith in immortality is therefore not so much new arguments for a debate as new experiences and a new assurance in living the kind of life that deserves to continue /**34**/ beyond this short and shifting mortal span.

These experiences and this assurance, however, are not given to most of us as a fixed and inalienable possession. Like the sunshine once more, they are profoundly affected by all the clouds and storms that sweep across our spiritual sky; they pass through many a night of doubt and sorrow, when the immortal hope burns low or dips beyond our horizon in some long arctic darkness. There are times in personal experience—perhaps there are even periods in human history—when religious men have to dig in and wait long for the return of the sun, like Byrd and his mates through their antarctic winter at Little America.[30]

It should be pointed out, too, that this assurance comes first to many of us, like our first experience of religion itself, not so much as an achievement of our own but rather through contact with some other person whose character or quality of life calls forth this conviction within us, as the sun calls forth life upon the earth. Faith in immortality, that is, is a claim that we make on behalf of some other people much /**35**/ earlier and much more strongly than we venture to make for ourselves. This fact has an important bearing on the charge so often made of late, that belief in immortality is an egregious case of wishful thinking, a projection beyond the grave of the elemental will to live, of the instinct of self-preservation. While it may, indeed, be so in some cruder cases, the nobler and deeper faith in immortality has more often been the conviction that some other person, honored or beloved, is too worthful to perish with the body in a universe that has produced and sustained moral values and spiritual capacities like these. Faith in immortality is thus a correlative, or rather a consequence, of faith in God.

It is natural, therefore, that the great personalities of our own and of other days—especially those whose greatness has been moral and spiritual—should become not only centers but strong supports for this faith. The New Testament abounds in evidence that this was true of Jesus to a very exceptional degree. The Book of Acts reports Peter as using in his first public address, on the Day of Pentecost, a striking /36/ phrase of which Professor Goodspeed in his American Translation has given us a highly significant rendering: "Death could not control him."[31] It is a revealing insight that lays bare with singular accuracy some of the main roots of a living faith in immortality then and now.

One of these is its plain disclosure of the motives and the perspectives that dominated Jesus' own living. Death could not control his *actions*. He himself once put his own attitude into words that have echoed down the centuries: "Be not afraid of them that kill the body, but are not able to kill the soul: but rather fear him who is able to destroy both soul and body."[32] And when his own life was threatened and finally taken from him, he turned his own counsel into a steadfast assurance and expectancy that dominated both his life and his death. This life was for him a room opening out into another and larger room, and death was the door leading through and beyond.[33]

In sharp contrast with this attitude and perspective there stands another, for which death is the decisive and controlling fact about human life. Not its /37/ high capacities and possibilities, which Jesus took as central for his estimate of human destiny, but rather its brevity and mortality, give us on this view the true clue to its significance. This attitude has always been with us down the centuries; but in our own disillusioned time it has become more than ever explicit and thoroughgoing. Man, says a modern novelist, is "only a bundle of cellular matter upon its way to become manure."[34] "Ours is a lost cause," writes a modern essayist, "and there is no place for us in the natural universe."[35] And a modern philosopher has painted in somber colors the ultimate shadow that from his viewpoint lies inexorably across all our human scene: "Brief and powerless is man's life; on him and all his race the slow, sure doom falls pitiless and dark."[36] These are attitudes toward life which death obviously can and does control: no wonder they have been named "Futilitarian."

Most modern men, of course, are not quite so explicitly hopeless. The more prevalent attitude in these matters does not thus magnify the fact and meaning of death, but rather ignores /38/ and tries to forget it. If death, as we often complain, was much overemphasized by our great-grandfathers in their theology and our great-grandmothers, on their samplers, we their descendants have certainly swung to the opposite extreme. We have left their gruesome skulls and crossbones and their solemn "Memento mori" out of everything from our tombstones to our ethics and our expectations; we have shortened their long funeral services to a minimum consistent with decency, have covered the raw edges of our open graves with banks of hothouse flowers, and have turned our faces the other way as soon as our hearts will let us. Death does not control us, largely because we do not face up to it.

Jesus' attitude contrasts sharply with both of these. Unlike most of us, he had faced up both to the fact and the imminent prospect of death. Unlike our modern Futilitarians, he had found in life itself something worthful enough and powerful enough to overcome it. "Death could not control him."[37] And down through the centuries he has quickened in countless hearts a courage and faith in the face of death /39/ like to his own. Death could not control *them*.

Thirty years ago the visiting preacher at an eastern university took as his text one Sunday morning, "I am the door."[38] He pictured life as a series of rooms leading one into the other: home to school, school to college, college out into active life; each room a preparation for the next. What then of life itself? His closing words no hearer could easily forget. Life itself, he said, seems to bring us at last only to a blank wall. But there stands One saying down the generations, "I am the door."

Twenty-five years later that same preacher's younger son had become one of the foremost medical scientists of his generation. In the mid-forties he was stricken with inoperable cancer, and his colleagues gave him six months to live. For eighteen months he looked death in the face with level eyes; and one of his medical colleagues has put into memorable words an experience shared during those months by many of his friends:

> The proof of a man's life—how much has been the living of a formula and how much an inward light—may often be found in the manner of his /40/ facing death. For courage is still, as it has always been, a thing of great beauty, that springs, whatever its form and expression, from an inner source of moral power. We wish, for ourselves and the ordinary human being, a swift and merciful death, which is most easily supported with dignity and composure. For him we would not have had it other than it came. Those who were fortunate in seeing him during those eighteen months when he and death sat face to face—who dreaded their first visits and came out gladly and inspired with a new faith in the nobility and courage to which rare men can attain—these know that the ugliness and cruelty of death were defeated. Death had no triumph, and he died as he had lived—with patience and love and submission in his heart, with the simple faith of a trustful child, and the superb gallantry of a great soul.[39]

But even these words fail to convey the full measure of his spiritual victory. Only a few weeks before his death, the doctor-son said to his preacher-father, speaking of the spirit of unity and serenity that pervaded the entire household, "If you want to see what the Kingdom of God is like, come over to 12 Irving Street." When his father told a friend of it, he added, "Every time I come into contact with the clarity of his mind and the serenity of his spirit, I am reminded of that great saying in the New Testament, 'Death hath no more dominion over him.'"[40] /41/

It is lives like these that have been our most persuasive "intimations of immortality" and have themselves constituted religion's largest contribution to the immortal hope and faith. Men looking on have intuitively said of such personalities, "These must surely be the valuable and enduring structure. The bodies that helped build them can have been only a temporary scaffolding. The scaffolding may have perished as scaffoldings must: but the structure shall endure. Death cannot control them."

At a memorial service held last year in the University of Chicago Chapel for Mrs. Henry W. Cheney, who had been the representative of that district in the Illinois legislature, Lorado Taft, the distinguished sculptor, was speaking.[41] Most men and women, he said, depressed and weakened his faith in immortality. But there had been a few in his acquaintance, Mrs. Cheney among them, who had greatly enlarged and reinforced this faith. They planned their lives on so large a scale, and lived them in such large dimensions, made such demands on life and cherished such expectations of it, that he could /42/ only think of them as hereafter going on to greater things. In the great phrase of Paul, they fought the good fight of faith, and laid hold on the life eternal.[42]

Nor is the quickening power of such lives limited to the time of their immediate passing. Of their *influence* also it may be truly said that death could not control them. This was obviously true of Jesus himself in most extraordinary measure. He became a far greater force in the lives of his friends and followers in the spirit than he had been when among them in the flesh: that is a plain fact of Christian history, which the New Testament not only states but by its very existence evidences. And this was not simply because he was a martyr who had been done to a violent and unjust death: "Men betrayed are mighty, and great are the wrongfully dead."[43] There was and is more in it than that. On behalf of all that have shared his attitude and spirit, our intuitive conviction deepens that the universe itself would somehow betray both them and us if it were to let them cease to be. They too would then be in some deeper sense "wrongfully /43/ dead"—martyrs in a cosmic persecution. Professor George H. Palmer has put that conviction unforgettably in the closing sentence of his biography of his wife, Alice Freeman Palmer: "Though no regrets are proper for the manner of her death, who can contemplate the fact of it and not call the world irrational, if out of deference to a few particles of disordered matter it exclude so fair a spirit?"[44]

The degree to which our beloved and honored dead not only keep, but even deepen and increase, their influence over us, in a way strangely independent of any physical presence, is a fact that soon or late falls intimately within the experience of us all. Their dear memory can touch even a common day with infinite significance, can speak with quickening power to the best within us, and lift our eyes to horizons and heights that stretch far beyond this mortal present. At the annual commemoration service of Columbia University two years ago, Dean Darrach of its medical faculty said:

> The continued influence of those departed this life, and the sense of reality of the continuing existence of their personalities, has been strong /44/ enough to remove for me any doubt as to some form of life after death. What it is or in what form I care not. I believe that they continue to exist and I believe that we can be influenced by them.[45]

Dean Darrach chose his words well. The faith in immortality cannot be proved in any conclusive sense, because the data for its demonstration (with all respect for and interest in modern psychical research) are still lacking; and even if the evidence of psychical research were much more conclusive than it seems to many of us as yet to be, the important question of the worthfulness of the life to come would still remain. Christian faith in immortality starts from the discoveries it has personally made of the worthfulness of life here and now when lived in Jesus' way, and moves on to the confidence that death can neither control nor interrupt lives that are lived that way. While it cannot be proved, it can be *lived*; and it is so lived whenever men measure their living not by time but by quality. It is therefore always an adventure of faith, but at the same time becomes a "moral certainty." As the Swiss boy put it when a traveler asked him where Kandersteg was: he /45/ did not know, but there was the way to Kandersteg![46]

When Dean Willard L. Sperry last came to the University of Chicago as visiting preacher, he said that he had learned to look forward expectantly to his first glimpse of Lake Michigan from the westbound Michigan Central train.[47] The sand dunes would first appear, with their hint of water beyond, but still shutting in the view with a wealth of interesting detail in the foreground. Presently, however, between the sand-hills, a sudden far vista of shoreless blue, . . . all too quickly gone . . . then for a moment glimpsed again. So, he said, is God in our human experience. And so, he might likewise have said, is "that great water in the West Termed Immortality"—stretching beyond this present that hints and yet hides it; but with the important difference that there is no railroad for this ultimate journey. Here we are each and all pioneers upon a great adventure, who cannot make a detour around this water but must cross it for ourselves. And the only boat that will carry us over is a kind of living that we can begin to build here and now.

## The Religion of a Geologist

Kirtley Fletcher Mather (1888–1978)

> The geologic eras succeeded each other for untold millions of years before there emerged from brute consciousness the self-conscious human being; but in a comparatively brief interval of time, self-consciousness is beginning to give place to an emerging world-consciousness such that men are daring to look all their fellows in the face with the eyes of a brother, and to act as if all men everywhere, regardless of color or intellect or nationality, are members of a single family. In this emerging ideal of brotherhood rests the hope for the world. It is the present high-water mark of the flood of evolution.
>
> —*Kirtley F. Mather,* Old Mother Earth

### Introduction

Given the amount of attention paid to evolution by fundamentalists during the 1920s, it is somewhat surprising that the AISL did not have a pamphlet by a geologist until September 1931, when they printed 27,000 copies of *The Religion of a Geologist* by Kirtley Fletcher Mather of Harvard University.[1] That would be the only printing; nor was any other title in the "Science and Religion" series printed again after that point, amid the deepening Depression and declining support from the Rockefeller Foundation.

Mather was an obvious choice for this topic. A direct descendent of Richard Mather (the father of Increase Mather and grandfather of Cotton Mather), he grew up in Chicago, where he had been a very active Baptist layman and was probably first influenced by the social gospel movement that shaped so much of his subsequent religious life and thought. As an undergraduate at the University of Chicago, Mather had taken a Bible course taught by Shailer Mathews, of whom he later said, "His approach was that of a 'modernist' rather than a 'fundamentalist' and I found it highly informative and fascinating." As a graduate student at the same university, Mather and his wife briefly joined Mathews's church, Hyde Park Baptist (now Hyde Park Union Church), a hotbed of modernism. Even though they soon left Chicago, the Mathers formed a lifelong friendship with their pastor, Charles Whitney Gilkey, who contributed to the AISL pamphlet, *Life After Death* (1930).[2]

Mather began writing about science and religion in 1918, just two years after completing his doctorate, when the *Atlantic Monthly* published an essay about the meaning of natural history. That subject would become increasingly contentious following the Great War, when Bryan and others underscored links between German militarism and the teaching of evolution that Stanford University biologist Vernon Kellogg had written about in *Headquarters Nights* (1917).[3] Mather went right to the heart of the matter. "Underneath the ancient warfare between theology and science" lies "an unvoiced, but very real, fear that in the last analysis the doctrine of the survival of the fittest in the struggle for existence is diametrically opposed to the conception of the brotherhood of man; that evolution according to Darwin and [Hugo] Devries and [August] Weissmann is the antithesis of Christianity according to Christ and John and Paul."[4] Assuaging that fear "was a theme with which I was concerned throughout many decades thereafter," he later wrote.[5]

To get that message across, Mather found it necessary to oppose fundamentalist views of the Bible in relation to science. Before moving to Harvard, he taught at Denison University, a Baptist college in Granville, Ohio. Simultaneously he taught a class designed for Denison students at Granville Baptist Church, introducing them to modernist views on miracles and divine action, while disparaging "the outworn science, the archaic philosophy, the man-made creeds and dogmas of traditional Christianity."[6] He continued this type of activity after coming to Harvard in 1924, teaching his famous "Mather class" at the Newton Centre Baptist Church for more than three decades.

The following summer, the Scopes trial brought Mather national attention. Concerned with the dynamics of a trial pitting Clarence Darrow against fundamentalism, Mather felt it would be important for the American Civil Liberties Union (ACLU) to have testimony from a few prominent scientists who were also religious believers. When he suggested this strategy to Roger Baldwin, the executive director of the ACLU, an invitation came for Mather himself to fill that role. Once on the ground in Dayton, Tennessee, he befriended Darrow. He not only helped him prepare for the famous cross-examination of Bryan, after the trial he also helped him sort through the mail he received during the trial. For the rest of his life, Mather spoke and wrote extensively about his experiences in Dayton. Stephen Jay Gould, his colleague at Harvard many years later, customarily invited Mather to lecture about it in his class. Gould's famous model of science and religion as "non-overlapping magisteria" sounds a great deal like Mather, whom he recalled as "perhaps the finest man I have ever known."[7]

Three years after the trial, in August 1928, Mather's book *Science in Search of God* (1928) was named a monthly selection for the Religious Book Club, an enterprise that had recently been launched in imitation of the Book-of-the-Month Club (which had begun only in 1926).[8] His book was chosen by an editorial committee of five modernist religious leaders, including Fosdick, and quickly proved an excellent decision. More members of the Club kept this book than any other selection—only three percent were returned, fewer even than two other very popular books about science and religion, *The New Reformation* (1927) by Pupin and *The Nature of the Physical World* (1928) by Arthur Eddington.[9] Perhaps the notoriety this brought Mather was a further reason the AISL enlisted him for what turned out to be the final installment of the "Science and Religion" series. The text of the pamphlet was first presented as a lecture at Phillips Brooks House, Harvard University, on March 8, 1931, and published in the *Harvard Alumni Bulletin* three months later, but clearly it had been conceived from the start as a publication for the AISL.[10]

The pamphlet and Mather's many other publications on science and religion present a characteristically modernist conception of God. Indeed, Mather was somewhat reluctant to use the word "God," because he wanted to avoid so many of the images with which it was traditionally associated. For him, it is "a symbolic term used to designate those aspects of the administration of the universe that affect the spiritual life and well being of mankind." God is "a creative and regulatory power operating within the natural order," who "is immanent, permeating all of nature, unrestricted by space or time," yet "transcendent only in that His spirit transcends every human spirit, possibly the sum total of all human spirits melded together. He is not supernatural in the sense of dwelling above, apart from, or beyond nature."[11] The term Mather preferred—the "administration of the universe," which (he noted) is singular and not capitalized (although it is capitalized once in his pamphlet), and not synonymous with "administrator"—he took with due acknowledgment from one of his teachers at Chicago, the great geologist Thomas Chrowder Chamberlin. Nevertheless, his god was still in some sense personal: it had produced us, and we are undoubtedly personal beings. "Because we recognize personality as inherent in the human species," he wrote, "we attribute personality to the motive powers which have produced mankind."[12] As he said on another occasion, "The emergence of personality in the evolutionary process is an event of transcendent importance," and it could have happened only "in response to personality-producing forces in the universe. It is to these particular portions of cosmic energy that I would apply the term God." Thus, for Mather, "God is the motive power which tends to produce a fine personality in a human being."[13] Mathews had expressed an almost identical view in his own pamphlet, where he identified an immanent but personal God with those "elements within the universe" that ultimately account for the "rational and purposeful activity which in the course of evolution results in personal life."[14]

Two of Mather's closest friends, the Unitarian scholar Ralph Wendell Burhoe and Harvard astronomer Harlow Shapley, held notions of God highly similar to Mather's "administration of the universe." When they (and others) founded the Institute on Religion in an Age of Science (IRAS) in 1954, they were in effect helping to bring Mather's modernist ideas from the 1920s into the modern "dialogue" of science and religion that IRAS did so much to create in the 1950s and 1960s and that has continued down to our own day. Given that influential voices in the contemporary dialogue espouse identical views, Mather's pamphlet has not yet lost its relevance.

---

Annotated Text

## *The Religion of a Geologist*

By Kirtley F. Mather

/3/ No individual man of science can speak for all scientists, nor can he represent science. As a matter of fact, there is no such entity as "Science." There is rather a large group of individual scientists who have ideas, who hold principles, and who develop beliefs, some of which they share in common and some of which are held by individuals more or less uniquely here and there. I cannot, therefore, speak as the authorized representative of all scientists or indeed of science as a system.

I presume, however, that all scientists will agree concerning certain fundamental principles from which each individual scientist must start as he tries to discover religious values and ideals for himself. In the quest for religion the scientist has certain advantages, or perhaps some would say certain handicaps. Whatever may be your attitude toward them, there are four basic principles with which every true scientist must start his quest. /4/

### Uniformity of Nature

There is first the principle of uniformity of nature. The results of all our investigations and observations indicate that there is, in the world of which we are a part, some organizing principle, some determining factor, which reveals itself in the logical, law-abiding sequence of cause and effect. I am quite well aware of the fact that in the last few years certain students of atomic physics have been led to the conclusion that the principle of cause and effect is not truly universal, that it breaks down when we study the world within the atom. The principle of indeterminism has been much discussed during the last two or three years. Unfortunately, some persons have misinterpreted that principle and have concluded that scientists have abandoned the old conception of a sequential relationship between cause and effect. As a matter of fact, however, the principle of indeterminism applies simply to the activities of the unit particles of matter and of energy within the atom. All our knowledge concerning the transformations of matter and of energy in larger units than atoms or quanta, indicates that the old conception /5/ of a causal relationship holds true wherever we have adequate experience with nature. The prediction of the movements of the planets in their courses, of eclipses of the sun, of the effects of heredity, of all the transformations of matter and of energy in the practical affairs of everyday life, is just as reliable today, if not more reliable, than it was before the days of Einstein and Heisenberg.[15]

The world in which we live is not only law-abiding; every event is produced by causes and forces resident within the universe. We have, from our point of view of science, absolutely no suggestion of any outside interference whatsoever. In other words, the scientist discovers no evidence of the supernatural, in the old orthodox concept of the term. For me personally, the supernatural is simply that portion of the whole which I know

I do not understand, and the natural, using the term in its older sense, is that portion of the whole which I think I understand. The first principle, then, with which we must start our quest is the principle of the uniformity of nature, involving as it does the abandonment of the old concept of a supernatural reality outside, /6/ above, beyond, or behind the universe.

### Dependence upon Human Intellect and Emotions

The second principle which the scientist must use is a frank and thoroughgoing dependence upon human intellect and human emotions for guidance in the search. Facts must be described; events must be observed. There is no such a thing as a true or a false fact. A fact simply is. It would probably be better to call it an event, and say that an event occurs. But the description of the fact or event may be true or false. Into every scientific problem the obvious fallibility of the human mind projects itself. All that we can know about our world is filtered through our senses and our intellects to our consciousness. We are limited by the human intellect and the human emotions.

Similarly in religion, the individual has certain experiences. That mysterious entity, myself, responds to something outside of myself; I experience something. There is no such thing as a true or a false experience. An experience simply is, or is not. But, to be useful /7/ to the individual, the experience must be interpreted. In the interpretation of experience, once more the human intellect and human emotions enter, with all the fallibility which that involves. It is in the interpretation of experiences that we find what used to be called revelation. A revelation, in the religious sense of the term, is simply an interpretation of an experience. We may describe the experience correctly or incorrectly; we may interpret it rightly or wrongly. But the interpretation of experience is the very essence of revelation.

In other words, the scientist seeks for authority in facts, or events, and experiences. He is skeptical always of descriptions, of inferences, and of interpretations. He trusts certain individuals correctly to describe and rightly to interpret facts and experiences only after those individuals have demonstrated, in some way which can be tested by others, their ability to do these truly difficult things.

To the scientist, then, the final court of authority is never found in the words or statements of human beings. It matters not with what degrees the speaker's name may be adorned, or /8/ what position he holds in church or school or state; it matters not upon which mountain top the words were engraved upon what tablets of stone, nor within which sacred and revered documents they are entombed. The scientist reserves to himself the right to challenge the interpretations placed upon facts and experiences. Although thus dependent upon the human intellect and the human emotions for guidance, he finds a sufficient if not an absolute authority in the facts and experiences of life.

### Knowledge Relative

That suggests the third basic principle from which the scientist must start. We have no knowledge and can have no knowledge of anything which is truly absolute or accurately described as ultimate. All knowledge is merely an approximation to the truth. We are dealing with the relative; we talk only in terms of probabilities. Truly absolute and actually ultimate realities are beyond the ability of our minds to comprehend. We observe a process of which we are a part. That process—the transformation of matter and of energy, going on, as far as we /9/ can tell, without any limit in time or space—is reality. The only reality which man has actually discovered is the reality of change, of process.

### Behavior the Clue

The fourth principle pertains to the nature of the process. The characteristics of reality are to be discovered by observing behavior. Behavior is the clue to any adequate understanding of the transformation of matter and of energy. It is the behavior of human beings which gives us knowledge concerning the nature of man. It is the behavior of cosmic forces which gives an insight into the real characteristics and qualities of the motive power of the universe.

These are the four posts upon which the scientific platform is constructed. Upon them we may stand and from them we may launch out into the unknown. Only thus may a scientist discover a religion which he can accept as satisfactory and sufficient for himself. Those four corners of the scientist's platform are, to repeat, the uniformity of nature, the dependence upon human intellect and human emotions, the abandonment of the search for the absolute /10/ or the ultimate, and the use of behavior as giving a sufficient and satisfactory clue to the real nature of cosmic forces.

Starting from the foundation, thus established foursquare, I want to present to you something of my own personal views about religion, the religion of a scientist based upon these principles. Other scientists starting from the same platform might have different announcements to make. I speak only for myself, and I recognize the possibility, in fact the probability, that other scientists could present their own personal views in quite different terms, and perhaps with even greater truth.

### Belief about God

The religion of a scientist must involve, first and foremost, certain beliefs about what we have been taught to call God. Every religion has its God, and every man has his religion. What is the nature of the God which the scientist can accept as satisfactory in the light of all that he knows about the universe?

For me, God is the Administration of the Universe, the power which determines the logical sequence of cause and /11/ effect, the energy which fills and thrills the universe, motivating the transformations of matter and of energy.[16]

We may discover the nature of cosmic energy only by observing what it has produced in the past and how it is operating at the present time. This involves the description of facts, the observation of events, and the interpretation of our own experiences. The geologist has a distinct advantage here over most other scientists because the geologist has learned to think in terms of vast intervals of time; he has necessarily adopted a world-view; he stands on the mountain top and looks far into the past; he surveys a universe, limitless both in time and in space. To him there comes no suggestion of any real beginning, no hint of any final end.[17] He is dealing simply with transformations of matter and of energy from previously existing forms and states into those which now exist. And he knows that the existing states of matter and of energy will in the future be transformed into still other states.

In this process he recalls the time when upon the earth the sensible display of cosmic energy was limited to the mineral kingdom. The only objects /12/, available at that time for study, were inanimate. These need to be analyzed. When we study crystals, rocks, minerals, and stones we find that they are worthy of our great respect. One comes to a pretty definite feeling of admiration for the power that could organize particles of matter—molecules, if you please—in the marvelous architectural designs of crystals. But that is simply a tribute to the law-abiding, consistent operations of cosmic energy.

The geologist would summon to his aid the physicist and ask him to carry the analysis down to the world within the atom. At first glance the atomic structure and molecular organization seem to be quite mechanical. For several decades the results of scientific investigation appeared to be leading directly toward a mechanistic explanation of the nature of cosmic energy. All that has changed in the last few years. We now know that the latest results of the analysis of material objects, when we penetrate as far as we may into the secret of the nature of things, give us a wholly different impression from that which our fathers had a generation ago.

With deeper understanding and /13/ truer knowledge, we find that the cosmic energy which operates within the atom has the attributes and characteristics of mind rather than of mechanics, of pure mathematics rather than of applied mathematics. The announcements which have recently been made by Sir James Jeans are of great significance here. As he puts it, the world now appears to be more like a great thought than like a great machine.[18] The facts which have been observed, the events which have been noted, are explainable not as the operations of a mechanical device, but as the expressions of mentality. The nearest approach we have thus far made to the ultimate in our analysis of matter and of energy indicates that the universal reality is mind. The old dualism of mind and matter disappears; matter becomes simply an expression of mind.

In that study of the world within the atom, just as in the study of the stellar reaches beyond the earth, there appears only cold, calculating mind; nothing of sentiment, no suggestion of kindliness; just the cold calculations of a mathematical genius, if you please to put it in that way.

The geologist, however, recalls that /14/ from this world of inanimate matter there emerged, at one stage in the history of the earth, the first animals and plants. We do not understand the process by which the first living cell came into existence, but we accept the facts, as indeed we must. Life appeared upon the earth.

Living creatures are characterized by certain qualities or attributes which distinguish them unmistakably from all preceding expressions of cosmic energy. There are differences between cells and crystals; animals and plants differ from minerals and stones. True, those differences are not nearly so distinct and conspicuous as at first glance they would appear to most of us to be. Certain expressions of inanimate nature very nearly duplicate certain qualities of animate nature. Nevertheless, there is a real distinction; the living is different from the non-living. Probably the most significant difference is indicated by what we call consciousness. The behavior of animals and plants is different from the behavior of minerals and stones. If you prod an amoeba with a needle, it indicates by its behavior that it is aware of this untoward circumstance in the physical /15/ environment. If you drop acid upon the tail or head of an angleworm, its behavior indicates that it is aware of this physical stimulus. You may prod a crystal of quartz from now till doomsday or drop acid upon a stone, but the behavior of these objects indicates absolutely no awareness of the external stimulus.

Consciousness is an expression of cosmic energy. In a certain organization of matter, the motive power of the universe reveals itself as conscious of its environment. I presume this characteristic of consciousness explains the fact that cells have changed during geologic time, whereas crystals have remained the same throughout the entire history of the earth, two billion years, more or less.

The geologist also recalls that just yesterday in the development of life upon the earth there emerged from the host of animals one species which is characterized by certain qualities and attributes which it does not share, at least in degree, with other animals. Man is self-conscious; man is aware of non-physical stimuli. He responds occasionally to the beauty of the sunset and the glory of the dawn. He is moved /16/ by the call to duty. He is stirred by high ideals and noble aspirations. He has a keen desire to discover goodness, moral law, righteousness. These are some of the non-physical stimuli to which he responds.

The awareness of such stimuli makes a difference in his life. Anything that makes a difference in the behavior of an organism must be real. The behavior of human beings indicates, therefore, the reality of these expressions of cosmic energy. To sum up the whole story, the geologist simply reminds you that in the transformation of matter and of energy upon the surface of the earth, in obedience to law, in response to the conditions of time and place, there has emerged that which we do well to call personality. There must, therefore, be personality-producing forces in the universe. It is to these particular portions of cosmic energy that I would apply the term God. For me, God is everything in the universe which tends to produce a fine personality in a human being. Does God, as thus defined, deserve to be called kind or good? It all depends upon whether the personalities which are produced are kindly and good. If /17/ you believe that there are kindly personalities and good human beings, then you must attribute to cosmic energy the necessary power to produce those results.

The forces which have produced righteous, kindly personalities must be at least as valuable and significant as their product. Cosmic energy displays, therefore, in one particular space or part of space and during one stage of time—the earth in glacial and post-glacial geologic time—the characteristics of a kindly thought.

### Belief about Man

That represents my belief about God. It leads naturally to a statement of my beliefs about man. Personality is creative. Man is a creative power, a creative being, having certain prerogatives and rights in the midst of time and space. There are only two alternatives, one of which must be accepted when we try to explain man. Either man was produced by a manufacturing process, wholly without volition on the part of himself or his predecessors all the way back through the entire animal kingdom to the one-celled plant-animal from which all life evolved, or else we /18/ must attribute to the developing organisms a certain element of volition.

Accept the first alternative and we have the old orthodox view of God, the Jehovah who manufactured the world and molded man out of the dust of the ground. Evolutionary processes may have been used in producing the results, but unless organisms have somehow had a certain freedom for self-expression, the result must be attributed to an external manufacturing agent.

Personally, it seems to me much more logical to believe that human beings were developed by the reaction of vital energy to its surroundings, that life itself is creative, and that volition enters into the process all along the way. I would not ascribe volitional activity solely to human beings, nor do I know where volition first begins to display itself in the behavior of the lower animals and plants. But somewhere in the scheme of things I would recognize freedom of action, volition, as an essential element in the creative process. If that be true, then man is in a very real sense the determiner of his own destiny, just as the dinosaurs were, within limits, the determiners of their destiny, and lungfishes determiners of theirs. /19/

I confess my inability to explain precisely how volition operates in the scheme of things. I do not fully understand how the prerogative of freedom, which seems to be one of the characteristics of observable expressions of vital energy, is preserved and developed within a world of law. But that it is there, I do not for a moment doubt. This means that to man there is given not the certainty of future perfection, but the opportunity of choosing for himself from among a variety of possible paths that in which he shall move onward.

There is nothing inherent in the evolutionary process which makes it necessarily progressive; but we cannot deny that it has been progressive. The earth is better now than it was two billion years ago when no living creature was upon its surface. There are more expressions of the nature of cosmic energy today than there were then. Finer qualities of the motive power of the universe are manifest in the world of sense perception now than there were then, if we have any basis whatever for passing judgment upon such things. /20/

### Belief in Evolution

It is sometimes suggested that our belief in the success of the evolutionary processes is solely a result of the supreme egotism of mankind. We designate animals as "higher" or "better" simply because they approximate more closely the structure or attainments of man himself. All our standards of judgment are human standards and therefore they are not a valid basis for concluding that there has been any real progress in life-development after all. There is, however, one basis for evaluating attainments which has more than a suggestion of being impersonal and universal. In how many ways may the vital energy which actuates an individual display itself to other individuals in this world of sense perception? There is no query here as to which expression of creative personality is better and which is poorer. It is a purely quantitative or mathematical estimate. In how many ways can an individual express his personality? The earthworm and the man, the dinosaur and the elephant, the bird and the fish? Obviously, man displays more ways of expressing his own nature than any other organism known to us. /21/

That, I should say, is a sufficient basis for the conclusion that evolution has been progressive. Apparently, it has been progressive because individuals, here and there, improved opportunities. There can be no opportunity unless there is also responsibility. One cannot exist without the other. Today, a very definite responsibility rests upon man, just as yesterday certain specific responsibilities rested upon the ape-man, and day before yesterday still other responsibilities rested upon the anthropoid ape. To meet that responsibility, man must commit himself to some belief about his future.

### Immortality an Achievement

And the third element in any man's religion is his belief about the future. There is, first, the collective future of mankind. Man has the opportunity of displaying cosmic energy in ways that have never yet been manifest. He has an opportunity to attempt the experiment of building a social order based upon the preservation of individual initiative and personality. Other experiments in developing societies have been tried by the motive power of the universe, but the social order of the insect /22/ is not the ideal for man. In insect societies, individual initiative and personality are submerged. Man apparently is trying the more difficult experiment of building a satisfactory social order in which there shall be preserved the initiative of the individuals and the personalities of the members. Whether or not this experiment will succeed no one can tell today. It would not be an experiment if we knew what its outcome would be. The collective future of mankind is apparently to be determined largely by human success in this particular experiment.

There is also the personal future of the human individual. This, of course, involves what the man of religion calls immortality. Here the scientist finds himself hampered by very few facts and, therefore, free to speculate almost as he pleases. Science has little to say concerning the possibilities of the continuation of an individual's personality after the individual's body has decayed. We do not know enough about personality to give us much of a basis of fact from which to draw any very definite conclusions.

I shall therefore present my own temporary and personal view concerning /23/ the future of the individual. There is only one expression of nature which can be immortal. Matter is not immortal; not even germ plasm can persist forever. The earth is a temporary and local thing. The day must come, far in the distant future but nevertheless at some time, when the earth will no longer be a suitable abode for the material bodies of animals or plants. Immortality, in the real sense of the term, must therefore depend upon something which is not limited by time or space. That "something" we call the spiritual.

Man is becoming aware of such spiritual realities in the universe. Beauty and truth and goodness and love are such realities, at least in certain of their expressions. They are not limited by time and space. If man wishes to become immortal he should lay hold of these eternal realities, these values which cannot be measured by foot-rules or calipers, which have no relationship to the beat of the pendulum or the movement of the planets. He should build these into his life.

Immortality is an achievement well within the bounds of possibility. I do not know how probable it is. The fact that it sometimes seems difficult or improbable /24/ should not discourage us. In the past, individuals here and there have accomplished things that never before had been accomplished. It is a corollary of our modern ideas concerning evolution, that things new do appear under the sun, that new conditions permit the operation of laws and forces which have not previously been effective. It is perfectly rational for man to strive to accomplish that which has never before been accomplished. There is at least the possibility that man may build into his life enough eternal values to make himself, even his spiritual personality, so valuable that it becomes indestructible, that it will be truly eternal. That I should say is the hope in my religion.

But my religion involves more than hope. Although I am uncertain concerning many things about which one might wish to have certitude, it is nevertheless true that in the midst of uncertainty I have a very real sense of security. As I have studied the results of the operation of cosmic energy, I have come to the belief that the administration of the universe is extremely wise and notably kind, that it will do the best it can for me. And I am content /25/ to take what comes as being what I deserve. The best that I can do may lead to oblivion. If so, I shall take that as satisfying all that I can expect to have satisfied. The best that I can do may lead to the merging of my frail and poor and feeble personality with similar feeble and frail personalities in what is sometimes called a cosmic personality. If so, all well and good, that is satisfactory for me. On the other hand, it may be that I will deserve something better than that. If so, I will welcome it. But permeating my entire philosophy there is a profound sense of security based on my knowledge of the remarkable achievements which must be credited to the creative forces which are an essential part of the structure of the universe and to which I am intimately related.

### Standard Scale of Conduct Found in Jesus of Nazareth

If this philosophy of life which I have been formulating is to serve adequately as the basis for my own personal behavior, I must advance one step more. How shall I discover in the many expressions of cosmic energy, those which are significant of the highest /26/ purposes, the finest qualities, in the universal administration? That there is purpose in the administrative processes cannot be denied. Ever since animals began to display the will to live, ever since squirrels began to store up food for the coming winter, ever since human beings gathered themselves together to face co-operatively the problems of unemployment or of depression in the stock market, purposes have been manifest in this particular corner of the universe.[19] The purposes may be temporary but they are none the less real. How shall I discover among them those which are the highest, the most valuable, in order that I may accord my life in harmony with them, in the hope that I can build into my personality sufficient values, as thus estimated, to make my life of eternal rather than simply of local and temporary significance?

Obviously, I need a standard scale of values by means of which I can judge or appraise the events which I observe. For me the standard of judgment is found in my belief about Jesus of Nazareth. I have assumed that the teachings and the life of Jesus of Nazareth, as I believe them to have /27/ been, represent truly the finest qualities in the universe. Here is the standard by which I evaluate the results of the experimental process, by which I decide what events are most significant of the highest good. I merely ask the question, "Does this event, when correctly described, carefully studied, and rightly understood, promote or create in men the type of life which Jesus of Nazareth would have men live?" If so, I believe it represents the finest qualities of administrative energy.

I have assumed, in other words, that Jesus actually opened unto us the very heart of God, that he has displayed the finest qualities of universal energy that are available for my study. That assumption I can never prove by any process of logic, any more than I can prove by logic that any of the many assumptions of science are true. But just as I can justify and validate the assumptions of science by trying them out to see how they work, so I can justify or validate this religious assumption by trying it out to see how it works.

We are all in the midst of that experimental operation. If the attempt to put Christlike motives and ideals /28/ into practical operation among human beings succeeds in making life better, then we are justified in the assumption that we have made. Is the truly Christian way of life the finest and the best? There is only one method of answering the question, the scientific method—try it and see.

Thus my religion, based on the fourfold foundation provided by the scientific study of the world in which we live, leads to certain very definite conclusions in the realm of belief, ideas which are well calculated to motivate one's life, ideals to which one may commit one's self—a confidence in the kindly thoughtfulness of administrative energy, a sense of security in the midst of uncertainty, a belief in the creative powers of human beings, a hope for the future, individual and collective, an assumption concerning Jesus of Nazareth which is in process of being tested in the experience of myself and of my fellows. These, I submit, provide a highly satisfactory basis on which one may order one's own life.

APPENDIX ONE

# Publication Details for AISL Pamphlet Series "Science and Religion" and Related Publications

| Author | Earlier versions (if any) | AISL version | Later versions (if any) |
|---|---|---|---|
| Conklin | (1) "Bryan and Evolution," *New York Times*, March 5, 1922, VII-14. Contains some paragraphs omitted in the pamphlet and lacks several paragraphs added to the pamphlet.<br><br>(2) The preface (dated May 1, 1922) in Conklin, *The Direction of Human Evolution*, 2nd ed. (New York: Scribner's, 1922), includes several paragraphs from the newspaper article and some additional paragraphs that are in the pamphlet but not the newspaper article. | *Evolution and the Bible* (September 1922). | (1) In *Fundamentalism vs Modernism*, ed. Eldred C. Vanderlaan (March 1925), 263–71.<br><br>(2) Newspaper version reprinted in *Creation-Evolution Debates*, ed. Ronald L. Numbers, vol. 2 of *Creationism in Twentieth-Century America: A Ten-Volume Anthology of Documents, 1903–1961* (New York: Garland, 1995), 15–19. |
| Fosdick | (1) "A Reply to Mr. Bryan in the Name of Religion," *New York Times*, March 12, 1922, VII-2.<br><br>(2) "Mr. Bryan and Evolution," *The Christian Century*, March 23, 1922, 363–65. | *Evolution and Mr. Bryan* (September 1922). | (1) In *Science and Religion: Evolution and the Bible*, ed. Harry Emerson Fosdick and Sherwood Eddy (New York: Doran, 1924), chap. 3.<br><br>(2) In William M. Goldsmith, *Evolution or Christianity, God or Darwin?* (St. Louis: Anderson Press, 1924), chap. 16.<br><br>(3) In *Fundamentalism vs Modernism*, ed. Eldred C. Vanderlaan (March 1925), 282–90.<br><br>(4) The final chapter in Henshaw Ward, *Evolution for John Doe* (Indianapolis: Bobbs-Merrill, 1925), includes lengthy excerpts from the pamphlet.<br><br>(5) In *Evolution and Religion*, ed. Gail Kennedy (Boston: D. C. Heath, 1957), 30–34. |
| Mathews | | *How Science Helps Our Faith* (September 1922). | (1) "How Science Helps Our Faith," *The Baptist* 3.36 (October 7, 1922): 1108–9.<br><br>(2) *Ke xue dui yu zong jiao de gong xian*, trans. Shizhang Zhang (Shanghai, 1934). |

(*continued*)

| Author | Earlier versions (if any) | AISL version | Later versions (if any) |
|---|---|---|---|
| Millikan | (1) "Deny Science Wars Against Religion, Forty Scientists, Clergymen and Prominent Educators Attack 'Two Erroneous Views,'" *New York Times*, May 27, 1923, I-1. Thirteen government and business executives who signed are omitted in the pamphlet.<br><br>(2) "Science and Religion," *Science* 57.1483 (June 1, 1923): 630–31.<br><br>(3) "Deny Science Wars against Religion," *Review of Reviews* 68 (July 1923): 88–89.<br><br>(4) "Men of Science Also Men of Faith," *Literary Digest* 78 (July 14, 1923): 30–31, includes lengthy excerpts from the article.<br><br>(5) In Robert Andrews Millikan, *Science and Life* (Boston: Pilgrim Press, 1924), 86–90. | *A Joint Statement Upon the Relations of Science and Religion* (n.d.). Published no later than September 1926, but probably much earlier. See Georgia L. Chamberlin to H. R. Clissold, October 1, 1926, AISLR 19:2. | (1) In *Fundamentalism vs Modernism*, ed. Eldred C. Vanderlaan (March 1925), 294–96.<br><br>(2) In Robert Andrews Millikan, *The Autobiography of Robert A. Millikan* (New York: Prentice-Hall, 1950), 289–92. |
| Millikan | (1) Address on "Science and Religion," Los Angeles (March 1923); no manuscript survives.<br><br>(2) "Science and Religion," *Bulletin of the California Institute of Technology* 32, no. 98 (March 1923): 3–20. The date on the front cover (1922) contradicts the internal date (March 9, 1923) at the end.<br><br>(3) "A Scientist Confesses his Faith," *The Christian Century*, June 21, 1923, 778–83.<br><br>(4) "Science and Religion," *Christian Education* 6.10 (July 1923): 517–24. Abridged, lacking several paragraphs. | *A Scientist Confesses His Faith* (September 1923). | (1) In Robert Andrews Millikan, *Science and Life* (Boston: Pilgrim Press, 1924), 38–64.<br><br>(2) Bailey Millard, "A Scientist Who Believes in Religion: A Nobel Prize-Winner Who is a Minister's Son," *World's Work* 51 (April 1926): 662–66, includes lengthy excerpts from the pamphlet.<br><br>(3) Abridged version in *Vendata Monthly: Message of the East* 16 (La Crescenta, CA: Ananda-Ashrama, January 1927): 206–14.<br><br>(4) In *Challenging Essays in Modern Thought*, ed. Joseph M. Bachelor and Ralph L. Henry (New York: Century Company, 1928), 383–96. Includes discussion questions and commentary on "Religion and the College Student," by "H. L. S."<br><br>(5) *Yi ge wu li xue jia di zong jiao guan*, trans. Shizhang Zhang (Shanghai, 1934). |
| Frost | | *The Heavens are Telling* (September 1924). | (1) *The Heavens are Telling*, Astronomical Society of the Pacific Leaflets, vol. 4, no. 186 (August 1944): 278–96. |
| Schmucker | | *Through Science to God* (September 1926). | |

| Author | Earlier versions (if any) | AISL version | Later versions (if any) |
| --- | --- | --- | --- |
| Pupin | (1) "Creative Co-ordination," commencement address at the University of Rochester, June 20, 1927, MIPP 1. | *Creative Co-ordination* (May 1928). | |
| | (2) "Creative Co-ordination: A Message from Physical Science," *Scribner's Magazine* 82, no. 2 (August 1927): 142–53. Much longer than the pamphlet, but the main ideas are the same. | | |
| | (3) "Creative Co-ordination," chapter 7 in Michael Pupin, *The New Reformation* (New York: Scribner's, September 1927). Similar to the version from *Scribner's Magazine*, with some omissions and some additions. | | |
| | (4) "Creative Co-ordination," address at inauguration of new president, Lafayette College, October 20, 1927. Published in *School and Society* 26 (October 29, 1927): 543–47. Identical to the pamphlet, except first and last paragraphs of the address are omitted in the pamphlet. | | |
| | (5) "The Unity of Knowledge," address at Methodist Preachers' Meeting, New York City, January 9, 1928. Published in *Methodist Review* 111 (March 1928): 169–75. Another version, with the most openly Christian comments. | | |
| Fosdick | (1) "Religion's Indebtedness to Science: A Sermon Preached at the Park Avenue Baptist Church New York on February 27, 1927" (New York?, 1927). Nearly identical to the pamphlet, which omits the final prayer and has minor editorial changes. | *Religion's Debt to Science* (May 1928). | (1) *Zong jiao suo shou ke xue de en ci*, trans. Shizhang Zhang (Shanghai, 1934). |
| | (2) "Religion's Debt to Science," *Good Housekeeping* 86 (March 1928): 21, 220, and 223–26. Identical to the pamphlet. | | |
| | (3) "Religion's Debt to Science," *Reader's Digest* 6 (April 1928): 715–16. Abridged version of article in *Good Housekeeping*. | | |
| Compton, Mathews, and Gilkey | (1) Compton, "Immortality From the Point of View of Science," *The Presbyterian Banner*, January 9, 1930, 10–12. Very similar to his part of the pamphlet, and to item (2a) at right—unclear why it was printed twice. | *Life After Death* (September 1930). | (1) Compton and Mathews, "Immortality: Four Faculty Men View the Future," *The University of Chicago Magazine* 23 (November 1930): 5–17. Includes addresses by Thomas Vernor Smith and Anton Julius. |

*(continued)*

| Author | Earlier versions (if any) | AISL version | Later versions (if any) |
|---|---|---|---|
| | (2) Addresses by Compton and Mathews at Rockefeller Chapel, University of Chicago, from Easter 1930 symposium on "Immortality." Addresses by Thomas Vernor Smith and Anton Julius Carlson were not published by the AISL. Manuscripts do not survive.<br><br>(3) Sermon by Gilkey from Easter Sunday, April 20, 1930; manuscript does not survive. | | (2a) Compton, "Life after Death: From the Point of View of a Scientist," *The Presbyterian Banner* 117 (March 26, 1931): 10–11 and 14. Identical to his part of the pamphlet.<br><br>(2b) Gilkey, "Life after Death: From the Point of View of a Christian Minister," *The Presbyterian Banner* 117 (March 26, 1931): 12–14. Identical to his part of the pamphlet.<br><br>(3) Compton, "A Scientist Considers Immortality," *The Christian: A Liberal Journal of Religion*, March 28, 1931, 310–12, includes lengthy excerpts from his part of the pamphlet.<br><br>(4) Compton, "Life after Death: From the Point of View of a Scientist," *Christian Education* 15 5 (February 1932): 315–23.<br><br>(5) Portions reused in Arthur H. Compton, *The Freedom of Man* (New Haven, CT: Yale University Press, 1935).<br><br>(6) Compton, "Why I Believe in Immortality," *This Week,* Sunday supplement to *New Orleans Sunday Item-Tribune*, April 12, 1936, 5 and 12. Draws heavily on his part of the pamphlet and cites version in *The Freedom of Man*. |
| Mather | (1) "The Religion of a Scientist," lecture at Phillips Brooks House, Harvard University, March 8, 1931; manuscript does not survive.<br><br>(2) "The Religion of a Scientist," *Harvard Alumni Bulletin*, June 18, 1931, 1142–49. Identical to the pamphlet, except the pamphlet omits the opening paragraph. | *The Religion of a Geologist* (September 1931). | (1) "The Religion of a Scientist," *The Intercollegian* 49 (1932): 143–46, includes lengthy excerpts from the pamphlet.<br><br>(2) "The Religion of a Scientist," *The Epworth Herald*, March 19, 1932, 274–75, reprints version from *The Intercollegian*.<br><br>(3) *Yi ge di zhi xue jia de zong jiao guan*, trans. Shizhang Zhang (Shanghai, 1934). |

Print copies listed in WorldCat database as of February 16, 2024: Conklin = 11; Fosdick 1922 = 14; Mathews = 10 in English, 2 in Chinese; Millikan, "Joint Statement" = 4; Millikan = 15 in English, 2 in Chinese; Frost = 16; Schmucker = 8; Pupin = 1; Fosdick 1928 = 4 in English, 2 in Chinese; Compton = 2; Mather = 2 in English, 2 in Chinese.

APPENDIX TWO

# Publication Runs for AISL Pamphlets and the Millikan "Statement"

| Author | First AISL imprint: date/number | Later imprints: dates/numbers |
|---|---|---|
| Conklin, *Evolution and the Bible* | September 1922: 30,000 | October 1922: 30,000<br>November 1922: 30,000?<br>September 1931: 27,000 |
| Fosdick, *Evolution and Mr. Bryan* | September 1922: 30,000 | October 1922: 30,000<br>November 1922: 30,000? |
| Mathews, *How Science Helps Our Faith* | September 1922: 30,000 | October 1922: 30,000<br>November 1922: 30,000?<br>FY 26–27 (February 1927?): 50,000 |
| Millikan, *A Scientist Confesses His Faith* | September 1923: ? | February 1927 (Second Impression): 50,000 |
| Frost, *The Heavens are Telling* | September 1924: 50,000? | Unknown date: ?<br>September 1930 (Third Impression): ? |
| Millikan, "A Joint Statement Upon the Relations of Science and Religion" | September 1926 or earlier: ? | |
| Schmucker, *Through Science to God* | September 1926: 50,000 | |
| Pupin, *Creative Co-ordination* | May 1928: 25,000 | FY 28–29: 25,000 |
| Fosdick, *Religion's Debt to Science* | May 1928: 25,000 | FY 28–29: 25,000 |
| Compton, Mathews, and Gilkey, *Life After Death* | September 1930: 25,000 | |
| Mather, *The Religion of a Geologist* | September 1931: 27,000 | |

FY 28–29 = Fiscal Year 1928–1929 (using figures given in annual statements).

All first imprints have "First Impression" on the front cover. For later imprints, the impression number is given here only if I have seen a copy. The existence and sizes of many printings has been inferred from other information, but publication records in AISLR are incomplete.

APPENDIX THREE (A)

# Scientists Who Supported AISL Pamphlets, 1922–1928

Starring as found in *American Men of Science*, 3rd ed. (1933). For example, *2 means that the person first received that honor in the second edition.

| Donor, starred where applicable | Field | Institution in first year of support | 1922–1923 | 1923–1924 | 1924–1925 | 1925–1926 | 1926–1927 | 1927–1928 |
|---|---|---|---|---|---|---|---|---|
| Adams, Walter S. (*3) | astronomy | Mt. Wilson Observatory | | | | 10 | 10 | |
| Aitken, Robert G. (*1) | astronomy | Lick Observatory | | | 5 | | | |
| Alden, William C. (*3) | geology | US Geological Survey | 3 | | 5 | | | |
| Alexander, Jerome (*3) | chemistry | private consultant | | 2.50 | 2.50 | 2.50 | 2.50 | 2 |
| Allee, Warder C. (*3) | zoology | University of Chicago | | 1 | | | | |
| Allen, Bennet M. (*2) | embryology | UCLA | 5 | | | | | |
| American Association for the Advancement of Science | | | 100 | 60 | | 60 | 55 | 60 |
| Austin, Louis W. (*1) | physics | National Bureau of Standards | | 5 | 5 | 5 | 5 | 5 |
| Bailey, Solon I. (*1) | astronomy | Harvard University | | 1 | 1 | | 2 | |
| Bain, Henry F. | botany | US Department of Agriculture | | | 10 | | 10 | |
| Barbour, Thomas (*3) | zoology | Harvard University | | | 5 | | | |
| Bentley, Madison (*2) | psychology | University of Illinois | | 5 | | | | |
| Bigelow, Robert P. (*1) | zoology | Massachusetts Institute of Technology | | 5 | | | | 6.50 |
| Bingham, Walter V. (*3) | psychology | Personnel Research Federation | | | 10 | 10 | 10 | 10 |
| Birge, Edward A. (*2) | zoology | University of Wisconsin | | 20 | | | | |
| Blakeslee, Albert F. (*2) | botany | Carnegie Institution | | 5 | 5 | 5 | 5 | 5 |
| Bliss, William J. (*1) | physics | Johns Hopkins University | | 10 | 10 | | 10 | 10 |
| Bogert, Marston T. (*1) | chemistry | Columbia University | | 5 | | | 5 | 5 |
| Boothroyd, Samuel L. | astronomy | Cornell University | | | | 5 | 5 | 5 |
| Bowen, Robert H. (*4) | zoology | Columbia University | | | | 5 | | |
| Brackett, E. E. | engineering | University of Nebraska | 1 | | | | | |
| Bradley, H. C. perhaps Bradley, Harold C. or Bradley, Harry C. | chemistry | University of Wisconsin | 20 | | | | | |
| Brannon, Melvin A. | engineering | Massachusetts Institute of Technology | | | | 26 | | |
| Brigham, Albert P. (*1) | botany | State University of Montana | | | | | | |
| Brode, Howard S. | geology | Colgate University | 10 | 10 | | | | |
| Brownlee, Roy H. | zoology | Whitman College | | 10 | 10 | | | |
| Bruner, Henry L. | chemistry | Brownlee Lab | | | | | 20 | |
| Bucher, Walter H. (*5) | zoology | Butler College | | | | | 5 | 5 |
| Bumpus, Herman C. (*1) | geology | University of Cincinnati | | 3 | 3 | | 2 | 1 |
| Cady, Walter G. (*2) | biology | Woods Hole Marine Biological Laboratory | | 5 | 5 | | | |
| Caldwell, Otis W. (*2) | physics | Wesleyan University (CT) | | 1 | 1 | 1 | | 1 |
| Calvert, Philip P. (*1) | botany | Columbia University | 20 | | | 10 | | |
| Carlson, Anton J. (*1) | zoology | University of Pennsylvania | | 5 | 5 | | | |
| Cessna, Orange H. | physics | University of Chicago | | | | | | |
| | psychology | Iowa State College | | | | | | 5 |

(continued)

| Donor, starred where applicable | Field | Institution in first year of support | 1922–1923 | 1923–1924 | 1924–1925 | 1925–1926 | 1926–1927 | 1927–1928 |
|---|---|---|---|---|---|---|---|---|
| Chamberlin, Thomas C. (*1) | geology | University of Chicago | 20 | | | 10 | 5 | 5 |
| Chamberlin, Rollin T. (*3) | geology | University of Chicago | | | 5 | | | |
| Charles, Grace M. | botany | Austin High School, Chicago | | | | | 5 | 2 |
| Choate, Helen A. | botany | Smith College | | | | 5 | 2 | 5 |
| Clapp, Grace L. | botany | Milwaukee-Downer College | | | 2 | 2 | 2 | |
| Clark, Hubert L. (*1) | zoology | Harvard University | | 2 | 2 | 2 | 2 | 2 |
| Clarke, John G. | medicine? | *perhaps* a surgeon from Philadelphia | 20 | | | | | |
| Clarke, John M. (*1) | paleontology | State Museum of New York | 20 | | | | | |
| Coker, William C. (*2) | botany | University of North Carolina | | | | 10 | | |
| Compton, Arthur H. (*3) | physics | University of Chicago | | | | 10 | | |
| Coulter, John M. (*1) | botany | University of Chicago | 20 | | 10 | 10 | 10 | 10 |
| Coulter, Stanley (*1) | biology | Purdue University | | 10 | 5 | | | |
| Coville, Frederick V. (*1) | botany | US Department of Agriculture | | | | | 5 | |
| Cowles, Henry C. (*1) | botany | University of Chicago | 20 | | | | | |
| Cram, Eloise B. | parasitology | Bureau of Animal Industry | | | | | | |
| Curtis, Winterton C. (*1) | zoology | University of Missouri | 10 | 1 | 5 | 5 | 5 | 5 |
| Davenport, Charles B. (*1) | zoology | Carnegie Institution | | 10 | 1 | | 5 | 5 |
| Day, William S. (*1) | physics | Columbia University | | | 10 | | 1 | |
| Dellinger, John H. (*3) | physics | National Bureau of Standards | | 5 | | | 5 | 5 |
| Dickinson, Hobert C. (*3) | physics | National Bureau of Standards | | 5 | 3 | | 3 | |
| Dodge, Raymond (*1) | psychology | Wesleyan University (CT) | | | | | | |
| Donaldson, Henry H. (*1) | neurology | Wistar Institute | | 20 | 20 | 20 | 20 | 2 |
| Doubt, Sarah L. | botany | Washburn College | | | | | 5 | |
| Drew, Gilman A. (*1) | zoology | Woods Hole Marine Biological Laboratory | | 5 | | 10 | | 10 |
| Dyar, Harrison G. (*1) | entomology | US National Museum | | 5 | | 5 | | |
| Emerson, Fred W. | botany | Penn College (Iowa) | 5 | 5 | 5 | | | 5 |
| Erders, Howard E. | zoology | Purdue University | | | | | 2.10 | |
| Fenneman, Nevin M. (*3) | geology | University of Cincinnati | 20 | 20 | 20 | 20 | 25 | |
| Finley, Charles W. | biology | Lincoln School, Teachers College, Columbia | | 5 | 5 | | | |
| Fletcher, Harvey (*4) | physics | Western Electric Company | | 5 | | | | |
| Foberg, J. A. | mathematics | State Department of Public Instruction, Harrisburg, PA | | 2.50 | 2.50 | 2.50 | 5 | |
| Fox, Henry | biology | Mercer University | | 4 | 2 | | | |
| Frisby, Edgar (*1) | astronomy | US Navy | | 1 | 2 | | | |
| Fuller, George D. | botany | University of Chicago | | | | | 5 | 5 |
| Gage, Harry H. | inventor | University of Chicago | | | | | 1 | |
| Gage, Simon H. (*1) | histology | Cornell University | | 11 | 10 | 10 | | 10 |

| Donor, starred where applicable | Field | Institution in first year of support | 1922–1923 | 1923–1924 | 1924–1925 | 1925–1926 | 1926–1927 | 1927–1928 |
|---|---|---|---|---|---|---|---|---|
| Gale, Henry G. (*2) | physics | University of Chicago | 10 | | | | | |
| Gardiner, Harry N. (*1) | psychology | Smith College | | 5 | | | | |
| Geology Department, University of Iowa | geology | University of Iowa | 37 | | | | | |
| Goddard, Henry H. (*3) | psychology | Ohio State University | | | 5 | 5 | | |
| Gooch, Frank A. (*1) | chemistry | Yale University | | | | 5 | 5 | 5 |
| Gould, E. S. | science teacher | Galva, IL | | | | 2 | 1 | |
| Grave, Benjamin H. | zoology | Wabash College | | | | | | |
| Grave, Caswell (*1) | zoology | Washington University (MO) | | 10 | 10 | 10 | 10 | 10 |
| Green, Wyman Reed | embryology | University of Chattanooga | | | 2 | | | |
| Groves, James F. | botany | Ripon College | 1 | | | | | |
| Hale, George E. (*1) | astronomy | Mt. Wilson Observatory | | 20 | 20 | 20 | 20 | 20 |
| Hardesty, Irving (*1) | neurology | Tulane University | | 10 | | 10 | 5 | 5 |
| Hargitt, Charles W. (*1) | zoology | Syracuse University | 10 | | 5 | 5 | | |
| Harper, James | business | Registrar, Rush Medical College; brother of William Rainey Harper | | 5 | | | | |
| Harshberger, John W. (*1) | botany | University of Pennsylvania | | 10 | 2 | | | |
| Hawkes, Herbert E. (*1) | mathematics | Columbia University | | 5 | | 10 | | |
| Hektoen, Ludvig (*1) | pathology | University of Chicago | | | 5 | 5 | | |
| Henderson, Junius | paleontology | University of Colorado | | | | 5 | | |
| Herrick, Charles J. (*1) | neurology | University of Chicago | | 5 | 5 | | 5 | |
| Hessler, John C. | chemistry | Knox College | 2 | | | | | |
| Hibbard, Angus S. | inventor | retired executive, Chicago Telephone Company | | | | | 10 | |
| Hitchcock, Albert S. (*1) | botany | US Department of Agriculture | | 5 | | 5 | 5 | |
| Hobbs, William H. (*1) | geology | University of Michigan | 5 | | | | | |
| Hole, Allen D. | geology | Earlham College | 1 | 1 | | | | |
| Holgate, Thomas F. (*1) | mathematics | Northwestern University | | | 5 | | | |
| Holt, Luther E. | medicine | Johns Hopkins University | | | | | 10 | |
| Hood, Ozni P. | engineering | US Bureau of Mines | | | | 1 | | |
| Howe, Charles S. (*2) | astronomy | Case School of Applied Science | | | 5 | 5 | | |
| Howe, Herbert A. (*1) | astronomy | University of Denver | | 10 | 10 | 10 | | |
| Huber, G. Carl (*1) | anatomy | University of Michigan | 10 | | | | | |
| Hull, Albert W. (*3) | physics | General Electric Company | | | | | | |
| Humphreys, William J. (*1) | physics | US Weather Bureau | | | | | 5 | |
| Hussey, W. J. (*1) | astronomy | University of Chicago | | | 5 | | 5 | 5 |
| Isham, George S. | astronomy | University of Chicago | | | 25 | | | |
| Ivy, Andrew C. (*4) | physiology | Northwestern Medical School | | | | | | 5 |

(*continued*)

| Donor, starred where applicable | Field | Institution in first year of support | 1922–1923 | 1923–1924 | 1924–1925 | 1925–1926 | 1926–1927 | 1927–1928 |
|---|---|---|---|---|---|---|---|---|
| Jeffrey, Edward C. (*1) | botany | Harvard University | 10 | 10 | 10 | | | |
| Jillson, Willliard R. | geology | Kentucky Academy of Science | | 25 | | | | 10 |
| Jones, Adam L. | psychology | Columbia University | | | | 5 | 10 | 5 |
| Jones, George H. *perhaps* | | | | | | | | |
| Jones, George E. | psychology | University of Pittsburgh | | | | | | 1 |
| Jones, Lewis R. (*1) | botany | University of Wisconsin | | | | 2 | | |
| Jordan, Edwin O. (*1) | pathology | University of Chicago | 20 | 20 | | | | |
| Kay, George F. (*5) | geology | University of Iowa | 10 | | | 5 | | |
| Keen, William W. | surgery | Jefferson Medical College | 1 | 20 | 10 | 10 | 20 | 10 |
| Keitt, George W. | botany | University of Wisconsin | | | | 5 | 5 | |
| Kellogg, Vernon L. (*1) | zoology | National Research Council | 10 | 10 | 5 | 5 | 5 | 5 |
| Kent, Norton A. (*2) | physics | Boston University | | 1 | 1 | 1 | 2 | 2 |
| Keyser, Casius J. (*1) | mathematics | Columbia University | | 5 | | | | |
| Kingsley, J. S. (*1) | anatomy | University of California | | | 5 | | | |
| Kraus, Edward H. | geology | University of Michigan | | 10 | 10 | 10 | 10 | 10 |
| Kraybill, Henry R. | chemistry | Boyce Thompson Institute | | | | 1 | | 2 |
| Kummel, Henry B. (*1) | geology | State of New Jersey | | 10 | 10 | 5 | 5 | 5 |
| Lamb, Arthur B. (*3) | chemistry | Harvard University | 5 | | | | | |
| Lane, Alfred C. (*1) | geology | Tufts College | 20 | | | 5 | | 10 |
| LaVenture, Anna B. | mathematics | high school teacher, Oak Park, Il | | | | | | 11.5 |
| Leffmann, Henry (*1) | chemistry | Leffmann-Trumper Clinical Laboratory, Philadelphia | | | 5 | 5 | | |
| Leverett, Frank (*1) | geology | University of Michigan | 5 | 5 | | 5 | | 5 |
| Lewis, E. P. (*1) | physics | University of California | | | 5 | | | |
| Lewis, Frederick T. (*2) | embryology | Harvard University | | 5 | 5 | 5 | 10 | 5 |
| Lillie, Frank R. (*1) | zoology | University of Chicago | 20 | 20 | 20 | | 25 | 25 |
| Lindley, Ernest H. (*1) | psychology | University of Kansas | | 5 | | 5 | | |
| Linton, Edwin (*2) | zoology | Washington and Jefferson College | | | | 5 | | 5 |
| Longden, Aladine C. (*1) | physics | Knox College | | 5 | 5 | | | |
| Loud, Frank H. (*4) | astronomy | Colorado College | | 10 | | | | |
| Lovenhart, A. S. (*2) | physiology | University of Wisconsin | | 5 | | | | |
| Mariette, Ernest S. | medicine | University of Minnesota | | | | | 10 | |
| Marlatt, Charles L. (*1) | entomology | US Department of Agriculture | | 10 | 10 | 10 | 10 | 10 |
| Martin, Charles L. | radiology | Baylor College of Medicine | | 5 | | | | |
| Martin, John N. | botany | Iowa State College | | 20 | | | | |
| Mathews, Clarence W. | horticulture | University of Kentucky | | 2 | 2 | 1 | 2 | |
| Mathews, Edward B. (*2) | geology | Johns Hopkins University | 20 | | | | | |
| McClendon, Jesse F. (*3) | physiology | University of Minnesota | | 1 | 3 | | | |

| Donor, starred where applicable | Field | Institution in first year of support | 1922–1923 | 1923–1924 | 1924–1925 | 1925–1926 | 1926–1927 | 1927–1928 |
|---|---|---|---|---|---|---|---|---|
| McGregor, James H. (*2) | geology | Columbia University | 15 | | | | 10 | |
| Mckee, Ralph H. (*3) | engineering | Columbia University | 20 | 20 | 20 | 20 | 20 | 20 |
| Mead, Albert D. (*1) | zoology | Brown University | 20 | | | | | |
| Mendenhall, Walter C. (*2) | geology | US Geological Survey | | 10 | | | | |
| Merrill, Paul W. (*3) | astronomy | Mt. Wilson Observatory | | 2 | 2 | 2 | | |
| Meyer, Arthur W. (*2) | anatomy | Stanford University | | | | 10 | | 5 |
| Middleton, Austin R. | zoology | University of Louisville | | | | | 2 | |
| Miller, Carl D. | physics | Washington College (MD) | | | | | | |
| Millikan, Robert A. (*2) | physics | California Institute of Technology | 20 | | | 20 | 10 | 10 |
| Moody, Agnes C. (*1) | zoology | University of California | | 5 | 5 | 5 | | |
| Moore, E. H. (*1) | mathematics | University of Chicago | | 5 | | 5 | | |
| Moore, Raymond C. (*5) | geology | University of Kansas | 10 | | | | | |
| Moulton, Charles R. | chemistry | Institute of American Meat Packers | | 5 | | 5 | | |
| Nabours, Robert K. (*3) | zoology | Kansas State Agricultural College | | 5 | | | 3 | 2.50 |
| Nelson, Nels C. (*3) | anthropology | American Museum of Natural History | | 5 | 5 | | | |
| Nichols, Edward L. (*1) | physics | Cornell University | | 5 | | | | |
| Noguchi, Hideyo (*3) | pathology | Rockefeller Institute | | 5 | 5 | 5 | | |
| Osburn, Herbert (*1) | zoology | Ohio State University | 10 | | | | | |
| Osborn, Raymond C. (*2) | zoology | Ohio State University | 20 | 20 | | | | |
| Pack, Dean A. | botany | US Department of Agriculture | | 2 | 2 | | | |
| Parkhurst, John A. (*2) | astronomy | Yerkes Observatory | | 6 | | | | |
| Parr, S. W. (*2) | chemistry | University of Illinois | | 5 | 5 | | | |
| Patten, William (*1) | zoology | Dartmouth College | | 10 | 10 | | | |
| Peter, Alfred M. | chemistry | Kentucky Experimental Station | | 6 | | | | |
| Porter, Jermain G. (*1) | astronomy | Cincinnati Observatory | | 5 | | | | 5 |
| Pound, Roscoe (*1) | botany | Harvard University | | | 5 | | | |
| Proctor, Charles A. | physics | Dartmouth College | | | 5 | | | |
| Pupin, Michael I. (*1) | physics | Columbia University | | 10 | | | | |
| Pusey, Brown | physician | Northwestern University | | 10 | | | 10 | 25 |
| Rankin, Walter M. (*1) | biology | Princeton University | | | | | | 10 |
| Reese, Albert M. (*3) | zoology | West Virginia University | 5 | 5 | | 5 | 5 | 3 |
| Reighard, Jacob E. (*1) | zoology | University of Michigan | | 20 | 10 | | | |
| Rice, Edward L. (*1) | zoology | Ohio Wesleyan University | | 5 | 5 | 6 | 6 | 5 |
| Rice, William North (*1) | geology | Wesleyan University (CT) | | 5 | 5 | 5 | 5 | 5 |
| Richards, Arthur | engineering | City of Altoona, PA | | | | | | 5 |
| Richards, Aute | zoology | University of Oklahoma | | | | | | 5 |
| Ridgway, Robert | engineering | US National Museum | | | 5 | | 5 | 5 |

*(continued)*

| Donor, starred where applicable | Field | Institution in first year of support | 1922–1923 | 1923–1924 | 1924–1925 | 1925–1926 | 1926–1927 | 1927–1928 |
|---|---|---|---|---|---|---|---|---|
| Rogers, Fred T. | physiology | Baylor Medical College | | | | | | |
| Ross, Luther S. | biology | Drake University | 20 | 5 | 5 | 5 | 5 | 3 |
| Ross, W. D. *perhaps* | | | | | | | 3 | |
| Ross, William Horace | chemistry | US Department of Agriculture | | | | | | 2 |
| Runnels, Scott C. | physician | Indianapolis, IN | 5 | | | | | |
| Russell, Henry N. (*3) | astronomy | Princeton University | | | | | 5 | 5 |
| Sanderson, F. *perhaps* | | | | | | | | |
| Sanderson, Everett S. | bacteriology | Rockefeller Institute | | | | | | |
| Sanderson, Ezra Dwight | entomology | Cornell University | | 10 | | | | |
| Sanford, Edmund C. (*1) | psychology | Clark University | | 5 | | | | |
| Sanford, Martha | ? | perhaps a retired teacher from Niles, California | | | | | | |
| Sawyer, Mary L. | botany | Wellesley College | | | | | 2 | |
| Schmucker, Samuel C. | biology | West Chester State Normal School (PA) | | 5 | | | 5 | 5 |
| Schneider, Edward C. (*3) | physiology | Wesleyan University (CT) | | 2 | 2 | 2 | 2 | 5 |
| Schuchert, Charles (*1) | paleontology | Yale University | | 5 | | 5 | | 2 |
| Seashore, Carl E. (*1) | psychology | University of Iowa | | 5 | 5 | 5 | 5 | |
| Severinghaus, Willard L. | physics | Columbia University | 5 | | | | | |
| Shambaugh, George E. | anatomy | University of Chicago | 20 | | 5 | | | |
| Sharp, Lester W. (*4) | botany | Cornell University | | | | | | |
| Shear, Cornelius L. (*1) | mycology | US Department of Agriculture | | 5 | | 5 | 8.10 | 5 |
| Shimer, Hervey W. | paleontology | Massachusetts Institute of Technology | | 5 | 5 | 5 | 5 | |
| Shull, Charles A. (*4) | botany | University of Chicago | | 10 | | | | |
| Shull, George H. (*2) | botany | Princeton University | | | | | | |
| Sigerfoos, Charles P. | zoology | University of Minnesota | 20 | 10 | | | 10 | |
| Sippy, Bertram W. | physician | Rush Medical College | 10 | 10 | | 5 | | |
| Slocum, Frederick (*3) | astronomy | Wesleyan University (CT) | | 5 | | | | |
| Slosson, Edwin E. | chemistry | Science Service, Washington, DC | 20 | | | | | |
| Smith, Edgar F. (*1) | chemistry | University of Pennsylvania | | | 10 | | 10 | |
| Smith, Erwin F. (*1) | botany | US Department of Agriculture | | | | 10 | 25 | |
| Smith, George O. (*2) | geology | US Geological Survey | | 25 | | | | 25 |
| Snow, Laetitia M. | botany | Wellesley College | | | | 5 | 5 | 5 |
| Stenton, Timothy W. (*1) | paleontology | US Geological Survey | | | | | 5 | 5 |
| Starr, Anna M. S. | psychology | Mt. Holyoke College | | | | 5 | 2 | |
| Stetson, Harlan T. (*5) | astronomy | Harvard University | | | | | | |
| Stewart, George W. (*1) | physics | University of Iowa | 10 | | | | | 5 |

| Donor, starred where applicable | Field | Institution in first year of support | 1922–1923 | 1923–1924 | 1924–1925 | 1925–1926 | 1926–1927 | 1927–1928 |
|---|---|---|---|---|---|---|---|---|
| Stieglitz, Julius O. (*1) | chemistry | University of Chicago | 10 | 10 | 10 | 10 | 10 | 10 |
| Stokey, Alma G. | botany | Mt. Holyoke College | | | | 5 | 5 | |
| Stone, Ormond (*1) | astronomy | University of Virginia | | | | | 1 | |
| Strong, Oliver S. (*1) | zoology | Columbia University | | | | | | |
| Sundwall, John | hygiene | University of Michigan | | 5 | | | | |
| Taylor, Aravilla M. | botany | Lake Erie College | | 3 | | | | 5 |
| Taylor, Lloyd W. | physics | Oberlin College | | | | | | 2.50 |
| Thomas, Abram | geology | University of Iowa | | | | | | 1 |
| Thwing, Charles B. | physics | Manufacturing Pyrometers, Philadelphia | | 1 | 1 | 1 | 1 | |
| Tilton, John L. | geology | West Virginia University | 5 | 5 | | | | |
| Tomlinson, Charles W. | geology | Schermerhorn-Ardmore Company (OK) | | | | | 10 | 20 |
| Tuttle, Franklin E. | chemistry | University of Kentucky | | 2 | 2 | | | |
| Tyler, Harry W. (*1) | mathematics | Massachusetts Institute of Technology | | | | 2 | | 2 |
| Upson, Fred W. | chemistry | University of Nebraska | | 2 | | | | |
| Van Schoiack, T. W. | ? | retired Army officer, executive with Koken Companies, St. Louis, MO | | | | | | 2 |
| Van Slyke, Lucius L. (*2) | chemistry | New York Experimental Station | | 5 | 5 | 5 | | |
| Van Vleck, Edward B. (*1) | mathematics | University of Wisconsin | | 5 | | 5 | | 5 |
| Vaughan, Thomas W. (*1) | geology | US National Museum | | 5 | 5 | 5 | 5 | 5 |
| Walcott, Charles D. (*1) | geology | Smithsonian Institution | | 10 | 10 | 10 | 10 | 10 |
| Ward, Henry B. (*1) | zoology | University of Illinois | 20 | 10 | | | | |
| Watson, Floyd R. (*3) | physics | University of Illinois | | | | | 2.50 | 2.50 |
| Weller, Stuart (*1) | geology | University of Chicago | | 10 | 10 | 10 | 10 | |
| Wendt, Gerald L. (*3) | chemistry | Standard Oil Company of Indiana | | 10 | 5 | | | |
| Westgate, Lewis G. | geology | Ohio Wesleyan University | | 5 | 5 | 5 | | |
| Whitnall, Harold O. | geology | Colgate University | | | 5 | | | |
| Wieman, Harry L. | zoology | University of Cincinnati | | | | | 10 | |
| Wilder, Russell M. | medicine | Mayo Clinic | 10 | | | | | |
| Wood, Horatio C. | pharmacology | University of Pennsylvania | | | | | 5 | |
| Woods, Albert F. (*1) | botany | University of Maryland | | | | 10 | 5 | 5 |
| Wright, Frederick E. (*2) | geology | Carnegie Institution/Geophysical Laboratory | | 11 | 10 | 10 | 10 | 10 |
| Yerkes, Robert M. (*2) | psychology | National Research Council | 10 | | | | | |

APPENDIX THREE (B)

# Scientists Who Supported AISL Pamphlets, 1928–1934

Starring as found in *American Men of Science*, 3rd ed. (1933). For example, *2 means that the person first received that honor in the second edition.

| Donor, starred where applicable | Field | Institution in first year of support | 1928–1929 | 1929–1930 | 1930–1931 | 1931–1932 | 1932–1933 | 1933–1934 |
|---|---|---|---|---|---|---|---|---|
| Aitken, Robert G. (*1) | astronomy | Lick Observatory | 1 | 2 | 2 | 2 | 2 | 2 |
| Alexander, Jerome (*3) | chemistry | private consultant | 2.50 | 2.50 | | | | |
| Ashman, George C. | chemistry | Bradley Polytechnic Institute | | 2 | 2 | 2 | 2 | 2 |
| Austin, Louis W. (*1) | physics | National Bureau of Standards | 5 | 5 | 5 | 5 | | |
| Bain, Henry F. | botany | US Department of Agriculture | | 10 | 5 | | | |
| Bigelow, Robert P. (*1) | zoology | Massachusetts Institute of Technology | 5 | 5 | 3 | 5 | | |
| Bigelow, Samuel L. (*2) | chemistry | University of Michigan | 5 | | | | | |
| Bingham, Walter V. (*3) | psychology | Personnel Research Federation | 10 | | | | | |
| Bissonnette, Thomas H. | embryology | Trinity College (CT) | | 3 | 3 | | | 3 |
| Blakeslee, Albert F. (*2) | botany | Carnegie Institution | 5 | | | | | |
| Bliss, William J. (*1) | physics | Johns Hopkins University | 10 | 10 | | | | |
| Bogert, Marston T. (*1) | chemistry | Columbia University | 5 | 5 | 5 | | | |
| Boothroyd, Samuel L. | astronomy | Cornell University | | 5 | | | 5 | |
| Brown, Frank E. | chemistry | Iowa State College | | 5 | 5 | 5 | | |
| Brownlee, Roy H. | chemistry | Brownlee Laboratory, Pittsburgh | 5 | 5 | | | | |
| Bruner, Henry L. | zoology | Butler College | 5 | 5 | 5 | | | |
| Bucher, Walter H. (*5) | geology | University of Cincinnati | | | | | 2 | |
| Cady, Walter G. (*2) | physics | Wesleyan University (CT) | 1 | 1 | | 1 | 1 | 1 |
| Caldwell, Otis W. (*2) | botany | Columbia University | | 10 | 5 | | | 5 |
| Carman, Joel E. | geology | Ohio State University | 5 | 5 | 5 | 5 | 5 | 5 |
| Choate, Helen A. | botany | Smith College | 5 | 5 | 5 | 5 | 2 | 2 |
| Clark, Hubert L. (*1) | zoology | Harvard University | 2 | 2 | 2 | 2 | 6 | 10 |
| Compton, Arthur H. (*3) | physics | University of Chicago | | | | | 5 | 5 |
| Cram, Eloise B. | parasitology | Bureau of Animal Industry | 3 | 5 | 5 | 5 | | |
| Curtis, Winterton C. (*1) | zoology | University of Missouri | | 5 | | | | |
| Davenport, Charles B. (*1) | zoology | Carnegie Institution | | 1 | | | | |
| DeLee, Joseph Bolivar | obstetrics | University of Chicago | | | | | 2 | 3 |
| Dellinger, John H. (*3) | physics | National Bureau of Standards | 5 | 5 | 5 | 5 | | |
| Dickinson, Hobert C. (*3) | physics | National Bureau of Standards | | | 5 | 3 | | |
| Donaldson, Henry H. (*1) | neurology | Wistar Institute | | | | | 10 | |
| Doubt, Sarah L. | botany | Washburn College | 2 | | | | | |
| Emerson, Fred W. | botany | Penn College (IA) | 5 | | 5 | | | |
| Fenneman, Nevin M. (*3) | geology | University of Cincinnati | 25 | 25 | 25 | 25 | 25 | 25 |

(*continued*)

| Donor, starred where applicable | Field | Institution in first year of support | 1928–1929 | 1929–1930 | 1930–1931 | 1931–1932 | 1932–1933 | 1933–1934 |
|---|---|---|---|---|---|---|---|---|
| Fletcher, Harvey (*4) | physics | Western Electric Company | 2 | | | | | |
| Fowler, Henry T. *perhaps* | | | | 2 | | | | |
| Fowler, Henry W. | zoology | Academy of Natural Sciences, Philadelphia | | | 2 | 2 | 2 | |
| Fox, Henry | biology | Mercer University | | 3 | | | | |
| Fry, Wilfred *perhaps* | | | | | | | 10 | |
| Fry, Wilfred E. | physician | University of Pennsylvania School of Medicine | 5 | 5 | | 5 | | |
| Fuller, George D. | botany | University of Chicago | 1 | | | | | |
| Gage, Harry H. | inventor | University of Chicago | | | | | | |
| Gage, Simon H. (*1) | histology | Cornell University | 10 | 10 | 10 | | | |
| Gilbert, W.M. | ? | administrative secretary, Carnegie Institute | | 5 | 10 | 5 | | |
| Goddard, Henry H. (*3) | psychology | Ohio State University | 5 | 5 | 10 | 5 | | 5 |
| Gooch, Frank A. (*1) | chemistry | Yale University | 1 | | | 5 | | 1 |
| Gould, E. S. | science teacher | Galva, IL | 1 | 1 | 1 | 1 | 1 | |
| Grave, Benjamin H. | zoology | Wabash College | 10 | 10 | 10 | 10 | 10 | 10 |
| Grave, Caswell (*1) | zoology | Washington University (MO) | 5 | 5 | 5 | 2 | | 2 |
| Grawe, Oliver R. | geology | Missouri School of Mines and Metallurgy | 20 | | | | | |
| Hale, George E. (*1) | astronomy | Mt. Wilson Observatory | | | 5 | | | |
| Hardesty, Irving (*1) | neurology | Tulane University | | | | | | |
| Henderson, Junius | attorney | curator, University of Colorado Museum of Natural History | | 5 | | | | |
| Herrick, Glenn W. | entomology | Cornell University | | 2 | 2.50 | 2.50 | | 3 |
| Holgate, Thomas F. (*1) | mathematics | Northwestern University | | 2 | | | | |
| Hollick, Charles A. (*1) | geology | New York Botanical Gardens | | 2 | | | | |
| Humphreys, William J. (*1) | physics | US Weather Bureau | 5 | 5 | | | | |
| Ivy, Andrew C. (*4) | physiology | Northwestern Medical School | | 10 | 10 | 10 | 10 | 10 |
| Jeffrey, Edward C. (*1) | botany | Harvard University | | | | 5 | | |
| Jennings, John G. (listed with scientists by AISL) | business | president of Lamson & Sessions, Cleveland, OH | 10 | | 10 | | | |
| Jones, Adam L. | psychology | Columbia University | | 5 | 5 | 5 | 5 | 5 |
| Jones, George H. *perhaps* | | | | | | | | |
| Jones, George E. | psychology | University of Pittsburgh | 5 | 5 | | | | |
| Jones, Lewis R. (*1) | botany | University of Wisconsin | 1 | | 1 | 1 | | |
| Jordan, Edwin O. (*1) | pathology | University of Chicago | | 2 | | | 5 | |

| Donor, starred where applicable | Field | Institution in first year of support | 1928–1929 | 1929–1930 | 1930–1931 | 1931–1932 | 1932–1933 | 1933–1934 |
|---|---|---|---|---|---|---|---|---|
| Jordon, Frank C. (*3) | astronomy | University of Pittsburgh | 1 | 1 | | | 1 | |
| Kadesch, William H. | physics | Iowa State College | | 1 | | | | |
| Keen, William W. | surgery | Jefferson Medical College | 20 | | | | | |
| Kellogg, Vernon L. (*1) | zoology | National Research Council | | 10 | 5 | | | |
| Kent, Norton A. (*2) | physics | Boston University | 2 | 5 | 2 | 1 | | |
| Knox, John K. | geology | Philips Petroleum Company | 10 | 2 | 5 | 10 | 10 | 10 |
| Kraus, Edward H. | geology | University of Michigan | 10 | 10 | | 10 | 10 | |
| Kraybill, Henry R. | chemistry | Boyce Thompson Institute | 2 | | | | | |
| Kummel, Henry B. (*1) | geology | State of New Jersey | 5 | 5 | 5 | 5 | | 5 |
| Kunkle, Edward C. | ? | perhaps a Baptist minister | 2.50 | | | | | |
| Lane, Alfred C. (*1) | geology | Tufts College | 10 | 5 | 5 | 5 | 3 | 5 |
| LaVenture, Anna B. | mathematics | high school teacher, Oak Park, IL | 10 | 10 | 10 | 10 | | |
| Leverett, Frank (*1) | geology | University of Michigan | 5 | 5 | | | | |
| Lewis, Frederic T. (*2) | embryology | Harvard University | 5 | | 5 | 5 | 5 | 3 |
| Lewis, Warren K. (*3) | engineering | Massachusetts Institute of Technology | 5 | | | | | |
| Lillie, Frank R. (*1) | zoology | University of Chicago | 25 | 25 | 25 | 25 | 12.5 | |
| Linton, Edwin (*2) | zoology | Washington and Jefferson College | 5 | | | | | |
| Lyon, Elias P. (*1) | physiology | University of Minnesota | | | | | | 2 |
| Magness, John R. | botany | US Department of Agriculture | 3 | | | | | |
| Marlatt, Charles L. (*1) | entomology | US Department of Agriculture | 10 | 10 | 10 | 10 | 10 | 10 |
| Martin, John N. | botany | Iowa State College | | | 2 | | | |
| Marvin, Charles F. (*1) | meteorology | US Weather Bureau | 5 | 5 | 5 | 5 | 3 | 3 |
| Mather, Kirtley F. (*4) | geology | Harvard University | | | 5 | 5 | 5 | 5 |
| Mckee, Ralph H. (*3) | engineering | Columbia University | 20 | | | | | |
| McPherson, William (*1) | chemistry | Ohio State University | | | 5 | | | |
| Mead, Albert D. (*1) | zoology | Brown University | 5 | 5 | | | | |
| Meyer, Arthur W. (*2) | anatomy | Stanford University | 5 | 5 | | | | |
| Miller, Carl D. | physics | Washington College (MD) | | 1 | | | | |
| Millikan, Robert A. (*2) | physics | California Institute of Technology | 10 | 10 | 10 | 15 | 10 | 10 |
| Montague, William P. (listed with scientists by AISL) | philosophy | Columbia University | 5 | | | | | |
| Nabours, Robert K. (*3) | zoology | Kansas State Agricultural College | 2.50 | | | | | |

(continued)

| Donor, starred where applicable | Field | Institution in first year of support | 1928–1929 | 1929–1930 | 1930–1931 | 1931–1932 | 1932–1933 | 1933–1934 |
|---|---|---|---|---|---|---|---|---|
| Pack, Frederick J. | geology | University of Utah | | | | 1 | | |
| Patterson, Thomas L. | physiology | Detroit College of Medicine and Surgery | | 2 | | | | |
| Patton, William (*1) | zoology | Dartmouth College | | 2 | 2 | | | |
| Payne, Fernandus (*3) | zoology | Indiana University | | 2 | | | | |
| Pupin, Michael I. (*1) | physics | Columbia University | 25 | | | 10 | | |
| Pusey, Brown | physician | Northwestern University | 10 | 10 | | 5 | 10 | 5 |
| Rankin, Walter M. (*1) | biology | Princeton University | 3 | 3 | 3 | 3 | | |
| Reese, Albert M. (*3) | zoology | West Virginia University | | 2 | | | | |
| Rice, Edward L. (*1) | zoology | Ohio Wesleyan University | 5 | 5 | 5 | 5 | | 5 |
| Richards, Aute | zoology | University of Oklahoma | 2.50 | | 5 | | 3 | 2 |
| Ross, Luther S. | biology | Drake University | 1 | | | 2 | 1 | 1 |
| Ross, W. D. *perhaps* | | | | | | | | |
| Ross, William Horace | chemistry | US Department of Agriculture | 3 | | | | | |
| Russell, Henry N. (*3) | astronomy | Princeton University | 5 | | 10 | | 5 | 5 |
| Sawyer, Mary L. | botany | Wellesley College (MA) | 5 | | | | | |
| Schear, Edward W. E. | biology | Otterbein College | 10 | | | | | |
| Schmucker, Samuel C. | biology | West Chester State Teachers College (PA) | 5 | | | | | |
| Schneider, Edward C. (*3) | physiology | Wesleyan University (CT) | 2 | 2 | | 2 | | 2 |
| Sharp, William B. | medicine | University of Texas | 5 | | | 5 | | |
| Shear, Cornelius L. (*1) | mycology | US Department of Agriculture | 5 | 5 | 5 | 5 | | 3 |
| Sherrill, Mary L. | chemistry | Mt. Holyoke College | | 5 | | | | |
| Shull, Charles A. (*4) | botany | University of Chicago | | 5 | | | | |
| Smith, George O. (*2) | geology | US Geological Survey | 25 | 25 | 25 | | | |
| Smith, Rowland F. | ? | *unidentified person from Charleston, SC* | 1 | | | | | |
| Snow, Laetitia M. | botany | Wellesley College | 5 | | 5 | 5 | 5 | 5 |
| Stanton, Timothy W. (*1) | paleontology | US Geological Survey | | 5 | 5 | 5 | 5 | |
| Stieglitz, Julius O. (*1) | chemistry | University of Chicago | 10 | 10 | 10 | 5 | 5 | 5 |
| Stokey, Alma G. | botany | Mt. Holyoke College | 2.50 | | | | | |
| Strong, Oliver S. (*1) | zoology | Columbia University | | 5 | | | | |
| Sturgis, William C. (*1) | botany | Colorado College (retired) | | 5 | | | | |
| Thomas, Abram | geology | University of Iowa | 1 | | | | | |
| Tilton, John L. | geology | West Virginia University | | 2 | | | | |
| Tomlinson, Charles W. | geology | Schermerhorn-Ardmore Company (OK) | 25 | 25 | 25 | 20 | 10 | 12 |

| Donor, starred where applicable | Field | Institution in first year of support | 1928–1929 | 1929–1930 | 1930–1931 | 1931–1932 | 1932–1933 | 1933–1934 |
|---|---|---|---|---|---|---|---|---|
| True, Rodney H. (*1) | botany | University of Pennsylvania | | | | 5 | | |
| Turner, William D. | engineering | Columbia University | 3 | 3 | 2 | 2 | 2 | 2 |
| Tyler, Harry W. (*1) | mathematics | Massachusetts Institute of Technology | | 2 | | 2 | 3 | |
| Upson, Fred W. | chemistry | University of Nebraska | 2 | | | | | |
| Van Schoiack, T. W. | ? | retired Army officer, executive with Koken Companies, St. Louis, MO | | | | | | |
| Van Vleck, Edward B. (*1) | mathematics | University of Wisconsin | 5 | 5 | 5 | | 5 | 5 |
| Vaughan, Thomas W. (*1) | geology | US National Museum | 5 | 5 | | 5 | | |
| Visher, Stephen S. | geology | Indiana University | | 1 | 1 | 1 | 1 | 1 |
| Ward, Henry B. (*1) | zoology | University of Illinois | | | 5 | | | |
| Watson, Floyd R. (*3) | physics | University of Illinois | 2.50 | 2.50 | 2.50 | 1 | 2 | 2 |
| Webb, Hanor A. | chemistry | George Peabody College for Teachers | | 2 | | | | |
| Wieman, Harry L. | zoology | University of Cincinnati | | | | 5 | | |
| Wildman, Earnest A. | chemistry | Earlham College | | 2 | 2 | | | |
| Winchester, George | physics | Rutgers University | | 2 | | 2 | | |
| Wright, Frederick E. (*2) | geology | Carnegie Institution/Geophysical Laboratory | 10 | 10 | 10 | 10 | | |
| Wright, Harry N. | mathematics | City College of New York | | | | 2 | | 2 |

# NOTES

## Preface

1. Edward B. Davis, "A Whale of a Tale: Fundamentalist Fish Stories," *Perspectives on Science and Christian Faith* 43.4 (1991): 224–37; Edward B. Davis, "Fundamentalism and Folk Science between the Wars," *Religion and American Culture* 5 (1995): 217–48.

## Introduction

1. Portions of the introduction are adapted from Edward B. Davis, "Science and Religious Fundamentalism in the 1920s: Religious Pamphlets by Leading Scientists of the Scopes Era Provide Insight into Public Debates about Science and Religion," *American Scientist* 93 (May–June 2005): 254–60.
2. William Jennings Bryan, "God and Evolution," *New York Times*, February 26, 1922.
3. Edwin Grant Conklin, "God and Evolution," *New York Times*, March 5, 1922; Henry Fairfield Osborn, "Evolution and Religion," *New York Times*, March 5, 1922; Harry Emerson Fosdick, "Evolution and Mr. Bryan," *New York Times*, March 12, 1922.
4. See Kenneth N. Beck, "The American Institute of Sacred Literature: A Historical Analysis of an Adult Education Institution" (PhD diss., University of Chicago, 1968); "Guide to the American Institute of Sacred Literature Records 1880–1943," https://www.lib.uchicago.edu/e/scrc/findingaids/view.php?eadid=ICU.SPCL.AISL#idp2511048, accessed February 7, 2024; "A Statement of the Present Activities and Prospects of the American Institute of Sacred Literature," 1917, and "Bulletin of Information," September 1917, AISLR 10:7; Shailer Mathews, *Will Christ come again?* (Chicago: American Institute of Sacred Literature, 1917).
5. Edward J. Larson, *Summer for the Gods: The Scopes Trial and America's Continuing Debate over Science and Religion* (New York: Basic Books, 1997); James Gilbert, *Redeeming Culture: American Religion in an Age of Science* (Chicago: University of Chicago Press, 1997), 9.

## Chapter 1 · "Spiking Bryan's Guns"

1. Parts of this chapter are based on Edward B. Davis, "Fundamentalist Cartoons, Modernist Pamphlets, and the Religious Image of Science in the Scopes Era," in *Religion and the Culture of Print in Modern America*, ed. Charles L. Cohen and Paul S. Boyer (Madison: University of Wisconsin Press, 2008), 175–98.
2. Ian G. Barbour, *Religion and Science: Historical and Contemporary Issues* (San Francisco: HarperCollins, 1997), 74.
3. "Convention Side Lights," *Watchman-Examiner*, July 1, 1920, 834.
4. George M. Marsden, "Introduction," in *The Fundamentals: A Testimony to Truth*, ed. George M. Marsden (New York: Garland, 1988), vol. 1, n.p.

5. George M. Marsden, *Fundamentalism and American Culture: The Shaping of Twentieth Century Evangelicalism, 1870–1925* (New York: Oxford University Press, 1980); Ernest R. Sandeen, *The Origins of Fundamentalism: Toward a Historical Interpretation* (Philadelphia: Fortress Press, 1968); and Ernest R. Sandeen, *The Roots of Fundamentalism: British and American Millenarianism, 1800–1930* (Chicago: University of Chicago Press, 1970); Michael S. Hamilton, "The Interdenominational Evangelicalism of D. L. Moody and the Problem of Fundamentalism," in *American Evangelicalism: George Marsden and the State of American Religious History*, ed. Darren Dochuk, Thomas Kidd, and Kurt Peterson (Notre Dame, IN: University of Notre Dame Press, 2014), 230–80; Lincoln A. Mullen, "The Cultural History of American Fundamentalism: A Review Essay," https://lincolnmullen.com/blog/the-cultural-history-of-american-fundamentalism-a-review-essay/, accessed February 9, 2024; Matthew Avery Sutton, *American Apocalypse: A History of Modern Fundamentalism* (Cambridge, MA: Belknap Press of Harvard University Press, 2014); and Adam Laats, *Fundamentalism and Education in the Scopes Era* (New York: Palgrave Macmillan, 2010).

6. Ronald L. Numbers, *The Creationists*, expanded ed. (Cambridge, MA: Harvard University Press, 2006), 53; John Stackhouse, "Not Fundamentalist, not Conservative, and not Liberal: The Fundamentals and the Mainstream of American Evangelicalism," *Christian Scholar's Review* 52.1 (Fall 2022): 7–24.

7. Marsden, "Introduction," n.p.

8. Jon H. Roberts, "Conservative Evangelicals and Science Education in American Colleges and Universities, 1890–1940," *Journal of the Historical Society* 5.3 (2005): 297–329, quotation on 304; Michael N. Keas, "Darwinism, Fundamentalism, and R. A. Torrey," *Perspectives on Science and Christian Faith* 62.1 (March 2010): 25–51.

9. Edward B. Davis, "Science Falsely So Called: Fundamentalism and Science," in *The Blackwell Companion to Science and Christianity*, ed. J. B. Stump and Alan G. Padgett (Oxford: Blackwell, 2012), 48–60; Ronald L. Numbers, "Science Falsely So-Called: Evolution and Adventists in the Nineteenth Century," *Journal of the American Scientific Affiliation* 27 (March 1975): 18–23; and Ronald L. Numbers and Daniel P. Thurs, "Science, Pseudoscience, and Science Falsely So-Called," in *Wrestling with Nature: From Omens to Science*, ed. Peter Harrison, Ronald L. Numbers, and Michael H. Shank (Chicago: University of Chicago Press, 2011), 281–306. For a related discussion, see Richard England, "Scriptural Facts and Scientific Theories: Epistemological Concerns of Three Leading English-Speaking Anti-Darwinians (Pusey, Hodge, and Dawson)," in *Nature and Scripture in the Abrahamic Religions: 1700–Present*, ed. Jitse M. van der Meer and Scott H. Mandelbrote (Leiden: Brill, 2008), 1: 225–56, esp. 251–52.

10. Stephen Jay Gould, "William Jennings Bryan's Last Campaign," *Natural History* 96 (November 1987): 16–26; Kristy Maddux, "Fundamentalist Fool or Populist Paragon? William Jennings Bryan and the Campaign against Evolutionary Theory," *Rhetoric & Public Affairs* 16.3 (2013): 489–520.

11. William Jennings Bryan, *The Prince of Peace* (1904), quoted in Gould, "William Jennings Bryan's Last Campaign," 22.

12. Numbers, *Creationists*, 41–44; and Edward J. Larson, *Summer for the Gods* (New York: Basic Books, 1997), 37–59. On the misperception of Kellogg's attitude, see Elizabeth Watts and Ulrich Kutschera, "On the Historical Roots of Creationism and Intelligent Design: German *Allmacht* and Darwinian Evolution in Context," *Theory in Biosciences* 140 (2021): 157–68.

13. William Jennings Bryan, "The Menace of Darwinism," *The Commoner*, April 1921, 5–8.

14. Lawrence Levine, *Defender of the Faith: William Jennings Bryan, the Last Decade, 1915–1925* (New York: Oxford University Press, 1965), 264; William Jennings Bryan, *In His Image* (New York: Revell, 1922).

15. William Jennings Bryan, *The Menace of Darwinism* (New York: Revell, 1922), 3, 17, 15, 19, and 35, his italics; W. B. Riley, *The Menace of Modernism* (New York: Christian Alliance, 1917).

16. Bryan, *Menace of Darwinism*, 5.

17. Bernard Lightman, "Christian Evolutionists in the United States, 1860–1900," *Journal of Cambridge Studies* 4 (December 2009): 14–22.

18. Joseph Cook, *Biology, with Preludes on Current Events* (Boston: James R. Osgood, 1877), 6–7 and 9; Sir John W. Dawson, *The Origin of the World, According to Revelation and Science* (New York: Harper), 225; Davis, "Science Falsely So Called," 53–54; A. Wilford Hall, *The Problem of Human Life* (New York: Hall, 1877), 13, the first of several such passages in the book; Asa Gray, *Natural Science and Religion: Two Lectures Delivered to the Theological School of Yale College* (New York: C. Scribner's Sons, 1880), 81.

19. Bryan, *Menace of Darwinism*, 20–22 and 31–32, his italics.

20. Bryan, *Menace of Darwinism*, 33 and 36–39; Charles Darwin, *The Descent of Man, and Selection in Relation to Sex*, 2nd ed. (London: John Murray, 1874), 133–34.

21. Bryan, *Menace of Darwinism*, 40–41.

22. William Jennings Bryan, *Seven Questions in Dispute* (New York: Revell, 1924). According to *N. W. Ayer & Son's American Newspaper Annual and Directory* (Philadelphia: N. W. Ayer & Son, 1920–1929), circulation for the *Sunday School Times* during the 1920s averaged 96,000. On Trumbull, see *WWWA*, and Joel A. Carpenter, *Revive Us Again: The Reawakening of American Fundamentalism* (New York: Oxford University Press, 1997), 25–26. For more on Pace, see Davis, "Fundamentalist Cartoons."

23. Bryan to Trumbull, January 31, 1924, William Jennings Bryan Papers, Library of Congress Manuscript Division, Washington, DC, General Correspondence, container 40, file "1925 [sic] Jan." See James R. Moore, *The Future of Science and Belief: Theological Views in the Twentieth Century* (Milton Keynes: Open University Press, 1981), 40.

24. Bryan, *Menace of Darwinism*, 51.

25. Bryan, *Menace of Darwinism*, 52 and 54–56.

26. "Proposed Legislation against the Teaching of Evolution," *Science* 55 (24 March 1922): 318–20; Edward J. Larson, *Trial and Error: The American Controversy over Creation and Evolution*, 3rd ed. (New York: Oxford University Press, 2003), 48–81; Numbers, *Creationists*, 55; Willard B. Gatewood Jr., ed., *Controversy in the Twenties: Fundamentalism, Modernism, and Evolution* (Nashville: Vanderbilt University Press, 1969), 36.

27. Numbers, *Creationists*, 66–67. For a statement of Fleischmann's position, see Albert Fleischmann, "The Doctrine of Organic Evolution in the Light of Modern Research," *Journal of the Transactions of the Victoria Institute* 65 (1933): 194–214.

28. Dunnington to Mathews, January 30, 1925, AISLR 12:7, his italics. On Dunnington, see the obituary in *Chemical & Engineering News* 22.4 (1944): 275.

29. Matthew to Ernest D. Burton, March 20, 1923, AISLR 11:7.

30. Mather to Kraatz, November 6, 1925, DUA 12:5.

31. Conklin to Eugene S. McCartney, November 2, 1923; Conklin to John C. Mahon, December 16, 1924, EGCP Evolution.

32. Angell to Clarke, January 16, 1923, RAMP 32:37.

33. Rogers to Mathews, April 15 [1923?], AISLR 19:3.

34. William North Rice, *Christian Faith in an Age of Science* (New York: A. C. Armstrong & Son, 1903), 255, 318, and 321.

35. Edward L. Rice, "Darwin and Bryan—A Study in Method," *Science* 61.1575 (March 6, 1925): 243–50, quotations on 243–44.

36. Rice, "Darwin and Bryan," 248–49 and 247. Rice's father had also underscored Bryan's view of certainty and contingency when reviewing Bryan's book *In His Image* in *The Christian Advocate* (April 20, 1922): 478, a passage quoted by his son.

37. Rice, "Darwin and Bryan," 249–50, with embedded quotation from Bryan, *In His Image*, 106, Bryan's italics.

38. William M. Goldsmith, *Evolution or Christianity?* (St. Louis: Anderson Press, 1924), 41 and 45–46. In addition to Bryan, Goldsmith was thinking of Judson E. Conant, Alfred Fairhurst, Theodore Graebner, Howard Agnew Johnston, Philip Mauro, Alfred Watterson McCann, and Alexander Patterson. David Starr Jordan's brief but appreciative review of Goldsmith's book appeared in *Science* 61 (January 9, 1925): 45. The best source of information on Goldsmith is Ina Turner Gray, "Monkey Trial—Kansas Style," *Methodist History* 14 (July 1976): 235–51.

39. Goldsmith, *Evolution or Christianity?*, 7–10, bold type in the original. John W. Judd, *The Coming of Evolution: The Story of a Great Revolution in Science* (Cambridge: Cambridge University Press, 1910). Judd quoted Lyell and Darwin on the creator (12–13), comparing them favorably to the "crudeness" of Milton's *Paradise Lost*. Goldsmith saw Milton identically.

40. William M. Goldsmith, *The Laws of Life: Principles of Evolution, Heredity and Eugenics: A Popular Presentation* (Boston: Badger, 1922), 30–32 and 44–47.

41. Goldsmith, *The Laws of Life*, 48.

42. Goldsmith, *Evolution or Christianity?*, 33–39, bold type in the original.

43. Gray, "Monkey Trial."

44. James R. Moore, *The Post-Darwinian Controversies: A Study of the Protestant Struggle to Come to Terms with Darwin in Great Britain and America, 1870–1900* (Cambridge: Cambridge University Press, 1979), 217–51; Christine Rosen, *Preaching Eugenics: Religious Leaders and the American Eugenics Movement* (New York: Oxford University Press, 2004).

45. Goldsmith, *Evolution or Christianity?*, 103–6, bold type in the original.

46. Shailer Mathews, *New Faith for Old: An Autobiography* (New York: Macmillan, 1936), 227.

47. Harry Emerson Fosdick, *Evolution and Mr. Bryan* (Chicago: American Institute of Sacred Literature, 1922), 13–15.

48. Jacques Loeb, *The Mechanistic Conception of Life: Biological Essays* (Chicago: University of Chicago Press, 1912), 31. On the interpretation Loeb gave his scientific work, see Garland Allen, *Life Science in the Twentieth Century* (New York: John Wiley & Sons, 1975), 73–81.

49. Conklin to Fosdick, February 6, 1928, and Fosdick to Conklin, February 9, 1928, EGCP 8:2.

50. Fosdick, *Evolution and Mr. Bryan*, 17.

51. Fosdick, *Evolution and Mr. Bryan*, 16–20, quoting Henry Drummond, *The Ascent of Man* (New York: James Pott, 1894), 333.

52. Richard England, "Interpreting Scripture, Assimilating Science: British and American Christian Evolutionists on the Relationship between Science, the Bible, and Doctrine," in *Nature and Scripture in the Abrahamic Religions: 1700–Present*, ed. Jitse M. van der Meer and Scott H. Mandelbrote (Leiden: Brill, 2008), 1:183–223.

53. Kirtley F. Mather, *Science in Search of God* (New York: Henry Holt, 1928), 74, and "The Impact of Modern Science on Religion," from the four-page version marked "For Conference at Deerfield," KFMP Lectures, Box 1, "Material for lectures on Religion, Sciences, etc." For a manuscript version of a talk with the same title given in December 1957 at the First Congregational Church, Webster Groves, Missouri, see KFMP Speeches, Box 1, "Notes and Texts of Speeches, 1956–58."

54. For more Brown and Millikan, see Edward B. Davis, "Robert Andrews Millikan: Religion, Science, and Modernity," in *Eminent Lives in Twentieth-Century Science & Religion*, ed. Nicolaas A. Rupke, rev. ed. (Frankfurt am Main: Peter Lang, 2009), 253–74, and Robert H. Kargon, *The Rise of Robert Millikan: Portrait of a Life in American Science* (Ithaca, NY: Cornell University Press, 1982), 145–46.

55. Brown to Millikan, May 23, 1922, RAMP 32:37.

56. Brown to Millikan, June 17, 1922, RAMP 32:37.

57. "RESOLUTIONS," RAMP 32:37; Roberts, "Conservative Evangelicals and Science Education," 298.

58. "A Declaration of the Rights of Science and Religion," RAMP 32:37.

59. On Freeman and Kelman, see *WWWA*; on Gordon and Wise, see *ANB*.

60. Kargon, *Robert Millikan*, does not mention Millikan's reliance on Brown's statements.

61. "The Report of the Committee on Freedom of Teaching in Science," *Science* 61 (March 13, 1925): 276–77; "Resolutions of the Committee of the American Association for the Advancement of Science Relative to the Freedom of Teaching," EGCP 45: folder for 1925–29.

62. Edwin Grant Conklin, "Dismissal of Dr. Henry Fox from the Faculty of Mercer University," *Science* 61 (February 13, 1925): 176–78; "Scientists Uphold Evolution Theory," *New York Times*, December 27, 1922.

63. Noyes to Millikan, undated, with two-page typed attachment, RAMP 32:37.

64. "Deny Science Wars against Religion, Forty Scientists, Clergymen and Prominent Educators Attack 'Two Erroneous Views,'" *New York Times*, May 27, 1923.

65. For publication details, see appendix one.

66. Quoting Millikan's copy of the NAS "Statement" mailed to him by NAS home secretary David White, RAMP 6:1, italics in the original. For related correspondence, see White to Millikan, July 9, 1923, RAMP 6:1, James R. Angell to John M. Clarke, January 16, 1923, and W. W. Campbell to Millikan, March 26, 1923, RAMP 32:37.

67. Henry Fairfield Osborn, "Credo of a Naturalist," *The Forum* 73 (April 1925): 486–94, quotation on 493.

68. Angell to Millikan, January 31, 1923, RAMP 32:37. On Millikan's attitude toward Jews, see David Goodstein, "In Defense of Robert Andrews Millikan," *American Scientist* 89 (January–February 2001): 54–60.

69. Robert A. Millikan, "Albert A. Michelson," *Biographical Memoirs of the National Academy of Sciences* 19 (1938): 120–47.

70. Dorothy Michelson Livingston, *The Master of Light: A Biography of Albert A. Michelson* (New York: Scribner, 1973), 294. I have benefited from a conversation with the late Loyd S. Swenson Jr.

71. Millikan to Merriam, February 13, 1923; Osborn to Millikan, February 12, 1923, RAMP 32:37; Millikan to Conklin, March 21, 1923, and April 12, 1923; Conklin to Millikan, March 29, 1923, and April 16, 1923, EGCP "Philosophy and Theology."

72. Eddy to Conklin, November 1, 1923, and November 14, 1923, Eddy's italics; and Conklin to Eddy, November 9, 1923, EGCP Philosophy.

73. Floyd to Conklin, June 1, 1923, and Conklin to Floyd, June 7, 1923, EGCP Evolution. On Floyd, who signed the Humanist Manifesto in 1933, see the obituary in the *New York Times*, November 27, 1943.

74. Wiley to Mathews, March 20, 1924, AISLR 12:4.

75. "Dr. Millikan's remarks at the Neighborhood Church 10/18/27—conference on church unity," 2–3, RAMP 59:14. On Millikan's church activities, see Davis, "Robert Andrews Millikan."

76. Reese to AISL, April 18, 1930, and Mathews to Reese, April 21, 1930, AISLR 19:3, italics in the original.

77. Mathews to James E. Essex, February 6, 1922, SMP 16:4.

78. William Adams Brown, *Why I Do Not Believe in Materialism* (Chicago: American Institute of Sacred Literature, 1927), introduced a series called "Why I Do Not Believe." The author is described as "one of the leading theologians of America, who may be classed as a conservative liberal in his thinking." Compton's pamphlet is "promised" in the "Statement Concerning Pamphlet Work of the American Institute of Sacred Literature for the Year Ending June 30, 1930," AISLR 11:3. However, on page 57 of the AISL ledger book for 1930–1931, AISLR 16:1, it is listed only as "Requested."

79. Brown, *Why I Do Not Believe in Materialism*, 6.

80. Porter to Mathews, March 10, 1923, and Mathews to Porter, April 11, 1923. The secretary misheard Mathews, recording the title of Foster's book as *The Analogy of the Christian Religion*. Porter's letter to Bryan, showing that the problem of evil was the main reason for his unbelief, is dated Jan 9, 1923, AISLR 19:3. On Porter, see the obituary in *Chicago Daily Tribune*, July 31, 1939, and Duff Peterson, "Eliot and Fairfield Porter, American Artists from Winnetka," http://www.winnetkahistory.org/gazette/eliot-and-fairfield-porter-american-artists-from-winnetka/, accessed February 9, 2024.

81. H. H. Lane, *Evolution and Christian Faith* (Princeton, NJ: Princeton University Press, 1923); Conklin to Osborn, June 19, 1925, and June 27, 1925, EGCP Evolution. Conklin also suggested that Pupin and Edward L. Rice would be good witnesses at Dayton; see Conklin to Forrest Bailey, associate director of the ACLU, June 2, 1925, EGCP C1:Scopes Trial.

82. Henry Fairfield Osborn, "Evolution and Education in the Tennessee Trial," *Science* 62 (July 17, 1925): 43–45, italics in the original.

83. James F. Porter, "Evolution vs. Purpose," *The New Republic* 43.558 (August 12, 1925): 323.

84. "Extracts from Letters from Scientists," AISLR 13:6. Curtis spoke at the dedication of the Irving Porter Church Memorial Telescope of the Fuertes Observatory on June 15, 1923. Heber D. Curtis, "The Influence of Astronomy upon Modern Thought," *Popular Astronomy* 32 (January 1924): 4–11.

85. Samuel Christian Schmucker, *The Study of Nature* (Philadelphia: Lippincott, 1908), 42–44. At the end of the embedded quotation, Schmucker paraphrases Philippians 4:8. On Schmucker's theology, see Edward B. Davis, "Fundamentalism and Folk Science between the Wars," *Religion and American Culture* 5 (1995): 217–48, and Edward B. Davis, "Samuel Christian Schmucker's Christian Vocation." *Seminary Ridge Review* 10.2 (Spring 2008): 59–75.

86. Mather to "Joint Editors," June 18, 1928, and Ralph W. Brown to L. J. A. Mercier, September 22, 1928, KFMP Conference. The late Ronald L. Numbers directed me to this correspondence.

87. On Dumm, see *WWWA* (however, this source gives incorrect dates for his brief stint at George Washington). I am grateful to Amy Stempler for information from the George Washington University Archives. On Mercier, see the obituary (quoted here) in the *New York Times*, March 14, 1953. Louis J. A. Mercier, *The Challenge of Humanism: An Essay in Comparative Criticism* (New York: Oxford University Press, 1933), is a translation of Mercier's *Le mouvement humaniste aux États-Unis* (Paris: Hachette, 1928). Ralph Wilder Brown, *Twenty Paragraphs about the World Conference on Faith and Order* (Gardiner, ME, 1922) is the best source of information on the conference.

88. See Paul Henry Hanus to "Joint Editors," February 10, 1927, KFMP Conference. Hanus mentions a paper Mercier had given on this idea, undoubtedly, "Wanted: New Methods of Conference. If Not Conference, What Then?," *The Harvard Graduate's Magazine* 35 (1926), 48–53.

89. Form letter from "Joint Editors," [October] 1926, KFMP Conference.

90. Letters to "Joint Editors," by M. J. Scott, S.J., February 4, 1927, L. O. Hartman, November 29, 1926, P. E. More, February 3, 1927, and R. S. Lull, October 19, 1926, KFMP Conference. On L. T. More, see Numbers, *Creationists*, 72; on P. E. More, see Arthur Hazard Dakin, *Paul Elmer More* (Princeton, NJ: Princeton University Press, 1960), esp. 249–50 and 338.

91. Ward to "Gentlemen," October 12 [1926], KFMP Conference. Henshaw Ward, *Evolution for John Doe* (Indianapolis: Bobbs-Merrill, 1925), was also published as *Evolution for Everyone* (New York: Grosset & Dunlap, 1925). See also Henshaw Ward, *Charles Darwin: The Man and His Warfare* (Indianapolis: Bobbs-Merrill, 1927); Henshaw Ward, *The Proofs of Evolution* (New York: Evolution, 1929); and Henshaw Ward, *Exploring Nature* (New York: H. Holt, 1923). On Ward, see *WWWA*; and Constance Areson Clark, "Evolution for John Doe," *Journal of American History* 87.4 (March 1, 2001): 1275–1303, quotation on 1282–83.

92. Ward, *Evolution for John Doe*, 179–80.

93. Thomson to "Sirs," October 18, 1926, KFMP Conference. Thomson wrote numerous popular books, including John Arthur Thomson, *Science and Religion, Being the Morse Lectures for 1924* (New York: C. Scribner's Sons, 1925).

94. Cunningham to "Dear Sirs," October 18, 1926, KFMP Conference. Cunningham was the author of *Hormones and Heredity: A Discussion of the Evolution of Adaptations and the Evolution of Species* (London: Constable, 1921). Conklin to "Joint Editors," October 11, 1926, and Rice to "Joint Editors," October 7, 1926, KFMP Conference.

95. Letters to "Joint Editors," by A. S. Pier, February 14, 1927, P. H. Hanus, February 10, 1927, and E. C. Jeffrey, February 11, 1927, KFMP Conference. See also George Howard Parker, "The Evolution of Mind," *Harvard Alumni Bulletin* 28.35 (1926): 1045–54. Both Parker and Wheeler, ironically, contributed essays to Frances Mason, *Creation by Evolution: A Consensus of Present-Day Knowledge as Set Forth by Leading Authorities in Non-Technical Language That All May Understand* (New York: Macmillan, 1928), a volume intended to show how "science widens and exalts our outlook on life and our religious faith" (vii). For Lake's critique of the Resurrection, see Kirsopp Lake, *The Historical Evidence for the Resurrection of Jesus Christ* (New York: G. P. Putnam, 1907).

96. Letters to "Joint Editors" by E. M. East, February 7, 1927, C. W. Dodge, February 7, 1927, W. E. Hocking, February 10, 1927, W. Duane, February 3, 1927, L. C. Graton, February 4, 1927, and H. Shapley, February 4, 1927, KFMP Conference.

97. On Slosson's beliefs, see David J. Rhees, "A New Voice for Science: Science Service under Edwin E. Slosson, 1921–29" (MA thesis, University of North Carolina, 1979).

98. Edwin E. Slosson, *Sermons of a Chemist* (New York: Harcourt, Brace, and Company, 1925), vi; Edwin E. Slosson, *The Contributions of the New Physics to Religion: An Address Given in the Park Avenue Baptist Church, New York City* (Boston: Unitarian Laymen's League, 1928).

99. Slosson to "Joint Editors," October 18, 1926, KFMP Conference. I have not located the issue of *The Independent* in which a forum on evolution had appeared. Joseph McCabe, *The Conflict Between Science and Religion* (Girard, KS: Haldeman-Julius Publications, 1927). A founding member of the Rationalist Press Association, McCabe was editor of *A Rationalist Encyclopaedia: A Book of Reference on Religion, Philosophy, Ethics, and Science* (London: Watts, 1948). Metcalf's address may have been his essay "The Truth of Evolution," *Scientific Monthly* 21 (September 1925): 291–95, one of several papers on "Evidences for Evolution: Statements Prepared for the Defense Counsel, State of Tennessee, vs. John Scopes." Vernon L. Kellogg, *Evolution: The Way of Man* (New York: Appleton, 1924). On Dorlodot, author of *Darwinism and Catholic Thought*, trans. Ernest C. Messenger (New York: Kennedy, 1922), see Peter J. Bowler, *Reconciling Science and Religion: The Debate in Early-Twentieth-Century Britain* (Chicago: University of Chicago Press, 2001), 324–25.

100. James H. Leuba, *The Belief in God and Immorality: A Psychological, Anthropological and Statistical Study* (Boston: Sherman French, 1916).

101. Larson, *Summer for the Gods*, 71–72. Larson is certainly right that Darrow "called himself an agnostic, but in fact he was effectively an atheist" (p. 71) in his efforts to demonstrate the utter ignorance of religion.

102. Osborn, "Credo of a Naturalist," 487.

103. Frederick Lewis Allen, *Only Yesterday: An Informal History of the Nineteen-Twenties* (New York: Harper & Brothers, 1931), 197–98.

104. Larson, *Summer for the Gods*, 225–28. More than anyone else, Allen created a false image of *Scopes* as a defeat for fundamentalism, leading to the decline of religion in America. See also Paul M. Waggoner, "The Historiography of the Scopes Trial: A Critical Re-Evaluation," *Trinity Journal*, n.s., 5 (1984): 155–74; and Ronald L. Numbers, *Darwinism Comes to America* (Cambridge, MA: Harvard University Press, 1998), 85–89.

105. Osborn, "Credo of a Naturalist," 488 and 493–94. He had developed his thoughts on materialism at greater length in Henry Fairfield Osborn, "Evolution and Daily Living," *The Forum* 73.2 (February 1925): 169–77. On Osborn's version of evolution, see Peter J. Bowler, *Evolution: The History of an Idea*, rev. ed. (Berkeley: University of California Press, 1989), 269–70. Osborn cited John Dewey, *Human Nature and Conduct* (New York: Holt, 1922), and William McDougall, *Outline of Psychology* (New York: C. Scribner's Sons, 1923).

## Chapter 2 · Liberal Protestant Scientists and Clergy Join Forces

1. Parts of this chapter are adapted from Edward B. Davis, "Science and Religious Fundamentalism in the 1920s: Religious Pamphlets by Leading Scientists of the Scopes Era Provide Insight into Public Debates about Science and Religion," *American Scientist* 93 (May–June 2005): 254–60, and Edward B. Davis, "Fundamentalist Cartoons, Modernist Pamphlets, and the Religious Image of Science in the Scopes Era," in *Religion and the Culture of Print in Modern America*, ed. Charles L. Cohen and Paul S. Boyer (Madison: University of Wisconsin Press, 2008), 175–98.
2. William Jennings Bryan, "God and Evolution," *New York Times*, February 26, 1922.
3. Birge to Fosdick, March 16, 1922, HEFP Evolution.
4. Sigerfoos to Fosdick, October 30, 1922, HEFP Evolution.
5. "Mr. Bryan and Evolution," *The Christian Century* (March 23, 1922): 363–65. The other magazines were *The Congregationalist*, *Zion's Herald*, *The Continent*, *Reformed Church Messenger*, *The Baptist*, and *Christian Work*. See Fosdick's handwritten memo in HEFP Evolution.
6. Eaton to Fosdick, April 5, 1922, and Fosdick to Eaton, April 11, 1922, HEFP Evolution.
7. John Roach Straton, "Fancies of the Evolutionists," *Forum* 75.2 (February 1926): 245–51; Henry Fairfield Osborn, "Facts of the Evolutionists," *Forum* 75.6 (June 1926): 842–51. For context, see Constance Areson Clark, "Evolution for John Doe—Pictures, the Public, and the Scopes Trial Debate," *Journal of American History* (March 2001): 1275–1303, and Alexander Pavuk, "The American Association for the Advancement of Science Committee on Evolution and the Scopes Trial: Race, Eugenics and Public Science in the U.S.A.," *Historical Research* 91.251 (February 2018): 137–59. George William Hunter, *A Civic Biology: Presented in Problems* (New York: American Book Company, 1914), 196.
8. Scribner's reprinted Osborn's essay three times, first as a twenty-one-page pamphlet, *Evolution and Religion* (1923), then as a chapter in his book, *The Earth Speaks to Bryan* (New York: C. Scribner's Sons, 1925), and again in his *Evolution and Religion in Education: Polemics of the Fundamentalist Controversy of 1922 to 1926* (New York: C. Scribner's Sons, 1926). It was also reprinted in *Evolution or Christianity?*, ed. William M. Goldsmith (St. Louis: Anderson Press, 1924), 131–36, and in *Forum* 73.5 (June 1925): 796–803, one month before Bryan's essay "Mr. Bryan Speaks to Darwin," *Forum* 74.1 (July 1925): 101–7.
9. Conklin to Burton, June 12, 1922, AISLR 11:4, and July 10, 1922, AISLR 11:5; cf. Conklin to Baptist clergyman Elmer W. Powell, January 19, 1923, EGCP Evolution. E. A. Birge, *An Open Letter* (Madison? 1922). Birge sent Fosdick a copy right after he read Fosdick's editorial, but Fosdick had already read it in another form. Birge to Fosdick, March 16, 1922, and Fosdick to Birge, March 21, 1922, HEFP Evolution. On Birge, see *ANB*.
10. Irvin G. Wyllie, "Bryan, Birge, and the Wisconsin Evolution Controversy, 1921–1922," *Wisconsin Magazine of History* 35 (1951–1952): 294–301, and "University President Defends His Faith," *The Christian Century* 39 (March 9, 1922): 292.
11. Worcester to Bryan, October 2, 1921, cited in Wyllie, "Wisconsin Evolution Controversy," 299.
12. E. A. Birge, *Science* (Madison? 1925); cf. E. A. Birge, *Science and Wisdom* (New Orleans: Phi Beta Kappa Society, 1911), an address at Tulane University.

13. Apparently never published, perhaps this sermon formed the basis for his letter in the *Capitol Times*, September 22, 1921, or his article "Mr Bryan Tilts against 'Modernism'—and Is Unhorsed," *Wisconsin Congregational Church Life* 28 (March 1922): 7–8.

14. William W. Keen, "I Believe in Evolution and in God," *Philadelphia Public Ledger*, June 11, 1922; William W. Keen, M.D., *I Believe in God and in Evolution* (Philadelphia: Lippincott, 1922); and William W. Keen, "Surgical and Anatomical Evidence of Evolution," *Science* 55 (June 9, 1922): 603–10. His address was also published in the *Bulletin of the Crozer Seminary* (July 1922).

15. Keen to Shailer Mathews, May 22, 1922, Keen to Ernest D. Burton, July 4, 1922, and Burton to Keen, July 17, 1922, AISLR 11:4. Riley's pamphlet is reprinted in William V. Trollinger Jr., *The Antievolution Pamphlets of William Bell Riley* (New York: Garland, 1995), 119–38. A few comments from Keen about Riley did later appear as "The Scientific Accuracy of the Sacred Scriptures," *Science* 62.1615 (December 11, 1925): 543–44.

16. Keen's $10 contribution is listed in "Gifts received during 1921–22," AISLR 11:5; Keen to Burton, January 17, 1923, AISLR 11:7, Keen's italics.

17. William North Rice, "In His Image," *The Christian Advocate* 97.16 (April 20, 1922): 477–78. The article by Abbott was probably "The Use and Misuse of the Bible," *The Outlook* 130 (1922): 290, reprinted after his death in William M. Goldsmith, *Evolution or Christianity?* (St. Louis: The Anderson Press, 1924), 117–21. Cadman's article must be "Darwin's Theory of Natural Selection," *Homiletic Review* 83 (June 1922): 452–56, which was published immediately following a revised version of Bryan's editorial "God and Evolution." Cadman's reply was also reprinted by Goldsmith, *Evolution or Christianity?*, 122–30. On Cadman, see *ANB*.

18. On Van Dyke, see *ANB*.

19. Conklin to Burton, July 10, 1922, AISLR 11:5.

20. Robert Campbell MacCombie Auld, *The Breed That Beats the Record and Wins in the Race for Supremacy as the Most Economical Producer of the Primest Meat for the Million* (Detroit: Aldine, 1886), and *How to Breed the Human Race: Synopses of Human Welfare* (New York: Stewart, 1911). Auld to Conklin, June 27, 1922, EGCP Evolution. For information on Auld, see the obituary in the *New York Times*, April 23, 1937, and *National Cyclopedia of American Biography*. I have not located Cameron's editorial or Conklin's letter to Auld.

21. "A Statement by the American Association for the Advancement of Science on the Present Scientific Status in Regard to the Theory of Evolution," published in "Scientists Uphold Evolution Theory," *New York Times*, December 27, 1922, and *Science* 57.1465 (January 26, 1923): 103–4; see also *School and Society* 17 (January 20, 1923): 74. Several drafts are in EGCP Carton 45: folder for 1925–1929 (despite the later dates in that folder's name).

22. Conklin to Burton, June 12, 1922, AISLR 11:4; Conklin to M. M. Miller, July 10, 1922, EGCP Evolution.

23. "Popular Religion Pamphlets and Leaflets," AISLR 11:5; Chamberlin to Barnes, November 29, 1921, AISLR 11:2; Mathews to W. S. Richardson, July 25, 1921, RAC.

24. "The University of Chicago Statement of the American Institute of Sacred Literature Endowment," written in September or October 1922 and sent by university auditor N. C. Plimpton to John D. Rockefeller Jr. on October 12, 1922, RAC, quotation on 7.

25. Quoting one of three blank fundraising letters, all dated "March, 1922," AISLR 11:4, apparently intended for different target groups, or perhaps multiple drafts of a letter for the same group of "rich men and a few people in 1922." Some responses survive, so the mailing did happen. Also see "Campaign No. 1 of Pamphlet Distribution," January–February 1922, AISLR 11:5.

26. Chamberlin to Mrs. H. E. Goodman, March 2, 1922, AISLR 11:5.

27. Gerald Birney Smith, *Significant Movements in Modern Theology: A Professional Reading Course* (Chicago: American Institute of Sacred Literature, 1915). Smith's course was

serialized under the same title in *The Biblical World* 44 (October 1914): 274–82; 44 (November 1914): 341–49; 44 (December 1914): 409–16; and 45 (January 1915): 37–44.

28. John Merle Coulter, "The Truth about the Bible: Biblical Views on the Physical Universe," *The Institute* 7.4 (December 1922), 56–71, one of a series of articles reprinted in Alexander Reid Gordon, Ernest Findlay Scott, and Edward I. Bosworth, *The Truth about the Bible* (Chicago: AISL, 1923); John Merle Coulter, "Evolution and Its Explanations," *The Christian Century* 39.21 (May 25, 1922), 649. Coulter later wrote a pamphlet, *Evolution and Christianity* (Auburn, NY: The Jacobs Press, 1925), for the liberal Presbyterian ministers who wrote the Auburn Affirmation of 1924. He also wrote *Where Evolution and Religion Meet* (New York: Macmillan, 1925). On his Sunday school class, see *Fiftieth Anniversary Celebration: Hyde Park Presbyterian Church, May 1–8, 1910*, ed. John Merle Coulter and Henry H. Belfield (Chicago: The Church, 1910), 41–42.

29. Burton to Conklin, June 8, 1922, AISLR 19:3; Burton to Fosdick, June 8, 1922, HEFP Evolution.

30. Chamberlin to Burton, July 24, 1922, AISLR 11:5.

31. Conklin to Burton, July 10, 1922, Mathews to Burton, August 3, 1922, and Chamberlin to Burton, August 15, 1922, AISLR 11:5.

32. John D. Rockefeller Jr., *The Christian Church: What of Its Future?* (New York? 1918), 14, reprinted from the *Saturday Evening Post* 190.32 (February 9, 1918): 16 and 37.

33. Burton to Rockefeller, December 16, 1918, AISLR 10:8.

34. Chamberlin to Burton, August 2, 1919, AISLR 10:9.

35. I have not located this letter, mentioned in a letter from Chamberlin to Burton, July 25, 1921, AISLR 11:3. Woelfkin's obituary appeared in the *New York Times*, January 7, 1928.

36. Mathews to Woelfkin, May 26, 1921, Mathews to Richardson, July 25, 1921, and Richardson to Rockefeller, July 29, 1921, RAC.

37. "Contributors for last year upon whom we ought to count this year" (1917), AISLR 10:7; other names include Charles Gilkey and the wife of Chicago biologist Frank R. Lillie.

38. Faunce to "Richard[son]," holograph letter dated "July 17—[1921]," Richardson to Rockefeller, July 29, 1921, and Richardson to Mathews, August 12, 1921, RAC.

39. "American Institute of Sacred Literature," report from advisory committee, August 24, 1933, RAC. Rockefeller made one more donation to the AISL of $250 in August 1935, in response to a special request from Mathews.

40. "Amounts received from Mr. John D. Rockefeller, Jr.," listing gifts through the fiscal year 1930–31, AISLR 11:3.

41. Chamberlin to Burton, June 25, 1922, and July 24, 1922; Conklin to Burton, July 10, 1922, AISLR 11:5.

42. Burton to Chamberlin, August 10, 1922, and Chamberlin to Burton, August 15, 1922, AISLR 11:5.

43. Michael M. Sokal, "Stargazing: James McKeen Cattell, *American Men of Science*, and the Reward Structure of the American Scientific Community, 1906–1944," in *Psychology, Science, and Human Affairs: Essays in Honor of William Bevan*, ed. Frank Kessel (Boulder, CO: Westview Press, 1995), 64–86, and (for an insider's view of the system) Stephen Sargent Visser, *Scientists Starred 1903–1943 in "American Men of Science"* (Baltimore: Johns Hopkins University Press, 1947).

44. James McKeen Cattell, ed., *American Men of Science: A Biographical Directory* (New York: Science Press, 1906).

45. See the draft letters tailored to various groups of people in AISLR 12:3.

46. For example, see Henry Leffmann to AISL, January 20, 1925, AISLR 19:3; B. Laufer to AISL, March 25, 1925, and R. M. Pearce to Mathews, 17 March 17, 1924, AISLR 12:3; R. W. Shufeldt to Mathews, March 19, 1924, AISLR 12:5.

47. These are found in AISLR 11:6, 14:3, 15:5, 16:3, 17:4, 18:4, and 20:3.

48. Cattell to Burton, November 20, 1922, Burton E. Livingston (permanent secretary, AAAS) to Burton, December 19, 1922, and Livingston to Mathews, January 1, 1924, AISLR 12:5; see appendix three for the full list of donors.

49. Kellogg to Burton, November 5, 1922, AISLR 11:4; Kellogg to Burton, January 5, 1923, and John M. Clarke to E. H. Moore, January 9, 1923, AISLR 11:7.

50. Conklin to Chamberlin, October 28, 1922, AISLR 11:5.

51. Mathews to George Ellery Hale, January 9, 1924, and Mathews to Hale, January 30, 1925, George Ellery Hale Papers, Archives, California Institute of Technology, on reel 24 of the microfilm, ed. Daniel J. Kevles. The AISL assigned Hale's book *The New Heavens* (1922) in a reading course for ministers.

52. "Science and Religion," *Bulletin of the California Institute of Technology* 32.98 (March 1923): 3–20. The date on the title page (1922) contradicts the date Millikan placed at the end of the article (March 9, 1923). This was reprinted as "A Scientist Confesses his Faith," *The Christian Century* (June 21, 1923): 778–83, apparently at the suggestion of Murray Shipley Howland, pastor of the Lafayette Avenue Presbyterian Church in Buffalo and author of the modernist declaration known as the "Auburn Affirmation"; Howland to Mathews, April 22, 1924, AISLR 12:4.

53. Slosson to Burton, April 23, 1923, Burton to Slosson, April 30, 1923, and Chamberlin to Slosson, May 9, 1923, AISLR 11:7.

54. Millikan to Burton, November 1, 1922, AISLR 11:4.

55. Slosson to Mathews, January 17, 1924, AISLR 12:4; Ritter to Burton, February 23, 1923, AISLR 11:7; William E. Ritter, "The Religion of a Naturalist," *The Open Court* (1923): 30–36.

56. Shailer Mathews, ed., *Contributions of Science to Religion* (New York: Appleton, 1924). Mathews to Frost, April 23, 1924, and June 2, 1924; Frost to Mathews, May 28, 1924, EBFP; Very to Mathews, January 29, 1925, AISLR 12:7. The late Owen Gingerich provided information on Very.

57. Mathews to Schmucker, March 10, 1924, AISL 12:5.

58. For more on Schmucker's theology and activities, see Edward B. Davis, "Samuel Christian Schmucker's Christian Vocation," *Seminary Ridge Review* 10.2 (Spring 2008): 59–75.

59. Michael Pupin, *From Immigrant to Inventor* (New York: Scribner's Sons, 1923), 384–85. For details on related publications, see appendix one.

60. Charles Harvey Arnold, *God Before You and Behind You: The Hyde Park Union Church Through a Century, 1874–1974* (Chicago: Hyde Park Union Church, 1974), 98.

61. Pupin, *From Immigrant to Inventor*, 385. For a fuller account of the development of this idea, see Edward B. Davis, "Cosmic Beauty, Created Order, and the Divine Word: The Religious Thought of Michael Idvorsky Pupin (1858–1935)," in *Eminent Lives in Twentieth-Century Science & Religion*, ed. Nicolaas A. Rupke, rev. ed. (Frankfurt am Main: Peter Lang, 2009), 295–316.

62. For publication details, see appendix one. Pupin to Manning, December 8, 1924, Michael Idvorsky Pupin papers, Box 2, Rare Book and Manuscript Library, Columbia University Library. On Manning, see *ANB*. Mathews to Appleget, August 12, 1927, AISLR 11:3.

63. For publication details, see appendix one.

64. Harry Emerson Fosdick, *Religion's Debt to Science* (Chicago: American Institute of Sacred Literature, 1928), 3–4.

65. See the ledger book for 1930–31, AISLR 16:1, on 57. On Sprengling, see Nabia Abbott, "Martin Sprengling, 1877–1959," *Journal of Near Eastern Studies* 19.1 (1960): 54–55.

66. Douglas Clyde Macintosh, *Why I Believe in Immortality* (Chicago: American Institute of Sacred Literature, 1925).

67. "Statement Concerning Pamphlet Work of the American Institute of Sacred Literature for the Year Ending June 30, 1929," AISLR 11:3; cf. a Rockefeller Foundation memo dated October 25, 1928, RAC.

68. For an analysis of that debate, see Edward B. Davis, "Prophet of Science—Part Three," *Perspectives on Science and Christian Faith* 61.4 (December 2009): 240–53.

69. *The University of Chicago Magazine* 23 (November 1930): 5–17.

70. "Statement Concerning Pamphlet Work of the American Institute of Sacred Literature for the Year Ending June 30, 1929," and "Statement Concerning Pamphlet Work of the American Institute of Sacred Literature for the Year Ending June 30, 1930," AISLR 11:3; William Adams Brown, *Why I Do Not Believe in Materialism* (Chicago: American Institute of Sacred Literature, 1927).

71. These three are listed as "Titles possible & suggested" in the AISL ledger book for 1930–1931, AISLR 16:1, on 57; Chamberlin to Compton, June 1, 1931, AISLR 17:2.

72. For publication details, see appendix one.

73. See the order form accompanying a letter to Aline B. Carter of March 8, 1933, AISLR 18:2, and the sales forms sent with letters to various parties in 1935, AISLR 21:2.

74. See the list by category in AISLR 12:3.

75. More than 300,000 copies of the first three pamphlets alone had been distributed prior to the publication of Millikan's pamphlet in September 1923; see the undated draft letter to W. S. Richardson, AISLR 11:3. Unfortunately, information about the print runs is scattered widely and unpredictably in AISLR, and I have not located all the figures necessary to determine the total more precisely. The numbers in appendix two can be documented.

76. "Campaign No. 1 of Pamphlet Distribution, January–February 1922," AISLR 11:5.

77. J. C. Labaree to Chamberlin, September 13, 1922, italics in the original; Mrs. J. Horner Kerr to AISL, September 18, 1922; and John Lohriller to AISL, September 23, 1922, AISLR 11:4. The defaced Millikan pamphlets are in AISLR 11:8 and AISLR 11:7, respectively.

78. C. S. Knight to AISL, September 22, 1922, AISL 11:4, italics in the original.

79. This recommendation is from an AISL handbill on "Pamphlets and Leaflets" (ca. 1928) in the author's collection. Burton to Robert Grant Aitken (director of Lick Observatory), April 20, 1932, AISLR 19:2.

80. "Statement Concerning Pamphlet Work of the American Institute of Sacred Literature for the Year Ending June 30, 1930," and Mathews to Rockefeller staffer Arthur W. Packard, January 31, 1935, AISLR 11:3.

81. The layman's name was William T. Shepard; see Murray Shipley Howland to Mathews, April 22, 1924, AISLR 12:4.

82. Edwin Brant Frost, *An Astronomer's Life* (Boston: Houghton Mifflin, 1933), 218; Bernard Benfield to Chamberlin, December 22, 1924, and Frost to Mathews, October 10, 1924, AISLR 12:5.

83. Robert E. Mathews to Chamberlin, January 28, 1933, AISLR 18:3.

84. Chamberlin to Conklin, October 20, 1922, AISLR 11:5; cf. Chamberlin to Illinois zoologist Henry Baldwin Ward, October 20, 1922, AISLR 11:4. Chamberlin to Osburn, November 23, 1922, AISLR 11:4.

85. *120 Years of American Education: A Statistical Portrait*, ed. Thomas D. Snyder (Washington, DC: National Center for Education Statistics, 1993), 56. Data prior to 1929–1930 are not given.

86. Burton to Burton E. Livingston, December 23, 1922, AISLR 11:5. Galley proofs of several magazine advertisements are in AISLR 12:6.

87. Davenport to Burton, November 21, 1922, Burton to Davenport, November 29, 1922, and Burton to Cattell, December 12, 1922, AISLR 11:5.

88. Tilton to Burton, February 18 and 24, 1923, AISLR 11:7.

89. Rogers to Burton, February 10, 1923, AISLR 11:7; Rogers to Mathews, April 15 [1923?], AISLR 19:3; Grove Samuel Dow, *Introduction to the Principles of Sociology* (Waco, TX: Baylor University Press, 1920). For information on this controversy, see S. A. R., "The Teaching of Evolution in the Baptist Institutions of Texas," *Science* 55 (May 12, 1922): 515.

90. AISL report to the university president for 1923–1924, AISLR 12:2; cf. Chamberlin to E. E. Slosson, January 30, 1924, AISLR 12:4.

91. Mathews to Hughes, January 3, 1924, AISLR 12:4.

92. See the form letter dated May 2, 1925, AISLR 13:1.

93. See the list and sample cover letter in AISLR 14:2. It is unclear why Chicago alumni of that particular year were included.

94. See the form letter dated June 4, 1925, and enclosures, AISLR 13:1.

95. A. S. Ulm to AISL, June 8, 1925, H. McCormick Lintz to AISL, June 17, 1925, italics in the original; M. R. Cooper to AISL, June 4, 1925, and Skinner to AISL, June 6, 1925, AISLR 12:7; *Compton's Pictured Encyclopedia*, ed. Guy Stanton Ford (Chicago: F. E. Compton, 1922), 3:1208–11.

96. Kearney to AISL, February 5, 1926, AISLR 14:1.

97. Form letter sent to 1,085 scientists beginning on February 18, 1932, AISLR 18:1.

98. Conklin to Chamberlin, August 12 and 19, 1922, AISLR 11:5. About two dozen copies of Conklin's pamphlet remain among his papers in EGCP Carton 9.

99. Conklin to Lewis E. Linzell, October 5, 1923, Conklin to Edward Rynearson, October 16, 1923, Conklin to William W. Johnson, October 30, 1923, Conklin to Mathilde O. Moos, January 3, 1923, Conklin to Dr. D. A. Whittle, November 29, 1923, and Conklin to Kelly, January 2, 1923, EGCP Evolution.

100. Chamberlin to Keen, November 11, 1922, AISLR 11:4; Smallwood to AISL, February 6, 1923, and Lane to Chamberlin, March 16, 1923, AISLR 11:7; Sigerfoos to AISL, January 30, 1924, AISLR 12:4.

101. Herbert E. Evans to Chamberlin, January 19, 1932, AISLR 18:2.

102. Gregg Mitman, "Evolution as Gospel: William Patten, the Language of Democracy, and the Great War," *Isis* 81.308 (September 1990): 446–63, quotation on 447; William Patten, *The Grand Strategy of Evolution: The Social Philosophy of a Biologist* (Boston: R. G. Badger, 1920). On Patten, see *ANB*. According to Anne M. Ostendarp, former College Archivist, Dartmouth began requiring evolution with the Class of 1923, which entered in 1919. In the early 1920s more than 600 freshmen enrolled annually. Ostendarp to Edward B. Davis, November 20, 2001, private correspondence. Conklin voiced similar views on cooperation and social evolution in *The Direction of Human Evolution* (1921), and Mather did in "Parables from Palaeontology," *Atlantic Monthly* 121 (April 1918): 35–43.

103. William Patten, "Mr. Bryan at Dartmouth: A Reply from the Faculty to His Sweeping Denial of the Truth of the Theory of Evolution," *New York Times*, December 23, 1923. For a comparison of the effects of Bryan's speech with those of the course, see "Dartmouth Students and Evolution," *The World's Work* 48 (June 1924): 131–32.

104. Patten to Burton, April 8, 1923, AISLR 11:7; Patten to Mathews, January 7, 1924, AISLR 12:3; Hopkins to Chamberlin, May 22, 1923, AISL 11:8.

105. William Patten, *Why I Teach Evolution* (Hanover, NH: Dartmouth Press, 1924), reprinted in *The Scientific Monthly* 19 (December 1924): 635–47.

106. Patten, *Why I Teach Evolution*, 1–3, 5, 10, 13, 15; his italics.

107. Wieland to Mathews, March 19, 1924, AISLR 12:4.

108. C. A. Adams to AISL, April 5, 1926, transcribed in "Extracts from Letters from Scientists," AISLR 13:6.

109. Robert C. Cook to Fosdick, March 23, 1923, HEFP Evolution.

110. Slosson to Burton, November 8, 1922, and Millikan to Burton, November 1, 1922, AISLR 11:4.

111. Osburn to AISL, November 18, 1922, AISLR 11:4, and Osburn to Burton, February 8, 1923, AISLR 11:7. Osburn's pamphlet was reprinted from *The Ohio Journal of Science* 22.7 (May 1922): 173–92. For a response by a creationist theologian from Wittenberg College, see Leander S. Keyser, "The Misconceptions of an Evolutionist: Some Critical Remarks on Both His Manner and His Matter," *Lutheran Quarterly* (January 1923): 82–102.

112. Newman to Burton, November 24, 1922; W. W. Otey, *The Origin and Destiny of Man*, 2nd ed. (Austin, TX: Firm Foundation Publishing House, 1938), 167. On Otey, see Cecil Willis, *W. W. Otey, Contender for the Faith: A History of Controversies in the Church of Christ from 1860–1960* (Akron, OH: Cecil Willis, 1964).

113. Kay to Burton, December 6, 1922, AISLR 11:4.

114. Keen to Burton, January 7, 1924, AISLR 11:7.

115. Blanche E. Smith to Conklin, September 13, 1939, EGCP B 20:14; Conklin's reply does not survive, but this letter shows that the pamphlets were still being used many years later. S. L. Hammond to AISL, September 27, 1922, and AISL to Hammond, October 5, 1922, AISLR 11:4; Lincoln to Mathews, April 26, 1924, Mathews to Lincoln, May 10, 1924, AISLR 12:4.

116. Hornaday to Mathews, April 22, 1924, AISLR 12:4; Bolton to Mathews, January 12, 1925, transcribed in "Extracts from Letters from Scientists," AISLR 13:6; East to Burton, January 12, 1923, AISLR 11:8; Slosson to Burton, November 8, 1922, AISLR 11:4.

117. Reighard to Mathews, March 14, 1924, AISLR 12:3.

118. Patten, *Why I Teach Evolution*, 3, his italics.

119. C. M Williams to Conklin, July 17, 1924, and Conklin to Williams, July 23, 1924; Conklin to Edward Rynearson, October 16, 1923; Conklin to W. G. Eggleston, September 17, 1923, EGCP Evolution.

120. Rodeheaver to Conklin, November 18, 1922, and Conklin to Rodeheaver, November 23 and 29, 1922, EGCP Evolution; William Berryman Scott, *The Theory of Evolution, with Special Reference to the Evidence upon Which it Is Founded* (New York: Macmillan, 1917).

121. Conklin to Robinson, May 21, 1924; Strausbaugh to Robinson, November 24, 1923; Robinson to Conklin, January 29, 1924, EGCP Evolution; John L. Robinson, *Evolution and Religion* (Boston: Stratford, 1923).

122. Venable to AISL, March 25, 1924, AISLR 12:5.

123. Leffmann to AISL, January 20, 1925, and February 7, 1925; Chamberlin to Leffmann, January 29, 1925, AISLR 19:3.

124. Jeffrey to Mathews, March 13, 1925, AISLR 12:7; Turner to AISL, May 3, 1923, AISRL 11:7.

125. Darrow to Mathews, March 27, 1926, AISLR 19:2; "Rats" to Conklin, October 31, 1922, EGCP Evolution; Shufeldt to Mathews, March 19, 1924, AISLR 12:5; H. Berman to Conklin, September 7, 1922, EGCP Evolution.

126. "Implication of the Scope Trial," AISLR 11:5.

127. "Implication of the Scope Trial," AISLR 11:5.

128. Paraphrased from Galileo's *Letter to the Grand Duchess Christina*, in *Discoveries and Opinions of Galileo*, trans Stillman Drake (New York: Doubleday, 1957), 186.

129. Shailer Mathews, *New Faith for Old: An Autobiography* (New York: Macmillan, 1936), 225.

## *Chapter 3 · Science and Religion, Chicago Style*

1. A. Hunter Dupree, *Asa Gray: 1810–1888* (Cambridge, MA: Belknap Press of Harvard University Press, 1959), 19–22, 37–47, 135–39, 182, 220–21, 266, and 358–78; Ronald L. Numbers, "George Frederick Wright: From Christian Darwinist to Fundamentalist," *Isis* 79 (1988): 624–45.

2. Dupree, *Asa Gray*, 373.
3. Asa Gray, *Natural Science and Religion: Two Lectures Delivered to the Theological School of Yale College* (New York: C. Scribner's Sons, 1880), 7-9.
4. Gray, *Natural Science and Religion*, 106 and 107-9.
5. Bessey to Cyprian Leycester Drawbridge, April 1909, Charles E. Bessey Papers, Archives & Special Collections, University of Nebraska-Lincoln Libraries. This correspondence is cited by Richard A. Overfield, *Science with Practice: Charles E. Bessey and the Maturing of American Botany* (Ames: Iowa State University Press, 1993), 183. Drawbridge was compiling data for an augmented edition of Arthur H. Tabrum, *Religious Beliefs of Scientists, Including over One Hundred and Forty Hitherto Unpublished Letters on Science and Religion from Eminent Men of Science*, new and enlarged ed. (London: Hunter & Longhurst, 1913), but he did not publish Bessey's letter. On Bryan, Bessey, and the Round Table Club, see Jeffrey McDonald, "William Jennings Bryan, the Round Table Club, and Religious Freedom," in *The Gospel and Religious Freedom: Historical Studies in Evangelicalism and Political Engagement*, ed. David Bebbington (Waco, TX: Baylor University Press, 2023), 29-44, and Andrew Koszewski, "Career Differentiation: The Legal Community in Lincoln, Nebraska, 1880-1891," *Great Plains Research* 2.2 (August 1992): 281-300.
6. David N. Livingstone, *Darwin's Forgotten Defenders: The Encounter between Evangelical Theology and Evolutionary Thought* (Grand Rapids, MI: Wm. B. Eerdmans, 1987); David N. Livingstone and Mark A. Noll, "B. B. Warfield (1851-1921): A Biblical Inerrantist as Evolutionist," *Isis* 91 (2000): 283-304; Bradley J. Gundlach, *Process and Providence: The Evolution Question at Princeton, 1845-1929* (Grand Rapids, MI: Wm. B. Eerdmans, 2013).
7. This echoes James R. Moore, *The Post-Darwinian Controversies: A Study of the Protestant Struggle to Come to Terms with Darwin in Great Britain and America, 1870-1900* (Cambridge: Cambridge University Press, 1979), 73-74.
8. Henry Higgins Lane, *Animal Biology: An Introduction to Zoology for College and University Students* (Philadelphia: P. Blakiston's Son, 1929); H. H. Lane, *Evolution and Christian Faith* (Princeton, NJ: Princeton University Press, 1923); *Henry Higgins Lane (1878-1965), Biographical Data* (Lawrence: Museum of Natural History, University of Kansas, 1967).
9. Lane, *Evolution and Christian Faith*, 180-82, his italics.
10. Lane, *Evolution and Christian Faith*, 195-99, his italics.
11. Bliss to Mathews, March 28, 1924, AISLR 12:7.
12. Shailer Mathews, *New Faith for Old: An Autobiography* (New York: Macmillan, 1936), 219-21.
13. E. D. Cope and J. S. Kingsley, "Editors' Table," *The American Naturalist* 21 (January 1887): 59-61.
14. On Huxley, see Colin A. Russell, "The Conflict Metaphor and Its Social Origins," *Science and Christian Belief* 1 (1989): 3-26.
15. Mathews, *New Faith for Old*, 222; Henry Drummond, *Natural Law in the Spiritual World* (New York: A. L. Burt, 1880).
16. See "Bulletin of Information," September 1917, and "A Statement of the Present Activities and Prospects of the American Institute of Sacred Literature," ca. 1918, AISLR 10:7.
17. Mathews, *Will Christ come again?* (Chicago: American Institute of Sacred Literature, 1917), 3, 13-14, and 20-21, his italics.
18. Shirley Jackson Case, *The Millennial Hope: A Phase of War-Time Thinking* (Chicago: University of Chicago Press, 1918). I rely here on George Marsden, *Fundamentalism and American Culture: The Shaping of Twentieth-Century Evangelicalism, 1870-1925* (New York: Oxford University Press, 1980), 146-47.
19. MacInnis to Mathews, September 27, 1917, AISLR 10:7. MacInnis is quoting *The Outlook for Religion* (New York: Funk & Wagnalls, 1918), 194, by the English ecumenical

minister William Edwin Orchard. John Murdoch MacInnis, *Peter the Fisherman Philosopher: A Study in Higher Fundamentalism* (Los Angeles: Biola Book Room, 1927). For information about MacInnis, see the "Biographical Note" at the Presbyterian Historical Society, https://www.history.pcusa.org/collections/research-tools/guides-archival-collections/rg-266, accessed February 13, 2024; Dan D. Crawford, "Fundamentalism vs. Modernism: The Trial of John MacInnis at the Bible Institute of Los Angeles," *Journal of Presbyterian History* 100.1 (Spring/Summer 2022): 20–37; Daniel W. Draney, *When Streams Diverge: John Murdoch Macinnis and the Origins of Protestant Fundamentalism in Los Angeles* (Eugene, OR: Wipf & Stock, 2008); and Fred Sanders, "Peter the Fisherman Philosopher," https://scriptoriumdaily.com/peter-the-fisherman-philosopher/, accessed February 13, 2024.

20. McPherson to Mathews, April 4, 1918, AISLR 10:8. On McPherson, see G. W. McPherson, *A Parson's Adventures* (Yonkers, NY: Yonkers Book Company, 1925), esp. his critique of modernism, pantheism, and evolution on 262–67.

21. Victor Anderson, "Pragmatic Theology and the Natural Sciences at the Intersection of Human Interests," *Zygon* 37 (2002): 161–73.

22. On Burtt and the Humanist Manifesto, see Diane Davis Villemaire, *E.A. Burtt, Historian and Philosopher: A Study of the author of The Metaphysical Foundations of Modern Physical Science* (Dordrecht: Kluwer Academic Publishers, 2002), 194. On Smith, see W. Creighton Peden, *Religion of Democracy: An Intellectual Biography of Gerald Birney Smith, 1868–1929* (Newcastle upon Tyne: Cambridge Scholars, 2014); Anderson, "Pragmatic Theology"; Delwin Brown, "The Fall of '26: Gerald Birney Smith and the Collapse of Socio-Historical Theology," *American Journal of Theology and Philosophy* 11 (1990): 183–201; and Henry David Koss, "The Development of Naturalism at the Divinity School of the University of Chicago with Special Emphasis on the Doctrine of God" (PhD diss., Northwestern University, 1972), 88–114.

23. Gerald Birney Smith, "Significant Movements in Modern Theology: A Professional Reading Course" (Chicago: American Institute of Sacred Literature, 1915), 3–4, AISLR 10:5. The course had already been serialized as Gerald Birney Smith, "A Professional Reading Course on Significant Movements in Recent Theological Thought," *The Biblical World* 44 (October 1914): 274–82, (November 1914): 341–49, (December 1914): 409–16, and 45 (January 1915): 3–44, his italics.

24. Smith, "Significant Movements," 7 and 9–10.

25. Smith, "Significant Movements," 28–29. On LeConte's religious journey, see Ronald L. Numbers, *Science and Christianity in Pulpit and Pew* (Oxford: Oxford University Press, 2007), 77–79.

26. Smith, "Significant Movements," 29–30; Lyman Abbott, *The Theology of an Evolutionist* (Boston: Houghton Mifflin, 1897).

27. Smith, "Significant Movements," 30–32; James Y. Simpson, *The Spiritual Interpretation of Nature* (New York: Hodder & Stoughton, 1912). See the autobiographical memoir in James Y. Simpson, *The Garment of the Living God: Studies in the Relations of Science and Religion* (London: Hodder & Stoughton, 1934), 15–78.

28. Smith, "Significant Movements," 32–33; Francis Howe Johnson, *God in Evolution: A Pragmatic Study of Theology* (New York: Longmans, Green, 1911).

29. "Bulletin of Information," September 1917, AISLR 10:7; Edward Scribner Ames, *The Psychology of the Religious Experience* (Boston: Houghton Mifflin, 1910).

30. Robert L. Kelly, *Theological Education in America: A Study of One Hundred Sixty-One Theological Schools in the United States and Canada* (New York: George H. Doran, 1924), vii.

31. Kelly, *Theological Education in America*, 412–15.

32. Kelly, *Theological Education in America*, 229–30.

33. "To Wealthy Men in 1922 Campaign," and Lincoln to Burton, April 26, 1922, AISLR 11:4. None of the leaflets and pamphlets named by Lincoln is cataloged on WorldCat, and I did not locate any archival copies. I mention them without having seen them. Lincoln also decried "the 'Pastor Russell' & similar swindles," an apparent reference to Charles Taze Russell, who founded Zion's Watch Tower and the Watch Tower Society, commonly known as Jehovah's Witnesses.

34. Smith, "Significant Movements," 10.

35. Andrew Dickson White, *The Warfare of Science* (New York: Appleton, 1876); Andrew Dickson White, *A History of the Warfare of Science with Theology in Christendom* (New York: Appleton, 1896), 1:v–vi and xii. On White's liberal religious background and agenda, see James C. Ungureanu, *Science, Religion, and the Protestant Tradition: Retracing the Origins of Conflict* (Pittsburgh: University of Pittsburgh Press, 2019); Glenn C. Slatschuler, "From Religion to Ethics: Andrew Dickson White and the Dilemma of a Christian Rationalist," *Church History* 47 (1978): 308–24; and Richard Schaefer, "Andrew Dickson White and the History of a Religious Future," *Zygon* 50 (2015): 7–27.

36. White to Cornell, August 3, 1869, quoted by Morris Bishop, *The History of Cornell* (Ithaca, NY: Cornell University Press, 1962), 191. The larger story is skillfully laid out by James R. Moore, *The Post-Darwinian Controversies: A Study of the Protestant Struggle to Come to Terms with Darwin in Great Britain and America, 1870–1900* (Cambridge: Cambridge University Press, 1979), 19–40; David C. Lindberg and Ronald L. Numbers, "Beyond War and Peace: A Reappraisal of the Encounter between Christianity and Science," *Church History* 55 (1986): 338–54; and Lawrence M. Principe, "The Warfare Thesis," in *The Idea That Wouldn't Die: The Warfare between Science and Religion*, ed. Jeffrey D. Hardin, Ronald L. Numbers, and Ronald A. Binzley (Baltimore: Johns Hopkins University Press, 2018), 6–26.

37. John William Draper, *History of the Conflict Between Religion and Science* (New York: Appleton, 1874).

38. For detailed criticisms, see Colin A. Russell, "Some Approaches to the History of Science," in Colin A. Russell, R. Hooykaas, and David C. Goodman, *The "Conflict Thesis" and Cosmology* (Milton Keynes, UK: Open University Press, 1974), 5–50; Moore, *Post-Darwinian Controversies*, 40–122; Lindberg and Numbers, "Beyond War and Peace"; David C. Lindberg and Ronald L. Numbers, "Introduction," in *God & Nature: Historical Essays on the Encounter between Christianity and Science*, ed. David C. Lindberg and Ronald L. Numbers (Berkeley: University of California Press, 1986), 1–18; John Hedley Brooke, *Science and Religion: An Historical Perspective* (Cambridge: Cambridge University Press, 1991). Some of White's "facts" are refuted in *Galileo Goes to Jail, and Other Myths about Science and Religion*, ed. Ronald L. Numbers (Cambridge, MA: Harvard University Press, 2009).

39. White, *Warfare*, 1:127; Edward E. Rosen, "Calvin's Attitude toward Copernicus," *Journal of the History of Ideas* 21 (1960): 431–41.

40. Edwin Grant Conklin, *Evolution and the Bible* (Chicago: American Institute of Sacred Literature, 1922), 21; White, *Warfare*, 1:128; cf.1:148.

41. George Sarton, "L'histoire de la science," *Isis* 1.1 (1913): 3–46, translated as "The History of Science," *Monist* 26.3 (1916): 321–65. Robert K. Merton, *Science, Technology, and Society in Seventeenth-Century England*, 2nd ed. (New York: Harper & Row, 1970), xvi. On Sarton, Merton, and the Conflict thesis, see Ungureanu, *Science, Religion, and the Protestant Tradition*, 250–56.

42. White, *Warfare*, 1:vi and xii. Matthew Arnold, *Literature & Dogma: An Essay Towards a Better Apprehension of the Bible* (London: Smith, Elder, 1873), is full of similarly worded, but not identical, descriptions of God, for example, "an enduring Power, not ourselves, that makes for righteousness"; cf. 61, 192, 230, 323–24, and 331.

43. Robert Andrews Millikan, *Evolution in Science and Religion* (New Haven, CT: Yale University Press, 1927), 43, his italics.
44. Harry Emerson Fosdick, *The Living of These Days, An Autobiography* (New York: Harper & Brothers, 1956), 52.
45. Edwin Grant Conklin, "Edwin Grant Conklin," in *Thirteen Americans: Their Spiritual Autobiographies*, ed. Louis Finkelstein (New York: Harper, 1953), 47–76, quotation on 58.
46. Robert Millikan, *The Autobiography of Robert A. Millikan* (New York: Prentice-Hall, 1950), 287.
47. Shailer Mathews, *The Growth of the Idea of God* (New York: Macmillan, 1931), 226, his italics.
48. Kirtley F. Mather, "Sermons from Stones," in *Has Science Discovered God? A Symposium of Modern Scientific Opinion*, ed. Edward H. Cotton (New York: Thomas Y. Crowell, 1931), 3–19, quotation on 9.
49. Kirtley F. Mather, *The Permissive Universe* (Albuquerque: University of New Mexico Press, 1986), 172.
50. Edward B. Davis, "Altruism and the Administration of the Universe: Kirtley Fletcher Mather on Science and Values," *Zygon* 46 (2011): 517–35.
51. Kirtley F. Mather, *Science in Search of God* (New York: Henry Holt, 1928).
52. Edward B. Davis, "Prophet of Science—Part Two: Arthur Holly Compton on Science, Freedom, Religion, and Morality," *Perspectives on Science and Christian Faith* 613.3 (September 2009): 175–90, on 178.
53. Samuel C. Schmucker, *The Study of Nature* (Philadelphia: Lippincott, 1908), 43; Samuel C. Schmucker, *Through Science to God* (Chicago: American Institute of Sacred Literature, 1926), 21.
54. Robert A. Millikan, "Alleged Sins of Science," *Scribner's Magazine* 87 (February 1930): 119–29, quotation on 129; Robert A. Millikan, "The Significance of Radium," *Science* 44 (1 July 1921): 1–8, quotation on 8, his italics.
55. Edwin Grant Conklin, "Religion of Science very different from religion of tradition & Revelation EGCP C 14: "Religion and Science." He spoke far less specifically about the "religion of science and evolution," vis-à-vis "the organized religions of the civilized world," in Edwin Grant Conklin, *The Direction of Human Evolution* (New York: C. Scribner's Sons, 1921), 242–44. For a provocative reply, see *The Religion of Science* (New York: Macmillan, 1922), by Dartmouth College biblical scholar William Hamilton Wood.
56. Conklin, *Direction of Human Evolution*, 209, 216, 218, and 224–25; Lawrence Joseph Henderson, *The Fitness of the Environment* (New York: Macmillan, 1913).
57. Arthur Holly Compton, "A Modern Concept of God," in *Man's Destiny in Eternity: The Garvin Lectures* (Boston: Beacon Press, 1949), 3–20, quotation on 11; Compton, *Freedom of Man*, 109 and 139–40.
58. [Robert A. Millikan], "Deny Science Wars against Religion," *New York Times*, May 27, 1923.
59. Shailer Mathews, *How Science Helps Our Faith* (Chicago: American Institute of Sacred Literature, 1922), 11–12.
60. Phillip E. Johnson, *Darwin on Trial* (Downer's Grove, IL: InterVarsity Press, 1991), Phillip E. Johnson, *Defeating Darwinism by Opening Minds* (Downer's Grove, IL: InterVarsity Press, 1997), and Phillip E. Johnson, *The Wedge of Truth: Splitting the Foundations of Naturalism* (Downer's Grove, IL: InterVarsity Press, 2000); John Shelby Spong, *Why Christianity Must Change or Die: A Bishop Speaks to Believers In Exile* (New York: HarperOne, 1999), and John Shelby Spong, *Eternal Life: A New Vision: Beyond Religion, Beyond Theism, Beyond Heaven and Hell* (New York: HarperOne, 2010).

61. William Bell Riley and Henry B. Smith, *Should Evolution Be Taught in Tax-Supported Schools? Riley-Smith Debate* (n.p., n.d.), 2.

62. The most comprehensive, reliable discussion of all types of creationism is Ronald L. Numbers, *The Creationists: From Scientific Creationism to Intelligent Design*, expanded ed. (Cambridge, MA: Harvard University Press, 2006).

63. Jerry Coyne, *Faith vs. Fact: Why Science and Religion Are Incompatible* (New York: Viking, 2015); Lawrence M. Krauss, *A Universe from Nothing: Why There Is Something Rather Than Nothing* (New York: Free Press, 2012). For a helpful summary of New Atheist approaches to the history of science and religion and a survey of critical responses, see Ronald L. Numbers and Jeffrey D. Hardin, "The New Atheists," in Hardin, Numbers, and Binzley, *The Idea That Wouldn't Die*, 220–38. On the fundamentally religious nature of scientific atheism, see Karl Giberson and Mariano Artigas, *Oracles of Science: Celebrity Scientists versus God and Religion* (Oxford: Oxford University Press, 2006).

64. Francis Collins, *The Language of God: A Scientist Presents Evidence for Belief* (New York: Free Press, 2006); Charles H. Townes, *Making Waves* (Woodbury, NY: AIP Press, 1995); William D. Phillips, "Ordinary Faith, Ordinary Science," *Invenio* 7 (June 2004): 7–18.

## *Conklin (1922)* · Evolution and the Bible

1. No full biography of Conklin has yet been published. The information here is mostly from Kathy J. Cooke, "A Gospel of Social Evolution: Religion, Biology, and Education in the Life of Edwin Grant Conklin" (PhD diss., University of Chicago, 1994); Edwin Grant Conklin, "Edwin Grant Conklin," in *Thirteen Americans: Their Spiritual Autobiographies*, ed. Louis Finkelstein (New York: Harper, 1953), 47–76; E. Newton Harvey, "Edwin Grant Conklin," *Biographical Memoirs of the National Academy of Sciences* 31 (1958): 54–91; J. W. Atkinson, "E. G. Conklin on Evolution: the Popular Writings of an Embryologist," *Journal of the History of Biology* 18 (1985): 31–50; and W. Barksdale Maynard, "The Darwin of Guyot Hall: Once World-Famous, Biologist Edwin Grant Conklin Revealed the Mysteries of Cells and Fought to Defend Evolution," *Princeton Alumni Weekly* 108.8 (February 11, 2009).

2. Edwin Grant Conklin, "Bryan and Evolution," *New York Times*, March 5, 1922, reprinted in Ronald L. Numbers, *Creation-Evolution Debates* (New York: Garland, 1995), 15–19, with editorial comments on ix.

3. Conklin to Ernest D. Burton, July 10, 1922; Mathews to Burton, August 3, 1922, AISL 11:5.

4. Edwin Grant Conklin, *The Direction of Human Evolution*, 2nd ed. (New York: C. Scribner's Sons, 1922), v–xii. The preface is dated May 1, 1922.

5. Conklin to Georgia L. Chamberlin, August 12, 1922, and October 28, 1922, AISLR 11:5.

6. Conklin, "Edwin Grant Conklin," 60 and 56.

7. Edward B. Davis, "Samuel Christian Schmucker's Christian Vocation," *Seminary Ridge Review* 10.2 (Spring 2008): 59–75.

8. Conklin, "Edwin Grant Conklin," 52 and 55–56.

9. Conklin to Fosdick, December 18, 1928, EGCP Carton 1, folder on "Religion 1928—."

10. Conklin, "Edwin Grant Conklin," 52 and 55–56.

11. "Objected to Monkeys," *Saint Paul Globe*, November 23, 1897.

12. Edwin Grant Conklin, *Evolution and Revelation* ([Philadelphia?]: [publisher not identified], 1897), 13 and 10.

13. Edwin Grant Conklin, *The Direction of Human Evolution* (New York: Scribner's, 1921), 206.

14. Edwin Grant Conklin, "Biographical Memoir of William Keith Brooks, 1848–1908," *National Academy of Sciences: Biographical Memoirs* 7 (1913): 22–88, on 72 and 68; cf. Conklin, "Edwin Grant Conklin," 73. Here I follow Cooke, "A Gospel of Social Evolution," 51 and 90.

15. Conklin, "Edwin Grant Conklin," 72–73 and 57–58.
16. "Religion of Science very different from religion of tradition & Revelation," EGCP, Carton 14, folder on "Religion and Science." Compare his comments on "this religion of science and evolution" in *Direction of Human Evolution*, 237–47, quotation on 242.
17. Conklin spoke favorably of "Higher Pantheism" in a lecture headed "March 10 [1925?] Nature and Supernatural," EGCP, Carton 14, folder on "Nature and the Supernatural."
18. Conklin to Miller, May 28, 1928, as quoted by Cooke, "A Gospel of Social Evolution," 234; Theodore A. Miller, *The Mind behind the Universe: A Book of Faith for the Modern Mind* (New York: Frederick A. Stokes, 1928).
19. Conklin, *Direction of Human Evolution*, 223–25.
20. Conklin to P. O'Dea, October 24, 1923, EGCP "Philosophy and Theology."
21. Conklin, "Edwin Grant Conklin," 58.
22. Conklin, *Direction of Human Evolution*, 212–13.
23. Perhaps a reference to the World's Christian Fundamentals Association, founded in 1919 by William Bell Riley, pastor of the First Baptist Church of Minneapolis.
24. The first anti-evolution bill had been introduced in Kentucky in January 1922, in response to a speech by Bryan to a joint session of the legislature. At the time this pamphlet was printed in September 1922, however, no anti-evolution law had yet been passed. Oklahoma became the first state to enact such legislation, when Democratic Governor John C. Walton signed a bill on March 24, 1923, and Florida followed suit two months later. In all, twenty states considered at least one anti-evolution bill. See Edward J. Larson, *Trial and Error: The American Controversy of Creation and Evolution*, 3rd ed. (New York: Oxford University Press, 2003), 48–57.
25. Acts 5:38–39. I have corrected a typo, "Sandehrin" to "Sanhedrin."
26. English zoologist Thomas Henry Huxley said similar things on various occasions. See the introduction to Huxley's *Autobiography and Selected Essays*, ed. Ada L. F. Snell (Boston: Houghton Mifflin, 1909).
27. Quoting the English geneticist William Bateson, whose address was published as "Evolutionary Faith and Modern Doubts," *Science* 55 (January 20, 1922): 55–61, quotation on 61. The italics are Conklin's, not Bateson's.
28. Dutch botanist Hugo de Vries held that new species formed rapidly when numerous individual organisms mutated in a similar way. Columbia University geneticist Thomas Hunt Morgan initially supported de Vries in *Evolution and Adaptation* (New York: Macmillan, 1903), though he later changed his mind. See Peter J. Bowler, *Evolution: The History of An Idea*, rev. ed. (Berkeley: University of California Press, 1989), 276–80.
29. Darwin discussed this in *On the Origin of Species* (London: John Murray, 1859), 22–23.
30. Quoting Bryan's editorial in the *New York Times*.
31. Josh Billings was the pseudonym of American humorist Henry Wheeler Shaw; the reference is unclear.
32. For more on Conklin's view of human "races," see Edwin Grant Conklin, *The Direction of Human Evolution* (London: Oxford University Press, 1921), 25–53.
33. Quoting the English biologist George John Romanes, *The Scientific Evidences of Organic Evolution* (London: Macmillan, 1882), 40.
34. Preformationists believed that human embryos had been formed by God at creation and were suspended until their time of development; see Shirley A. Roe, *Matter, Life, and Generation* (New York: Cambridge University Press, 1981).
35. *Homo neanderthalensis* was discovered in Germany in 1856. "Java man" (*Pithecanthropus erectus*) was discovered by the Dutch anatomist Eugène Dubois in 1891. Subsequently, Java man was reclassified as *Homo erectus*, and the relationship between Neanderthals and modern humans is still a contested subject. A 1921 exhibit at the American Museum of

Natural History depicted *Pithecanthropus*, Neanderthal man, and Cro-Magnon man in a progressive sequence. See Edward J. Larson, *Evolution: The Remarkable History of a Scientific Theory* (New York: Modern Library, 2004), 141–47, and Bowler, *Evolution*, 231–32 and 322–23.

36. Gen. 2:7 and 1:3.

37. The ram-headed Egyptian god Khnum made humans from clay on a potter's wheel. Cyrus H. Gordon, "Khnum and El," in *Scripta Hierosolymitana: Egyptological Studies*, ed. Sarah Israelit-Groll, vol. 28 (Jerusalem: Magnes Press, 1982).

38. Heb. 2:10; Rom. 11:36.

39. A reference to a letter from Scottish geologist Charles Lyell to Thomas S. Spedding, May 19, 1863, in *Life, Letters and Journals of Sir Charles Lyell*, ed. Mrs. (Katharine M.) Lyell, 2 vols. (London: John Murray, 1881), 2:376. Lyell attributed this remark to "one of Darwin's reviewers." Conklin had already mentioned this in *Direction of Human Evolution*, 233.

40. The notion that medieval Christians believed in a flat earth and therefore opposed Columbus is a secular myth created mainly in the nineteenth century by Washington Irving and others. Jeffrey Burton Russell, *Inventing the Flat Earth: Columbus and Modern Historians* (New York: Praeger, 1991). Columbus was briefly imprisoned on charges that he abused his powers as governor of Hispaniola, but he was never in trouble with the Inquisition. However, he accepted a popular medieval interpretation of 2 Esdras 6:42, that land covers sixth-sevenths of the earth. Samuel Eliot Morison, *Admiral of the Ocean Sea: A Life of Christopher Columbus* (Boston: Little, Brown, 1946), 71n and 94.

41. Evangelist Wilbur Glenn Voliva was leader of the Christian Apostolic Church of Zion, Illinois, which held "flat earthism" as a doctrine. Robert J. Schadewald, "Flat Earthism," in *The History of Science and Religion in the Western Tradition: An Encyclopedia*, ed. Gary B. Ferngren et al. (New York: Garland, 2000), 359–61; on Voliva, see *WWWA*.

42. A reference to Mrs. Jameson (Estelle M. Hurll), *Sacred and Legendary Art* (London: Longmans, Green, and Co., 1870), 1:80.

43. For subtler analysis, see Donald H. Kobe, "Copernicus and Martin Luther: An Encounter between Science and Religion," *American Journal of Physics* 66 (March 1998): 190–96.

44. Conklin's reference to theologian Philip Melanchthon is based on Andrew Dickson White, *A History of the Warfare of Science with Theology in Christendom* (New York: Appleton, 1896), 1:26–27. Although initially Melanchthon was hostile to Copernicanism, later he encouraged its teaching as mathematical hypothesis at Lutheran universities. Robert S. Westman, "The Melanchthon Circle, Rheticus, and the Wittenberg Interpretation of the Copernican Theory," *Isis* 66 (1975): 164–93, and "The Copernicans and the Churches," in *God and Nature*, ed. David C. Lindberg and Ronald L. Numbers (Berkeley: University of California Press, 1986), 76–113, esp. 82–83.

45. Conklin quotes White, *History of the Warfare of Science*, 1:128. In a paragraph about theological opposition to Copernicus that also includes an imaginary quotation from John Calvin, White claims that "even John Wesley declared the new ideas to 'tend toward infidelity.'" White based this on a discussion of various theological opinions about life on other worlds—not the solar system—in Charles Woodruff Shields, *The Final Philosophy, or System of Perfect Knowledge Issuing from the Harmony of Science and Religion* (New York: Scribner, 1877), 61. According to Shields, in a sermon written "after [William] Derham and [Christiaan] Huyghens had associated a plurality of worlds with revealed truths," Wesley "termed that opinion the palmary argument of infidels, and declared he would doubt it, even though it were allowed by all the philosophers in Europe." Wesley did preach a sermon, "What Is Man?" (Psalm 8:4), where he rejects "the plurality of worlds, a very favourite notion with all those who deny the Christian revelation," partly on the basis of Huygens's observation "that the moon has no atmosphere," and therefore "no clouds, no rain, no springs, no rivers; and therefore no plants or animals." *The Works of John Wesley*, ed. Albert C. Coulter (Nashville:

Abingdon Press, 1984–), 3:454–63, quotation on 462. For more on Wesley's position, see Michael J. Crowe, *The Extraterrestrial Life Debate, 1750-1900* (Cambridge: Cambridge University Press, 1986), 92–96; on his overall attitude toward science, see J. W. Haas Jr., "John Wesley's Views on Science and Christianity: An Examination of the Charge of Antiscience," *Church History* 63 (1994): 378–92.

46. A reference to *Moses's Principia* (1724), by the English anti-Newtonian John Hutchinson. White accurately says that Hutchinson "assaulted the Newtonian theory as 'atheistic,'" but the words Conklin quoted are not found there. White, *History of the Warfare of Science*, 1:148.

47. James Dwight Dana's reconciliation of the Bible and geology in his *Manual of Geology* (1862) and elsewhere was viewed favorably by many conservative Protestants, including Bryan. Edward B. Davis, "The Word and the Works: Concordism in American Evangelical Thought," in *The Book of Nature in Early Modern and Modern History*, ed. Klaas van Berkel and Arjo Vanderjagt (Leuven: Peeters, 2006), 195–207.

48. James Ussher, *Annals of the Old Testament* (1650), calculated that the creation took place in 4004 B.C.

49. See note 24.

50. John Alexander Dowie founded the Christian Catholic Apostolic Church in 1896 and Zion City (north of Chicago), exclusively for members of his church, in 1899. The first Zion school opened in Chicago in 1899 with the goal of raising a "royal generation." Philip L. Cook, *Zion City, Illinois: Twentieth Century Utopia* (Syracuse, NY: Syracuse University Press, 1996), 125–30; on Dowie, see *ANB*.

## *Fosdick (1922)* · Evolution and Mr. Bryan

1. The standard biography is Robert Moats Miller, *Harry Emerson Fosdick: Preacher, Pastor, Prophet* (New York: Oxford University Press, 1985), which is frustrating to use owing to the absence of footnotes. Harry Emerson Fosdick, "Harry Emerson Fosdick," in *Thirteen Americans: Their Spiritual Autobiographies*, ed. Louis Finkelstein (New York: Harper, 1953), 106–20, is more helpful here.

2. For publication details, see appendix one.

3. Graves to Fosdick, February 20 and 24, 1922, HEFP Evolution; Harry Emerson Fosdick, "Evolution and Mr. Bryan," *New York Times*, March 12, 1922; Harry Emerson Fosdick, "A Reply to Mr. Bryan in the Name of Religion," HEFP Evolution.

4. Fosdick, "Harry Emerson Fosdick," 107–11, 113. For early influences, see Miller, *Fosdick*, 43–54.

5. Harry Emerson Fosdick, *Evolution and Mr. Bryan* (Chicago: American Institute of Sacred Literature, 1922), 15.

6. Harry Emerson Fosdick, *The Meaning of Faith* (New York: Association Press, 1917), 179, his italics. This book was printed dozens of times by numerous domestic and foreign publishers down to the early 1950s.

7. Here the pamphlet has a footnote: "This article appeared in New York Times, Sunday, March 12."

8. Fosdick quotes the reply to Bryan by Henry Fairfield Osborn (whose name Fosdick misspells), "Evolution and Religion," *New York Times*, March 5, 1922, 91. Osborn's essay was reprinted at least three times in the next four years, including once as a pamphlet, *Evolution and Religion* (New York: Charles Scribner's Sons, 1923). For a modern edition, see Ronald L. Numbers, *Creation-Evolution Debates* (New York: Garland, 1995), 9–14.

9. Fosdick quotes *Pansophia Mosaica e Genesi delineata* (Leipzig: Gleditsch, 1685), by the German Lutheran theologian August Pfeiffer, as found in Andrew Dickson White, *A History of the Warfare of Science with Theology in Christendom* (New York: Appleton, 1896), 2:312. Fosdick erroneously dates this about a century too early.

10. Fosdick quotes Martin Luther's *Table Talk*, as found in White, *History of the Warfare of Science*, 1:126. On Luther's criticism of Copernicus, see Donald H. Kobe, "Copernicus and Martin Luther: An Encounter between Science and Religion," *American Journal of Physics* 66 (March 1998): 190–96.

11. The Hungarian Jesuit Melchior Ildephonsus Inchofer served on the Vatican committee investigating Galileo's *Dialogue on the Two Chief World Systems* in 1632. His detailed report, a longer version of which is *Tractatus syllepticus* (Rome: Ludovicus Grignanus, 1633), charged that Galileo had argued for the literal truth of the earth's motion, violating an injunction issued by the Vatican. Fosdick again takes the quotation from White, *History of the Warfare of Science*, 1:139. Although White dated this passage to 1631, it is actually Galileo's own description of Inchofer's *Tractatus* from a letter to the French lawyer Elia Diodati of July 25, 1634. For a translation of Inchofer's report, see Maurice A. Finnocchiaro, *The Galileo Affair: A Documentary History* (Berkeley: University of California Press, 1989), 262–70; for a translation of the *Tractatus*, see Richard J. Blackwell, *Behind the Scenes at Galileo's Trial* (Notre Dame, IN: University of Notre Dame Press, 2006), 105–206. Also see William R. Shea, "Melchior Inchofer's 'Tractatus Syllepticus': A Consultor of the Holy Office Answers Galileo," in *Novità celesti e crisi del sapere*, ed. Paolo Galluzzi (Florence: Babèra, 1984), 283–92. William R. Shea and Lawrence M. Principe provided helpful information.

12. Quoting the English evangelist Henry Drummond, *The Ascent of Man* (New York: James Pott, 1894), 333.

## *Mathews (1922)* · How Science Helps Our Faith

1. No full-length biography of Mathews exists. Information comes from two of his works, *New Faith for Old: An Autobiography* (New York: Macmillan, 1936), and "Theology as Group Belief," in *Contemporary American Theology: Theological Autobiographies*, ed. Virgilius Ferm, 2nd ser. (New York: Roundtable Press, 1933), 2:163–93; W. Creighton Peden, "Shailer Mathews," in *Makers of Christian Theology in America*, ed. Mark G. Toulouse and James O. Duke (Nashville: Abingdon Press, 1997), 392–98; and *ANB*. For insightful analysis of Mathews's theology, placing him squarely in the radical milieu at the Divinity School, see David Henry Koss, "The Development of Naturalism at the Divinity School of the University of Chicago with Special Emphasis on the Doctrine of God" (PhD diss., Northwestern University, 1972).

2. Shailer Mathews, *How Science Helps Our Faith* (Chicago: American Institute of Sacred Literature, 1922), 13–14. Kristin Johnson, the author of *Imagining Progress: Science, Faith, and Child Mortality in America* (Tuscaloosa: University of Alabama Press, 2024), introduced me to the term "natural law theodicy" and its history. On the related idea of "meliorism," see Howard R. Murphy, "The Ethical Revolt against Christian Orthodoxy in Early Victorian England," *American Historical Review* 60.4 (July 1955): 800–817. Mathews associated meliorism with William James; see *Contributions of Science to Religion*, ed. Shailer Mathews (New York: D. Appleton, 1924), 409.

3. Mathews, *Contributions of Science to Religion*, 407–8 and 411.

4. Koss, "Development of Naturalism," 56–58.

5. Shailer Mathews, *The Growth of the Idea of God* (New York: Macmillan, 1931), 226, his italics.

6. Robert Millikan, *The Autobiography of Robert A. Millikan* (New York: Prentice-Hall, 1950), 287. It is not clear in context whether Millikan or someone else asked this question.

7. For more on his changing theology, see Kenneth Cauthen, *The Impact of American Religious Liberalism* (New York: Harper and Row, 1962), 147–68, and Koss, "Development of Naturalism," 76–87.

8. Mathews, *Contributions of Science to Religion*, 391.

9. Mathews, "Theology as Group Belief," 173.

10. Shailer Mathews, *The Spiritual Interpretation of History* (Cambridge, MA: Harvard University Press, 1920), 25.

11. Shailer Mathews, *Immortality and the Cosmic Process* (Cambridge, MA: Harvard University Press, 1933). Comprising forty-nine pages, set in large type with wide margins on small paper, this is the published version of his Ingersoll Lecture.

12. Address by Charles Gilkey in *Shailer Mathews, Born Portland, Maine, May 26, 1863, Died Chicago, Illinois, October 23, 1941; Selections from the Memorial Service Held in Joseph Bond Chapel, the University of Chicago, on Sunday, October 26, 1941* (Chicago: n.p., 1941), 17–29, on 25.

13. Arthur H. Compton, Shailer Mathews, and Charles W. Gilkey, *Life After Death* (Chicago: American Institute of Sacred Literature, 1930), 23.

14. Mathews, *Immortality and the Cosmic Process*, 27.

15. Psalm 139:8.

16. Most scholars today would agree with Mathews's assessment of how various Christian bodies responded to the heliocentric theory of Copernicus and the new physics of Galileo. For a careful analysis, see Robert S. Westman, "The Copernicans and the Churches," in *God and Nature*, ed. David C. Lindberg and Ronald L. Numbers (Berkeley: University of California Press, 1986), 76–113.

17. Contrary to Mathews, nearly all Christian theologians since the early church have accepted the sphericity of the earth. Jeffrey Burton Russell, *Inventing the Flat Earth: Columbus and Modern Historians* (New York: Praeger, 1991).

18. Mathews later wrote extensively about this topic in *The Growth of the Idea of God* (New York: Macmillan, 1931).

19. In a letter to Mathews from June 6, 1924, EBFPY, Frost pointed out that the word "discharge" should be "charged" in both instances. Physicist Ernest Rutherford discovered that atoms have positive nuclei, and in 1911 he proposed a new theory of atomic structure on this basis.

20. The idea that the Milky Way is only one of uncountable enormous galaxies was advanced especially by astronomer Edwin Hubble of Mount Wilson Observatory in 1925 and subsequently. Robert W. Smith, "Cosmology 1900–1931," in *Cosmology: Historical, Literary, Philosophical, Religious, and Scientific Perspectives*, ed. Norriss S. Hetherington (New York: Garland, 1993), 329–45, and Norriss S. Hetherington, "Hubble's Cosmology," in *Cosmology*, 347–69.

21. Bryan's *New York Times* editorial made this a central claim.

22. "Theistic evolution may be described as an anesthetic which deadens the pain while the patient's religion is being gradually removed, or it may be likened to a way-station on the highway that leads from Christian faith to No-God-Land." William Jennings Bryan, *The Menace of Darwinism* (New York: Revell, 1922), 5.

## *Millikan (1923)* · A Scientist Confesses His Faith

1. The information in this introduction comes mainly from Edward B. Davis, "Robert Andrews Millikan: Religion, Science, and Modernity," in *Eminent Lives in Twentieth-Century Science & Religion*, ed. Nicolaas A. Rupke, rev. ed. (Frankfurt am Main: Peter Lang, 2009), 253–74.

2. Maynard Shipley, *The War on Modern Science* (New York: A. A. Knopf, 1927), 239.

3. For details on related publications, see appendix one.

4. "Holds Science Is Aid to Religion: Dr. Millikan Gives Address at Oberlin's Commencement," *Cleveland Plain Dealer*, June 15, 1926.

5. Robert A. Millikan, "The Significance of Radium," *Science* 44 (July 1, 1921): 1–8, on 8, his italics.

6. Robert A. Millikan, *The Autobiography of Robert A. Millikan* (New York: Prentice-Hall, 1950), 279, his italics.

7. Robert A. Millikan, "Alleged Sins of Science," *Scribner's Magazine* 87 (February 1930): 119–29, on 129, his italics.

8. [Millikan], "Deny Science Wars against Religion," *New York Times*, May 27, 1923.

9. Robert A. Millikan, "Available Energy," *Science* 68 (September 28, 1928): 279–84, quotation on 283, and Robert A. Millikan, "Present Status of Theory and Experiment as to Atomic Disintegration and Atomic Synthesis," *Science* 73 (January 2, 1931): 1–5, quotation on 5.

10. Millikan misdates Galileo's work, which was published between 1610 and 1638.

11. St. Augustine of Hippo, *The Literal Meaning of Genesis*, 1.19.39.

12. Other versions of this text have 1922, the year in which Bryan's anti-evolution campaign came to Millkan's attention. It was apparently altered here to correspond to the publication date of September 1923. The second impression (February 1927) is identical to the first impression.

13. The fundamentalist–modernist controversy came to a head in the 1920s.

14. All five of these scientists were strongly religious, although their specific beliefs differed widely. See Frank Manual, *The Religion of Isaac Newton* (New York: Oxford University Press, 1974); Geoffrey N. Cantor, *Michael Faraday, Sandemanian and Scientist* (London: Macmillan, 1991); Phillip L. Marston, "Maxwell and Creation: Acceptance, Criticism, and His Anonymous Publication," *American Journal of Physics* 75 (August 2007): 731–40; Ivan Tolstoy, *James Clerk Maxwell: A Biography* (Edinburgh: Canongate, 1981); Silvanus P. Thompson, *The Life of Lord Kelvin* (New York: Chelsea, 1910), 2:1086–104; on English physicist John William Strutt, 3rd Baron Rayleigh, active in the Synthetic Society and the Society for Psychical Research, see Janet Oppenheim, *The Other World: Spiritualism and Psychical Research in England, 1850–1914* (Cambridge: Cambridge University Press, 1985), 330–33. For briefer, helpful comments on Maxwell, Kelvin, and Rayleigh, see David B. Wilson, "The Thought of Late Victorian Physicists: Oliver Lodge's Ethereal Body," *Victorian Studies* 15 (September 1971): 29–48.

15. The first quotation is from a letter Kelvin wrote in 1887; Thompson, *Life of Lord Kelvin*, 2:1103. The second quotation is from a short speech he offered during a series of lectures on "Christian Apologetics," conducted at University College, London in 1903 (2:1099; Millikan also quotes Thompson's description on 2:1089). For Kelvin's views on the age of the earth, see Joe D. Burchfield, *Lord Kelvin and the Age of the Earth* (New York: Science History Publications, 1975).

16. Graham Lusk, "Pasteur, the Man," *Science* 57 (February 2, 1923): 139–41, quotation on 141.

17. From the eulogy Pasteur delivered for the positivist philosopher Émile Maximilien Paul Littré, his predecessor at the Académie Française. The source for Millikan's version of the epitaph is unclear; for another version, see the introduction by William Osler to René Vallery-Radot, *The Life of Pasteur*, trans. R. L. Devonshire (Garden City, NY: Doubleday, Page, 1924), xvi.

18. Walcott was secretary of the Smithsonian Institution from 1907 until his death. Millikan corresponded with many scientists when he drafted his statement for the *New York Times* early in 1923 (see note 24). Some of the correspondence survives in RAMP 32:37, but Walcott's letter is not among them although he signed the statement.

19. Millikan misspelled the surname of Charles Greeley Abbot, director of the Smithsonian Astrophysical Observatory and assistant secretary (later secretary) of the Smithsonian Institution. Walcott, Osborn, Conklin, Merriam, Pupin, Coulter, Angell, and Arthur A. Noyes all signed Millikan's statement in the *New York Times* (see note 24).

20. John 8:32.

21. Following other versions, here I have corrected the pamphlet, which has "graduations."

22. Millikan quotes one of the many editions of John Wesley, *A Survey of the Wisdom of God in the Creation, Or, a Compendium of Natural Philosophy* (1763), which included a lengthy abridgement of Charles Bonnet, *The Contemplation of Nature* (1764). Following the publication

of this pamphlet, Syracuse University zoologist Charles Wesley Hargitt told Millikan that he had erred in attributing these words to Wesley rather than Bonnet; Hargitt to Millikan, January 10, 1924, RAMP 32:37. In the version of the pamphlet published in Robert A. Millikan, *Science and Life* (Boston: Pilgrim Press, 1924), the error is corrected in a footnote. Hargitt published at least two articles on Wesley, "John Wesley—Evolutionist," *Zion's Herald* (1925): 1061–88, and "John Wesley and Science: A Challenge from the Eighteenth Century," *Methodist Review* 110 (1927): 383–93. Owing to this passage being quoted out of context, Wesley had often been seen as an evolutionist since at least 1893; see Laura Bartels Felleman, "John Wesley's *Survey of the Wisdom of God in Creation:* A Methodological Inquiry," *Perspectives on Science and Christian Faith* 58 (March 2006): 68–73.

23. Scottish theologian Henry Drummond, author of *Natural Law in the Spiritual World* (1883); Henry Ward Beecher, first pastor of the Plymouth Congregational Church in Brooklyn; theologian Lyman Abbott, who succeeded Beecher at Plymouth Congregational Church; Fosdick, who in 1923 was pastor of the First Presbyterian Church in Manhattan; Chicago ethicist Theodore Gerald Soares, Millikan's pastor at Hyde Park Congregational Church, later hired by Millikan as professor of ethics at Caltech and pastor of the Neighborhood Church in Pasadena; theologian Henry Churchill King, president of Oberlin College from 1902 to 1927; Congregationalist pastor Robert Elliott Brown, married to Millikan's sister; Baptist theologian Ernest DeWitt Burton, president of the University of Chicago from 1923 until his death in 1925; and Mathews, dean of the University of Chicago Divinity School from 1908 to 1933. King, Brown, and Burton signed Millikan's statement in the *New York Times* (see note 24).

24. Here Millikan borrowed the definitions of science and religion that he had first given in a statement he drafted that was signed by forty prominent scientists, theologians, and politicians, "Deny Science Wars against Religion," *New York Times*, May 27, 1923.

25. Matt. 23:23.
26. Hab. 2:2.
27. Acts 17:28–29.
28. 1 Cor. 13:11.
29. Job 11:7.
30. See note 15 above, where Millikan quoted the same passage from Thompson, *Life of Lord Kelvin*, 2:1089.
31. Millikan borrows again from his statement, "Deny Science Wars against Religion."
32. Gal. 6:7.
33. Rom. 6:23. "Love begets love" might be quoting "Herein is Love," a sermon by Charles Haddon Spurgeon from January 19, 1896; see https://www.spurgeon.org/resource-library/sermons/herein-is-love-2/#flipbook/, accessed February 16, 2024.
34. Matthew Arnold, *Literature & Dogma: An Essay towards a Better Apprehension of the Bible* (London: Smith, Elder, 1873), is full of similarly worded, but not identical, descriptions of God, for example, "an enduring Power, not ourselves, that makes for righteousness"; cf. 61, 192, 230, 323–24, and 331. Similar language is found in an 1894 lecture by Columbia University ethicist Felix Adler, founder of the New York Society for Ethical Culture; see "The Religion of Humanity," *New York Times*, November 12, 1894.
35. See note 15 above.

## *Frost (1924)* · The Heavens are Telling

1. Information about Frost's religious life and beliefs is quite limited. Even in his autobiography, Edwin Brant Frost, *An Astronomer's Life* (Boston: Houghton Mifflin, 1933), references to religion are scarce and usually brief. Most of his personal papers were probably destroyed by his wife after his death. Only a few survive among his professional correspondence, and some of those formerly held at Yerkes Observatory are now missing.

2. Frost, *An Astronomer's Life*, 188–92.

3. This information comes from Kyle Cudworth, former Director of Yerkes Observatory; Cudworth to Davis, February 21, 2013.

4. Philip Fox, "Edwin Brant Frost," *Astrophysical Journal* 83 (1936): 1–9, quotation on 6.

5. James G. K. McClure to Frost, November 1, 1923; Frost to McClure, November 5, 1923, EBFP 1:8.

6. Edwin B. Frost, "The Structure of the Cosmos," in *Contributions of Science to Religion*, ed. Shailer Mathews (New York: D. Appleton, 1924), 58–104.

7. Frost, *An Astronomer's Life*, 212. Information about the Norton Lectures comes from Adam Winters, Processing Archivist at the Southern Baptist Theological Seminary; Winters to Davis, February 21, 2013.

8. Frost, *An Astronomer's Life*, 275–86. In the preface, Frost says, "I have been asked to include [this lecture] in the book," wherein it constitutes pp. 275–86. For his sixty-fifth birthday in July 1931, when 300 people gathered at Yerkes to celebrate, some newspapers ran a story with very similar remarks, supplemented with musings about the eternity of the universe and the likelihood of extraterrestrial life. A lengthy excerpt from "Fragments of Cosmic Philosophy" was printed many years later: Edwin Brant Frost, "The Unity of the Universe," *Science of Mind* 29 (June 1956): 24–28.

9. Frost, *An Astronomer's Life*, 285 and 284; Edwin Brant Frost, *The Heavens are Telling* (Chicago: American Institute of Sacred Literature, 1924), 31.

10. Frost, *An Astronomer's Life*, 281 and 211–12.

11. Mathews to Frost, April 23, 1924, EBFPY.

12. Chamberlin to Frost, August 4, 1924, EBFP 2:3.

13. The title of this pamphlet implicitly cites Psalm 19:1 ("The heavens declare the glory of God"), but directly quotes the opening line ("The Heavens are telling the glory of God") of the most famous chorus in Joseph Haydn's oratorio, "The Creation." On the front cover there are no quotation marks, but the title is enclosed in quotation marks at the top of page 3. I am grateful to the late Owen Gingerich for pointing this out and for advice about several other notes to this pamphlet.

14. John Milton, *Paradise Lost*, book IV, lines 604–6, and book VII, lines 580–81.

15. Quoting the conclusion of Immanuel Kant, *Critique of Practical Reason* (1788).

16. It is not clear where Frost's numbers were coming from. Two years before, a paper by the Dutch astronomer Jacobus Cornelius Kapteyn, "First Attempt at a Theory of the Arrangement and Motion of the Sidereal System," *Astrophysical Journal* 55 (1922): 302–27, estimated that there are forty-seven billion stars in the known universe, a region smaller than the Milky Way galaxy is now known to be.

17. Methods to date the earth and the sun were in flux when Frost wrote this. In the latter half of the nineteenth century, several physicists tried various ways of calculating the ages of the earth and the sun, especially Lord Kelvin. In 1862, he found the earth between 20 and 400 million years old, and the sun as probably less than 100 million years old and almost certainly less than 500 million years old. At the end of the nineteenth century, the earth was widely believed to be about 100 million years old, but within a dozen years new methods using radioactive elements were yielding estimates of one billion years or more—the number Frost gives. No one yet knew the real source of the sun's energy, although Frost was right to suggest that nuclear processes were involved. See Joe D. Burchfield, *Lord Kelvin and the Age of the Earth* (New York: Science History Publications, 1975); Kelvin, "On the Age of the Sun's Heat," *Macmillan's Magazine* 5 (March 5, 1862): 288–93; and Lawrence Badash, "The Age-of-the-Earth Debate," *Scientific American* (August 1989): 90–96.

18. In 1920 and 1921, physicist Albert Abraham Michelson and astronomers John A. Anderson and Francis Gladheim Pease measured the angular diameters of Betelgeuse and

Antares with an interferometer attached to the 100-inch telescope at Mount Wilson. A. A. Michelson and F. G. Pease, "Measurement of the Diameter of Alpha-Orionis by the Interferometer," *Proceedings of the National Academy of Sciences of the United States of America* 7 (May 15, 1921): 143–46.

19. Harvard astronomer Harlow Shapley calculated the distance to the Andromeda galaxy to be about one million light years; Shapley, "Novae in Spiral Nebulae," *Publications of the Astronomical Society of the Pacific* 29 (1917): 213–17. According to Owen Gingerich (private communication), Frost's number of perhaps one million spiral galaxies probably came from Heber D. Curtis of Lick Observatory.

20. Physicist Ernest Rutherford put forth the planetary model of the atom in 1911, with the atomic nucleus analogous to the sun and the electrons analogous to the planets.

21. 2 Cor. 5:1.

22. Psalm 8:3–4.

23. After astronomers began studying the spectral lines of stars and nebulae in the mid-nineteenth century, a number of lines were found in various objects that did not match the known spectral lines of terrestrial elements. Eventually these were all satisfactorily explained, but when Frost wrote this, two very strong green lines in nebular spectra had not yet been accounted for and were being attributed to the putative element "nebulium." A few years later, Caltech astronomer Ira Sprague Bowen showed that they originated in a highly ionized state of oxygen, under conditions of extremely low density found in certain nebulae. Bowen, "The Origin of the Nebular Lines and the Structure of the Planetary Nebulae," *Astrophysical Journal* 67 (1928): 1–15.

24. Alexander Pope, *An Essay on Man*, Epistle I, lines 23–28.

25. Luke 10:25–37.

26. In the nineteenth century, physicists almost universally believed that light and other electromagnetic waves traveled through a subtle material substance called ether that pervaded all of space. The special theory of relativity, first proposed by Albert Einstein in 1905, eventually made the ether hypothesis redundant, but for many years his work was often seen as a modification to the ether theory. John Stachel, "The Theory of Relativity," in *Companion to the History of Modern Science*, ed. R. C. Olby et al. (London: Routledge, 1990), 442–57, esp. 447–48.

27. Perhaps this should read: "length or extent," but the third impression (September 1930) has the same reading as the first.

28. Nicolaus Copernicus proposed that the earth goes around the sun, Johannes Kepler showed that the planets move in elliptical orbits, not circular orbits, and Isaac Newton proved that planetary motion in elliptical orbits is caused by universal gravitation.

29. French biologist Louis Pasteur, who helped establish the germ theory of disease.

30. Tycho Brahe, *De nova et nullius ævi memoria prius visa stella* (1573), analyzed the supernova he had seen in Cassiopeia in November 1572; Kepler, *De stella nova in pede Serpentarii* (1606), studied a supernova in Ophiuchus that first appeared in October 1604. For scientific and historical information on both, see David H. Clark and F. Richard Stephenson, *The Historical Supernovae* (New York: Pergamon, 1977), 172–206. Frost's reference to two first-magnitude events "in the past quarter of a century" is puzzling, since no supernovae brighter than magnitude 5.8 have been observed since 1885 and no supernovae have reached the first magnitude since Kepler's time. See "List of Supernovae," from the International Astronomical Union, http://www.cbat.eps.harvard.edu/lists/Supernovae.html, accessed February 16, 2024.

31. Matt. 26:39, Mark 14:36, or Luke 22:42.

32. John Milton probably visited Galileo in September 1638, while Galileo was under house arrest and after he had gone blind, writing about this in *Areopagitica; A speech of Mr. John Milton for the Liberty of Unlicenc'd Printing, to the Parlament of England* (London, 1644).

### Schmucker (1926) · Through Science to God, The Humming Bird's Story, An Evolutionary Interpretation

1. The information in this introduction comes mainly from Edward B. Davis, "Samuel Christian Schmucker's Christian Vocation," *Seminary Ridge Review* 10.2 (Spring 2008): 59–75.

2. Samuel C. Schmucker, *The Meaning of Evolution* (New York: Macmillan, 1913); Samuel C. Schmucker, *Man's Life on Earth* (New York: Macmillan, 1925). See the curriculum for the Chautauqua Literary and Scientific Circle, https://www.chq.org/wp-content/uploads/2021/05/2020-CLSC-Historic-Booklist.pdf, accessed February 20, 2024.

3. Mathews to Schmucker, March 10, 1924, AISLR 12:5, acknowledging a gift of five dollars to the AISL and indicating that he looked forward to seeing him at Chautauqua.

4. Samuel C. Schmucker, *Through Science to God: The Humming Bird's Story, An Evolutionary Interpretation* (Chicago: American Institute of Sacred Literature, 1926), 20–22.

5. Samuel C. Schmucker, *Heredity and Parenthood* (New York: Macmillan, 1929), 207.

6. Schmucker, *Man's Life on Earth*, 285–86.

7. Schmucker, *Heredity and Parenthood*, vi and 13–14.

8. Carl Linnaeus devised the modern system of binomial nomenclature, which assigns to each organism a pair of Latin names for the genus and species to which it belongs.

9. Despite the reference to Isaac Newton, Schmucker's physical and theological interpretations of gravitation are directly opposed to Newton's. In a famous letter to Richard Bentley of January 17, 1692, Newton wrote, "You sometimes speak of gravity as essential and inherent to matter. Pray do not ascribe that notion to me." H. S. Thayer, *Newton's Philosophy of Nature: Selections from His Writings* (New York: Hafner Press, 1953), 53. Indeed for much of his life Newton believed that God was probably the direct cause of gravitation. Edward B. Davis, "Newton's Rejection of the 'Newtonian World View': The Role of Divine Will in Newton's Natural Philosophy," in *Facets of Faith and Science, vol. 3: The Role of Beliefs in the Natural Sciences*, ed. Jitse M. van der Meer (Lanham, MD: University Press of America, 1996), 75–96.

### Pupin (1928) · Creative Co-ordination

1. The information in this introduction comes mainly from Edward B. Davis, "Michael Idvorsky Pupin: Cosmic Beauty, Created Order, and the Divine Word," in *Eminent Lives in Twentieth-Century Science & Religion*, ed. Nicolaas A. Rupke, rev. ed. (Frankfurt am Main: Peter Lang, 2009), 295–316. According to Serbian sources, Pupin was born in 1854. Most American sources have 1858, a date that Pupin and his daughter provided on several occasions for reasons too complicated to explain here, but the earlier date is almost certainly accurate.

2. Marcel LaFollette, *Making Science Our Own: Public Images of Science, 1910–1955* (Chicago: University of Chicago Press, 1990), 50–53.

3. Michael Pupin, *From Immigrant to Inventor* (New York: Scribner's Sons, 1923), *The New Reformation* (New York: C. Scribner's Sons, 1927), and *Romance of the Machine* (New York: C. Scribner's Sons, 1930).

4. Erin A. Smith, "The Religious Book Club: Print Culture, Consumerism, and the Spiritual Life of American Protestants Between the Wars," in *Religion and the Culture of Print in Modern America*, ed. Charles L. Cohen and Paul S. Boyer (Madison: University of Wisconsin Press, 2008), 217–42.

5. Michael Pupin, "From Chaos to Cosmos," *Scribner's Magazine* 76 (July 1924): 3–10, and "Creative Co-ordination: A Message from Physical Science," *Scribner's Magazine* 82 (August 1927): 142–53.

6. Michael I. Pupin, "Creative Co-ordination," *School and Society* 26 (October 29, 1927): 543–47.

7. According to Charles Harvey Arnold, *God Before You and Behind You: The Hyde Park Union Church through a Century, 1874-1974* (Chicago: Hyde Park Union Church, 1974), it was Chamberlin who suggested that Pupin be approached as a pamphlet author (on 98).

8. Michael Pupin, "The Spiritual Influence of a Noted Scientist," *Columbia Alumni News* (November 2, 1923): 66-67.

9. These themes are developed in Davis, "Michael Idvorsky Pupin."

10. Pupin, *From Immigrant to Inventor*, 382.

11. The original pamphlet has a footnote after the title: "From *School and Society*. Reprinted by permission of author and publishers."

12. Psalm 19:1.

13. The pre-Socratic philosopher Democritus and the Roman poet Lucretius are often seen as founders of the atomic theory.

14. Newton explained the motions of bodies in terms of forces and three laws of motion, Scottish physicist James Clerk Maxwell formulated four equations describing electromagnetic radiation, and French engineer Nicolas Léonard Sadi Carnot explained heat engines and laid the foundation for the science of thermodynamics.

15. Matt. 22:37 and 22:39.

## *Fosdick (1928)* · Religion's Debt to Science

1. The original sermon is Harry Emerson Fosdick, "Religion's Indebtedness to Science," February 27, 1927, HEFP, Series 1A, 12:2. For publication details, see appendix one.

2. Fosdick, *Religion's Debt to Science* (Chicago: American Institute of Sacred Literature, 1928), 1.

3. John William Draper, *History of the Conflict between Religion and Science* (New York: D. Appleton, 1874); Andrew Dickson White, *A History of the Warfare of Science with Theology in Christendom*, 2 vols. (New York: D. Appleton, 1896).

4. Harry Emerson Fosdick, *For the Living of These Days, An Autobiography* (New York: Harper & Brothers, 1956), 52.

5. Harry Emerson Fosdick, "Science and Mystery," *Atlantic Monthly* 112 (October 1913): 520-30, quotation on 528-29.

6. Fosdick, *Religion's Debt to Science*, 4.

7. Harry Emerson Fosdick, "The Real Point of Conflict of Science & Religion," February 18, 1940, HEFP Series 1A, 11:39, quoting the version published the following year in Harry Emerson Fosdick, *Living under Tension: Sermons on Christianity* (New York: Harper & Brothers, 1941), 140-48, quotation on 143-44. The original source of the borrowed words may have been a sermon by Smith College economist William Aylott Orton at Skidmore College in 1929; see "Dr. Orton Vesper Speaker: The Role of Religion in Modern Life," *Skidmore News*, April 19, 1929, 1.

8. Harry Emerson Fosdick, "Science Demands Religion," October 14, 1945, HEFP, Series 1A, 12:29. This was published as "Science Demands Religion," *The Christian World*, January 24, 1946, but I have not seen a copy.

9. Harry Emerson Fosdick, *Dear Mr. Brown: Letters to a Person Perplexed about Religion* (New York: Harper & Brothers, 1961), chaps. 5 and 6. At the height of his interest in this topic during the decade of the Scopes trial, Fosdick published at least five articles about Christianity and science. In addition to those already cited, these were "Evolution and Religion," *Ladies' Home Journal* 42 (September 1925): 12, 180, 183, and 185, which has four notes referencing White; "Science and Religion," *Harper's Monthly Magazine* 152 (February 1926): 296-300; and "Will Science Displace God?," *Harper's Monthly Magazine* 152 (August 1926): 362-66.

10. The pamphlet has a footnote after the title: "Printed by permission of *Good Housekeeping* (March, 1928)."

11. Twenty states considered at least one anti-evolution bill during the 1920s and early 1930s. Edward J. Larson, *Trial and Error: The American Controversy of Creation and Evolution*, 3rd ed. (New York: Oxford University Press, 2003), 48.

12. Fosdick probably based this claim on a passage in White, *History of the Warfare of Science*, 1:183–84. White in turn cited Leopold Friedrich Prowe, *Nicolaus Coppernicus* (Berlin: Weidmann, 1883–1884), but he did not give a more specific reference. The relevant passage can only be a lengthy footnote about comets in vol. 1, part 2, 269–70, where Prowe says without further documentation, "Noch am Ende des 17. Jahrhunderts wurde dem Professor der Astronomie in manchen Ländern Europa's ein Eid abverlangt, dass er mit den Grundsätzen des Aristoteles, und ausdrücklich auch mit dessen Ansichten über die Kometen, überein- stimme." This hardly supports Fosdick's claim. Furthermore, according to the late Owen Gingerich (private communication), Prowe's statement is itself "weak and unsubstantiated."

13. Alfred Watterson McCann, *God—or Gorilla; How the Monkey Theory of Evolution Exposes Its Own Methods, Refutes Its Own Principles, Denies Its Own Inferences, Disproves Its Own Case* (New York: Devin-Adair, 1922).

14. The sermon diverges slightly here: "In particular this morning I speak to you on the positive indebtedness of religion to science and if you wish a text remember the words of our Lord, so amazingly confirmed in this realm: 'Ye shall know the truth, and the truth shall make you free'" (John 8: 32).

15. Probably a reference to a periodic comet originally discovered in 1812 by French astronomer Jean-Louis Pons, which returned in January 1884.

16. Gen. 11:1–9.

17. The sermon cites *The Outline of Science: A Plain Story Simply Told*, ed. J. Arthur Thompson, 4 vols. (New York: G. P. Putnam's Sons, 1922), 4:11.

18. The sermon cites Paul De Kruif, *Microbe Hunters* (New York: Harcourt, Brace, 1926), chap. 1. Contrary to Fosdick, the Dutch cloth merchant Antoni van Leeuwenhoek used simple microscopes with a single lens, not compound microscopes; and he was hardly the first person to use microscopes of any type. Gerald L'E. Turner, "Microscope, Optical (Early)," in *Instruments of Science: An Historical Encyclopedia*, ed. Robert Bud and Deborah Jean Warner (New York: Garland, 1998), 387–90.

19. William Wordsworth, "Tintern Abbey," lines 94–102.

20. Thomas Hood, "I Remember, I Remember," lines 25–32.

21. Psalm 19:1.

22. Samuel Walter Foss, "A Greater God," lines 17–24. Fosdick had already quoted this in *The Meaning of Prayer* (New York: Association Press, 1915), 98–99.

23. The typescript sermon and the first printed version have, "In the second place, we as Christians are unpayably indebted to science for knowledge of the reign of law." Harry Emerson Fosdick, *Religion's Indebtedness to Science: A Sermon Preached at the Park Avenue Baptist Church New York on February 27, 1927* (New York?, 1927).

24. Probably a reference to what is said about the English anti-Newtonian writer John Hutchinson and others in White, *History of the Warfare of Science*, 1:148.

25. The sermon has a footnote here: "Quoted in personal conversation by Thomas Gladding, N. Y. chemist, as from a chemistry professor in his youth." Thomas Gladding was a founding partner of the New York firm Stillwell and Gladding. "Thomas Stantial Gladding," *The Analyst* 65 (July 1940): 391–92.

26. Edmund Halley predicted that the comet he observed in 1682 would return around 1758, and it was actually observed in late December of that year, although it did not reach perihelion until March 1759. Peter Lancaster-Brown, *Halley & His Comet* (Dorset, UK: Blandford Press, 1985), 78–86. The source of this dubious story about a "fearful crisis" over Halley's comet and its predicted return is unclear. Regardless, Fosdick could be confusing

Cotton Mather with his father, Increase Mather. Both saw comets as divine omens, but neither expressed concerns about the return of Halley's comet (which neither lived to see) or denied that comets have natural causes. Indeed, they saw comets both as providential signs and as natural causes themselves of certain calamitous terrestrial events. Comets appeared over Boston in November 1680 and August 1682; the second was Halley's comet. Increase Mather wrote two sermons, *Heaven's Alarm to the World* (Boston, 1681) and *The Latter Sign Discoursed* (Boston, 1682), and a treatise, *Kometographia, or, A Discourse Concerning Comets* (Boston, 1683), in which he interpreted them as divine warnings to a sinful world. Cotton Mather's second publication, *The Boston Ephemeris: An Almanack for MDCLXXXIII* (Boston, 1683), included a brief astronomical description of Halley's comet and an advertisement for his father's forthcoming book, but nothing of an interpretive nature. He said more many years later in the anonymously published *A Voice from Heaven. An Account of a Late Uncommon Appearance in the Heavens* (Boston, 1719), where he saw comets as fiery abodes for sinners, veritable heavenly hells, while drawing on naturalistic theories. Significantly, in *The Christian Philosopher* (London, 1721) he reported Halley's prediction without any anxiety, cited Isaac Newton to support the suggestion that "the Appearance of *Comets* is not so dreadful a thing, as the *Cometomania*, generally prevailing, has represented it," while quoting the Scottish physician George Cheyne, "that these frightful Bodies are the Ministers of *Divine Justice*, and in their Visits lend us *benign* or *noxious* Vapours, according to the Designs of Providence." A later version of the section on comets, augmented with expanded quotations from Cheyne, appeared as *An Essay on Comets, their Nature, the Laws of their Motions, their Cause and Magnitude of their Atmosphere, and Tails; With a Conjecture of their Use and Design* (Boston, 1744). See Cotton Mather, *The Christian Philosopher*, ed. Winton U. Solberg (Urbana: University of Illinois Press, 1994), xxxiv, lxxvi, and 50–54; Winton U. Solberg, "Science and Religion in Early America: Cotton Mather's 'Christian Philosopher,'" *Church History* 56 (1987): 73–92; Michael G. Hall, *The Last American Puritan: The Life of Increase Mather, 1639–1723* (Middletown, CT: Wesleyan University Press, 1988), 158–71; and Sara Schechner Genuth, "From Heaven's Alarm to Public Appeal: Comets and the Rise of Astronomy at Harvard," in *Science at Harvard University: Historical Perspectives*, ed. Clark A. Elliott and Margaret W. Rossiter (Bethlehem, PA: Lehigh University Press, 1992), 28–54. I am grateful to the late Ronald L. Numbers for advice.

27. John 8:32. The chapter on comets in White's book closes by quoting the same biblical text to make the same overall point about science providing "emancipation from terror and fanaticism." White, *History of the Warfare of Science*, 1:208.

28. Scottish Presbyterian John Gibson Paton, English Congregationalist Robert Morrison, American Baptist Adoniram Judson, and Scottish Congregationalist physician David Livingstone.

29. Physicians James Carroll, Walter Reed, and Jesse William Lazear were all assigned to The United States Army Yellow Fever Commission, which Reed directed. In a successful effort to discover the cause of the disease, Carroll and Private John R. Kissinger allowed themselves to be bitten by mosquitoes that had previously bitten infected persons. Carroll and Kissinger recovered, although Kissinger's health suffered permanent damage. In a separate incident, Lazear contracted the disease, perhaps from self-experimentation, and died a few days later.

30. John 15:13.

31. John J. Moran also volunteered to be bitten, but he did not develop yellow fever.

32. In October 1846, dentist William Thomas Green Morton successfully demonstrated the anesthetic properties of ether, when he administered it to a patient undergoing the removal of a neck tumor by physician John Collins Warren. English surgeon Edward Jenner discovered that the effects of smallpox could be reduced or prevented by vaccinating people with matter taken from the sores of people afflicted with cowpox, a much less dangerous disease.

33. John 14:12.

34. Alice Griffith Carr, a nurse with the American Red Cross during the Great War, remained in Europe after the war, working for the Red Cross and later for the Near East Foundation. The Greek government honored her multiple times for her service. Clara D. Noyes, "Department of Red Cross Nursing," *American Journal of Nursing* 23.12 (September 1923): 1036–39; Mabell S. C. Smith, "American Public Health Methods in the Near East," *American Journal of Nursing* 28.5 (May 1928): 463–65; and the introduction to the Alice Griffith Carr Papers, MS-135, Wright State University Libraries, Dayton, Ohio, https://wright.libraryhost.com/repositories/2/resources/280, accessed February 20, 2024. I am grateful to the late Owen Gingerich for identifying Carr.

35. The sermon cites George Freeland Barbour, *The Life of Alexander Whyte, D. D.* (London: Hodder and Stoughton, 1923), 210.

36. The sermon cites the Journal of English biologist Thomas Henry Huxley, as quoted by Leonard Huxley, *Life and Letters of Thomas Henry Huxley* (New York: D. Appleton, 1900), 1:162.

37. From the final paragraph of philosopher Bertrand Russell's essay "The Free Man's Worship," December 1903. Fosdick evidently considered Russell a scientist.

### *Compton, Mathews, and Gilkey (1930)* · Life After Death

1. The information here comes mainly from Edward B. Davis, "Prophet of Science—Part One: Arthur Holly Compton on Science, Freedom, Religion, and Morality," *Perspectives on Science and Christian Faith* 61.2 (June 2009): 73–83; "Prophet of Science—Part Two," *Perspectives on Science and Christian Faith* 61.3 (September 2009): 175–90; and "Prophet of Science—Part Three," *Perspectives on Science and Christian Faith* 61.4 (December 2009): 240–53.

2. "Statement Concerning Pamphlet Work of the American Institute of Sacred Literature for the Year Ending June 30, 1929," AISLR 11:3.

3. For publication details, see appendix one.

4. This information comes from his son, the late John J. Compton. See Edward B. Davis, "Prophet of Science—Part Two: Arthur Holly Compton on Science, Freedom, Religion, and Morality," *Perspectives on Science and Christian Faith* 613.3 (September 2009): 175–90, on 178.

5. Arthur Holly Compton, *The Idea of God as Affected by Modern Knowledge* (Boston: American Unitarian Association, 1940), 13 and 5; reprinted as "A Modern Concept of God," in *Man's Destiny in Eternity: The Garvin Lectures* (Boston: Beacon Press, 1949), 3–20, quotations on 11 and 5 (the wording of the second quotation is slightly different in the book).

6. George W. Gray, "Compton Sees a New Epoch in Science," *New York Times*, March 13, 1932.

7. Arthur Holly Compton, *The Freedom of Man* (New Haven, CT: Yale University Press, 1935), 109 and 115.

8. Compton, *Freedom of Man*, 110 and 112.

9. Compton, "The Effect of Social Influences on Physical Science," in *The Cosmos of Arthur Holly Compton*, ed. Marjorie Johnston (New York: Alfred A. Knopf, 1967), 81–100, quotation on 96.

10. Compton, *Freedom of Man*, 129–30, 139–40, and 147.

11. This Editor's Note, printed on an unnumbered page, precedes the text of the pamphlet.

12. Compton cites (but incorrectly names) a book by the English astrophysicist James Hopwood Jeans, *The Universe around Us* (Cambridge: Cambridge University Press, 1929), 324–27.

13. A reference to *Human Immortality: Two Supposed Objections to the Doctrine* (Boston: Houghton, Mifflin, 1898), by psychologist and philosopher William James. I am grateful to Jon H. Roberts for this information.

14. For comments on brain ablation, see *Brain, Mechanism and Intelligence* (Chicago: University of Chicago Press, 1929), by Chicago psychologist Karl Spencer Lashley; on Lashley, see *DSB*.

15. Henry Louis Bergson and Herbert Wildon Carr, *Mind-Energy* (London: Henry Holt, 1920); on Bergson, see *DSB*.

16. Canadian geologist J. D. MacKenzie, "The Experience of Dying," *Atlantic Monthly* 131 (May 1923): 585–90.

17. A reference to his older brother Karl T. Compton, a physicist at Massachusetts Institute of Technology who was later president of MIT.

18. Geneticist Clarence Cook Little, president of the University of Michigan from 1925 to 1929. The source of this quotation has not been identified.

19. According to orthogenesis, a form of neo-Lamarckian evolution that was popular among American scientists of the early twentieth century, biological variation is directed toward fixed goals by forces within an organism. Its leading proponent was paleontologist Henry Fairfield Osborn of the American Museum of Natural History. Peter J. Bowler, *Evolution: The History of An Idea*, rev. ed. (Berkeley: University of California Press, 1989), 268–70.

20. A reference to *Emergent Evolution: The Gifford Lectures, Delivered in the University of St. Andrews in the Year 1922* (London: Williams and Norgate, 1923), by psychologist C. Lloyd Morgan, who emphasized the emergence of new qualities at higher levels of organization within nature, especially life itself and mental processes.

21. Arthur Stanley Eddington, *The Nature of the Physical World* (Cambridge: Cambridge University Press, 1928), 178. With the previous sentence, the passage reads: "I do not think that the whole purpose of the Creation has been staked on the one planet where we live; and in the long run we cannot deem ourselves the only race that has been or will be gifted with the mystery of consciousness. But I feel inclined to claim that *at the present time* our race is supreme; and not one of the profusion of stars in their myriad clusters looks down on scenes comparable to those which are passing beneath the rays of the sun." Compton admired Eddington's defense of human freedom and consciousness. For an analysis of Eddington's Gifford Lectures in historical context, see Matthew Stanley, *Practical Mystic: Religion, Science, and A. S. Eddington* (Chicago: University of Chicago Press, 2007), 196–204, and Peter J. Bowler, *Reconciling Science and Religion: The Debate in Early-Twentieth-Century Britain* (Chicago: University of Chicago Press, 2001), 114–20.

22. The pamphlet has a footnote here: "I have borrowed this parable of the light and the candle flame from our distinguished American astronomer, Henry Norris Russell." Henry Norris Russell, *Fate and Freedom* (New Haven, CT: Yale University Press, 1927), 141–51.

23. From "Sinaloa," a poem by biologist David Starr Jordan, president of Stanford University, from his autobiography *The Days of a Man: Being Memories of a Naturalist, Teacher, and Minor Prophet of Democracy*, 2 vols. (New York: World Book Company, 1922), 1:528. After the poem, a blank page (20) precedes Mathews's address, which starts on page 21.

24. Mathews's good friend William Rainey Harper, first president of the University of Chicago, died of cancer at age 49 in 1906 and was buried in the chapel where Mathews gave this address; on Harper, see *ANB*.

25. Terminal punctuation here is missing in the original, where a small space at the end of a line is followed by "We" at the start of the next line.

26. Mathews quotes an unidentified work by the American idealist philosopher Josiah Royce, in which Royce apparently made a verbal play on 1 Cor. 15:53. Mathews used the identical quotation in at least four other writings: "Immortality and Morality," in *Twenty-first Annual Session of the Baptist Congress for the Discussion of Current Questions* (New York: Baptist Congress Publishing Company, 1904), 130–40, quotation on 134; *The Church and the Changing Order* (New York: Macmillan, 1912), 75; *The Gospel and the Modern Man* (New York: Macmillan, 1913), 219; and "The Reasonable Hope of the Christians," in *My Belief in Immortality*, ed. Aaron Avery Gates (New York: Doubleday, Doran, 1928), 3–14, quotation on 11.

27. Job 14:14.

28. Perhaps a garbled version of a sentence in Henry David Thoreau, *Walden* (1854): "Morning is when I am awake and there is a dawn in me."

29. Mal. 4:2 and Prov. 4:18.

30. Naval aviator and explorer Richard Evelyn Byrd established a base camp called "Little America" in Antarctica; see Byrd, *Little America* (New York: G. P. Putnam's Sons, 1930).

31. Acts 2:24, from Edgar J. Goodspeed, *The New Testament: An American Translation* (Chicago: University of Chicago Press, 1923).

32. Matt. 10:28.

33. Several passages on 35–37 were quoted in *The Instructor*, published by Deseret Sunday School Union, 73.1 (January 1938), 33–35, https://archive.org/details/instructor731dese/page/34/mode/2up, accessed February 23, 2024.

34. The reference is unclear.

35. Quoting the closing lines of Joseph Wood Krutch, *The Modern Temper: A Study and a Confession* (New York, Harcourt, Brace, 1929).

36. From the final paragraph of Bertrand Russell's December 1903 essay "The Free Man's Worship."

37. Acts 2:24.

38. John 10:9. The next paragraph indicates that this "visiting preacher" was the Unitarian minister Francis Greenwood Peabody, who had several positions at Harvard University, including dean of the Divinity School from 1901 to 1906. See *ANB*.

39. Quoting Harvard bacteriologist Hans Zinsser, describing the death of physician Francis W. Peabody in the *Harvard Graduates' Magazine* 36.141 (September 1927): 242–45, quotation on 245; also see Simeon Burt Wolbach, "Hans Zinsser (1878–1940)," *Biographical Memoirs of the National Academy of Sciences* 24 (1947): 323–60, quotation on 346. On Peabody's extraordinary life and career, see Oglesby Paul, *The Caring Physician: The Life of Dr. Francis W. Peabody* (Boston: Boston Medical Library in the Countway Library of Medicine, 1991).

40. Rom. 6:9.

41. Flora Sylvester Cheney, wife of physician Henry W. Cheney, was a member of Hyde Park Baptist Church, where Gilkey had been pastor before becoming dean of the Rockefeller Chapel. Very active in Democratic politics, she was elected to the Illinois legislature in November 1928. See the obituary in the *New York Times*, April 29, 1929. On Lorado Taft, whose studio was near the University of Chicago, see *ANB*.

42. 2 Tim. 4:7.

43. From "The Story of Sigurd the Volsung," a retelling of the Icelandic saga by the Pre-Raphaelite poet William Morris.

44. George Herbert Palmer, *The Life of Alice Freeman* (Boston: Houghton Mifflin, 1908), 327. Alice Freeman Palmer was president of Wellesley College from 1887 to 1902; on G. H. Palmer and A. F. Palmer, see *ANB*.

45. Surgeon William Darrach, dean of the Columbia University Medical School from 1919 until 1930.

46. Kandersteg, a village in the Bernese Oberland, can be reached by a road that snakes across the mountains; perhaps this is the basis for the story.

47. Congregationalist minister and educator Willard Learoyd Sperry, dean of Harvard Divinity School from 1922 to 1953; see *ANB*.

## *Mather (1931)* · The Religion of a Geologist

1. The information in this introduction comes mainly from Edward B. Davis, "Altruism and the Administration of the Universe: Kirtley Fletcher Mather on Science and Values," *Zygon* 46 (2011): 517–35.

2. Kirtley Fletcher Mather, "Geologist at Large," 23 and 70, DUA 1:5.

3. Vernon L. Kellogg, *Headquarters Nights: A Record of Conversations and Experiences at the Headquarters of the German Army in France and Belgium* (Boston: Atlantic Monthly Press, 1917).

4. Kirtley F. Mather, "Parables from Palaeontology," *Atlantic Monthly* 121 (April 1918): 35–43, quotation on 35.

5. Mather, "Geologist at Large," 84.

6. Kirtley F. Mather, *Christian Fundamentals in the Light of Modern Science* (Granville, OH: Times Press, 1924), 83.

7. Stephen Jay Gould, "Nonoverlapping Magisteria," *Natural History* 106 (March 1997): 16–22; Stephen Jay Gould, *Hen's Teeth and Horses' Toes* (New York: W. W. Norton, 1983), 273.

8. Kirtley F. Mather, *Science in Search of God* (New York: Henry Holt, 1928).

9. Erin A. Smith, "The Religious Book Club: Print Culture, Consumerism, and the Spiritual Life of American Protestants Between the Wars," in *Religion and the Culture of Print in Modern America*, ed. Charles L. Cohen and Paul S. Boyer (Madison: University of Wisconsin Press, 2008), 217–42.

10. Kirtley F. Mather, "The Religion of a Scientist," *Harvard Alumni Bulletin* (June 18, 1931): 1142–49.

11. Kirtley F. Mather, *The Permissive Universe* (Albuquerque: University of New Mexico Press, 1986), 171–72.

12. Mather, *Science in Search of God*, 71.

13. Kirtley F. Mather, "Sermons from Stones," in *Has Science Discovered God? A Symposium of Modern Scientific Opinion*, ed. Edward H. Cotton (New York: Thomas Y. Crowell, 1931), 3–19, quotation on 9.

14. Shailer Mathews, *How Science Helps Our Faith* (Chicago: American Institute of Sacred Literature, 1922), 12.

15. Werner Heisenberg formulated his famous uncertainty principle in 1927, and discussion of its larger meaning and significance began immediately. As Heisenberg put it himself, "the resolution of the paradoxes of atomic physics can be accomplished only by further renunciation of old and cherished ideas. Most important of these is the idea that natural phenomena obey exact laws–the principle of causality." Werner Heisenberg, *The Physical Principles of the Quantum Theory*, trans. Carl Eckart and Frank C. Hoyt (Chicago: University of Chicago Press, 1930), 62. Contrary to what Mather implies, however, Einstein never abandoned a strictly causal picture of nature.

16. For more on Mather's conception of God, see Davis, "Altruism and the Administration of the Universe."

17. An implicit quotation of a famous passage from James Hutton: "The result, therefore, of our present enquiry is, that we find no vestige of a beginning,—no prospect of an end." Hutton, *Theory of the Earth, With Proofs and Illustrations* (Edinburgh: William Creech, 1795), 1:200.

18. Sir James Jeans, *The Mysterious Universe* (Cambridge: Cambridge University Press, 1930), 148.

19. Mather delivered the first version of this text as a talk in March 1931, less than seventeen months after the great stock market crash of October 1929.

# INDEX

Abbot, Charles Greeley, 150
Abbott, Lyman, 22, 52, 54–55, 152; *The Theology of an Evolutionist*, 68, 91–92
Abraham, 202
Adam, 117–18, 120
Allen, Frederick Lewis, 47
American Association for the Advancement of Science, xiii, 5, 29, 54, 60, 61, 82; efforts to defend evolution, 20–21, 30, 50–53; starred scientists in *American Men of Science*, 57–58; support for AISL pamphlets, 57–59
American Civil Liberties Union, 236
American Institute of Sacred Literature, xiii, 7, 9, 18, 31, 49, 51, 78, 80, 90, 95, 99–100; alternative ideas for pamphlets, 35, 37, 45, 51–55, 60, 62–63, 72; dissemination of pamphlets, 5, 54–55, 64–73, 75, 284n75; financial support for, 35, 45, 52, 55–59, 72, 95, 259–71; first pamphlets published by AISL, 55, 86; goals of, 36; history of, 4; reception of pamphlets, 34–38, 64–66, 68–69, 71–77, 88–89
Ames, Edward Scribner, 93
Anderson, Victor, 89
Angell, James Rowland, 19–20, 29, 32, 150
Apostles' Creed, 9, 82, 100, 211
Appleget, Thomas Baird, 62
Aristotle, 48, 109
Arnold, Matthew, 98–99, 289n42, 298n34
Augustine, Saint, 21, 146–47, 151–52
Auld, Robert Campbell MacCombie, 53

Babbitt, Irving, 40
Baldwin, Roger, 236
Barbour, Ian G., 9

Baronio, Cesare (cardinal), 79
Bartley, James, xiii
Bateson, William, 114
Beck, P. W., 25
Beecher, Henry Ward, 85, 152
Bergson, Henri, 42; *Creative Evolution*, 68; *Mind-Energy*, 216
Bessey, Charles E., 82
Bible: German enemies of, 12–13, 89; inspiration or inerrancy of, 10–11, 21, 33, 73, 75, 90, 92, 155, 198; not a science book, 42, 78–79, 81, 83, 120–21, 124, 127–29, 137, 154–55, 198
Billings, Josh. *See* Shaw, Henry Wheeler
Birge, Edward Asahel, 49, 52, 55
Birkhoff, George David, 32
Blaine, John J., 52
Bliss, William J. A., 84
Bolton, Thaddeus L., 73–74
Bonnet, Charles, 297n22
Brahe, Tycho, 175
Breasted, James H., 150
Brooks, William Keith, 109–10
Brown, Ralph Wilder, 40
Brown, Robert Elliott, 28–31, 33, 59, 143, 152; concerns about youth and education, 28–29
Brown, William Adams, 37, 63, 277n78
Browne, Borden Parker, 124
Bryan, William Jennings, 101–2, 107–8, 125; blamed evolution for modernism, 16–17; concerns about youth and education, 16–17, 130–31; efforts to ban evolution, 12–18; "God and Evolution," 3, 115, 123, 125–33; against "hypotheses" in science, 3, 14–15, 126; *The Menace of Darwinism*, 9, 12–15; objections to

Bryan, William Jennings (*cont.*)
  materialism and reductionism, 15–17; objections to social Darwinism, 12, 15, 17–18; objections to theistic evolution, 14, 18, 127, 131, 296n22; Round Table Club, 82; seen as "medieval," 30, 111, 122, 132
Burbank, Luther, 33
Burhoe, Ralph Wendell, 238
Burton, Ernest DeWitt, 31, 51–53, 54–60, 67, 70–71, 73, 95, 152
Burtt, Edwin Arthur, 80, 90
Byrd, Richard Evelyn, 228

Cadmen, S. Parkes, 22, 29, 53
Calvin, John, 21, 97, 293n45
Campbell, William Wallace, 32
Carlson, Anton Julius, 63, 211
Carnot, Nicolas Léonard Sadi, 196
Carr, Alice Griffith, 207–8, 305n34
Carroll, James, 206–7
Carus, Paul, 60
Case, Shirley Jackson: *The Millennial Hope*, 88
Cattell, James McKeen, 46–47, 191
Chamberlin, Georgia L., 54–57, 60–61, 63, 66, 75, 100, 191
Chamberlin, Thomas Chrowder, 100, 150, 237
Chautauqua Institution, 39, 56, 61, 177–78
Cheney, Flora Sylvester, 232
Clarke, John M., 19, 31
Collins, Francis, 102–3
Columbus, Chrisopher, 23, 120, 293n40
Communism, 64, 69
Compton, Arthur Holly, 7, 58, 83, 89, 211; conception of God, 71, 211–12; on intelligent design, 71, 101, 211–12; *Life After Death: From the Point of View of a Scientist*, 5, 63, 213–22; modernist faith of, 71, 100, 211, 213; objections to materialism and reductionism, 37, 63, 210, 215–17
Compton, Karl T., 217, 306n17
Conant, Judson E., 276n38
conflict view of science and religion, 6–8, 23, 34, 37, 85–86, 97–99, 197–99, 200, 207–8. *See also* Draper, John William; White, Andrew Dickson
Conklin, Edwin Grant, 22, 26, 28, 31–34, 38–39, 41–43, 46, 50–53, 55, 57, 59, 78, 83, 150; and academic freedom, 30; acceptance of hypotheses in science, 15, 113–15; conception of God, 71, 100–101, 110; conversion to modernism, 99, 109–11; *The Direction of Human Evolution*, 34, 74, 110; editorial in *New York Times*, 3, 49, 108; *Evolution and the Bible*, 4, 53, 64–65, 67–70, 72, 75, 77, 97, 107, 111–22, 285n98; frustration with anti-evolutionists, 19, 43, 108–9, 112–13, 120; objections to materialism and reductionism, 110; religion of science, 100, 107, 110
Cope, Edward Drinker, 85–86
Copernicus, Nicholaus, 127–28, 138, 174, 193, 200, 227
Cook, Joseph, 14
Cornell, Ezra, 96
Coulter, John Merle, 22, 54, 150
Coyne, Jerry, 102
Cunningham, Joseph Thomas, 42
Curie, Marie, 144
Curtis, Heber Doust, 39, 48

Dana, James Dwight, 79, 82, 121
Darby, John Nelson, 10
Darrach, William, 233
Darrow, Clarence, 46, 76, 236
Darwin, Charles, 3, 21, 89, 236; *Autobiography*, 15; *The Descent of Man*, 15; *On the Origin of Species*, 108, 113, 115, 126, 200
Davenport, Charles, 30, 50, 58, 67
Dawson, John W., 14, 81
deism, 13–15, 21, 27, 39, 43, 131–32
Delbrueck, Hans, 135
Democritus, 48, 193
Descartes, René, 46–47
de Vries, Hugo, 41, 115, 236
Dewey, John, 46–47
dispensationalism, 10
Dodge, Carroll William, 43, 46
Dods, Marcus, 208
Dorlodot, Henricus Henry de (canon), 46
Dow, Grove Samuel, 67
Dowie, John Alexander, 122, 294n50
Draper, John William: *History of the Conflict between Religion and Science*, 6, 96, 98–99, 110, 197; influence on Conklin, 110–11
Drawbridge, Cyprian Leycester, 287n5
Drummond, Henry, 125, 151–52; *The Ascent of Man*, 27, 132; *Natural Law in the Spiritual World*, 86, 91
Duane, William, 44
Dumm, Alfred Benjamin, 40

Dunn, Gano, 31
Dunnington, Francis Perry, 18–19

earth: age of, 102, 121, 144, 148, 167, 242, 244, 299n17
East, Edward Murray, 43, 47, 74
Eddington, Arthur, 191, 227; *The Nature of the Physical World*, 220, 237, 306n21
Eddy, J. Arthur, 33–34
Edison, Thomas, 32
Einstein, Albert, 28, 144, 161, 173, 200, 239
ether (medicine), 304n32
ether (physics), 144, 173, 216, 300n26
eugenics, xiii, 4, 9, 17–18, 23–25, 39, 50, 53, 68, 94, 177, 179
evolution: aristogenesis, 47; believed universally by scientists, 19–20, 32, 45, 113, 122; of birds, 184–88; compared with epigenesis, 118–19; compared with special creation, 115, 117, 119–20; creation by evolution, 21, 109, 177; efforts by states to ban its teaching, 18, 67–68, 111–12, 121, 200, 292n24; emergent evolution, 219; of humans, 23, 36–38, 45, 50, 72, 77, 79, 115–19, 130, 151, 160, 186; orthogenesis, 219; progressive evolution, 30–31, 101, 110, 145, 157, 159, 178–79, 188, 212, 218–19, 246–47; and purpose, 38, 49, 92, 100–101, 110, 141, 160, 210; and race, 23–24, 50–51, 117, 186; sexual selection, 61, 178; social evolution, 87, 172. *See also* theistic evolution
extraterrestrial life, 97, 171–72, 220, 306n21

Fairhurst, Alfred, 276n38
Faraday, Michael, 148
Fisher, George Park, 81
flat earth, 19, 31, 87, 120, 128, 137–38, 293n40
Fleischmann, Albert, 18
Flexner, Abraham, 32
Floyd, William, 34
Fosdick, Harry Emerson, 5, 29, 35, 43, 45–46, 56, 108, 152, 237; acceptance of hypotheses in science, 15, 125–27; *Christianity and Progress*, 68; conception of God, 14, 27, 125; concerns about youth and education, 26–27, 50, 124, 130–31; conversion to modernism, 99, 123–24, 197–98; editorial in *New York Times*, 4, 22, 49, 51, 55, 107, 123; *Evolution and Mr. Bryan*, 4, 55, 64–65, 67, 69–70, 72, 107, 111, 123, 125–33, 134; frustration with anti-evolutionists, 200–201; objections to materialism and reductionism, 26–27, 46, 110, 124–25, 129–31, 136; *Religion's Debt to Science*, 4, 62–63, 197, 200–209; on science and religion, 36, 62, 134, 136, 197–99, 200–201, 208–9; "Shall the Fundamentalists Win?," 125
Foss, Samuel Walter, 204
Foster, George Burman: *The Finality of the Christian Religion*, 38
Fox, Henry, 30
Fox, Philip, 160
free will, 38, 100, 212, 216–17, 246
Freeman, James Edward, 29
Frost, Edwin Brant, 296n19; acceptance of hypotheses in science, 173; conception of God, 161–62; *The Heavens are Telling*, 4, 60–61, 66, 76, 161–62, 162–76
fundamentalist or fundamentalism, 5, 12, 17–19, 21, 34–35, 43–44, 50, 55–56, 60, 62, 73, 75, 83, 89, 95, 143, 148, 192, 235–36; defined, 10–11; no tolerance for other beliefs, 35; and six-day creation, 102, 144; views driving people out of churches, 125
*Fundamentals, The*, 11, 14, 52, 69. *See also* World's Christian Fundamentals Association

Galilei, Galileo, 121, 128, 138, 146–47, 175, 193, 202, 227
Gamaliel, 112
Gilbert, James, 5–6
Gilkey, Charles Whitney, 63, 136, 210–11, 213, 236; *Life After Death: From the Point of View of a Christian Minister*, 5, 227–34, 236
Gilkey, Langdon, 210
God: as administration of the universe, 71, 237, 242, 249, 251; did not create natural laws, 178, 188–89; freedom of, 198; immanence of, 9, 21, 27, 39, 87, 90–93, 108, 111, 123, 125, 132, 135, 140–42, 144–45, 178–79, 189, 192, 238; imperfect conceptions of, 154–56, 170, 203–4; important to believe in, 139; imprisoned by natural law, 204–6; and miracles, 10–11, 83–84, 91, 161, 175–76, 198; omnipotence of, 93, 161, 169, 176, 177; omniscience of, 161, 169–70, 176; as personality-producing forces, 38, 135, 141, 225–27, 238, 245; and scientific law, 142; as soul of the universe, 156; transcendence of, 9, 90–92, 125, 191
"god of the gaps," 20, 27, 125, 132
Goldsmith, William Marion: *Evolution or Christianity?*, 22–25; *The Laws of Life*, 23
Goodspeed, Edgar J., 229

Gordon, George Angier, 29
Gould, Steven Jay, 237
Graebner, Theodore, 276n38
Graton, Louis Caryl, 44
Graves, Ralph H., 123
Gray, Asa, 79–81, 83–86, 90, 102–3, 109; *Darwiniana*, 81–82; *Natural Science and Religion*, 14, 81–82, 84
Great Depression, 4, 235, 250
Great War, 4, 6, 9, 10–12, 31, 55, 70, 83, 86–89, 111, 159, 207, 236
Guyer, Michael F., 60

Haeckel, Ernst, 12, 89
Hale, George Ellery, 29, 31–32, 59, 72
Hall, Alexander Wilford, 14
Halley's Comet, 205, 303n26
Hanus, Paul Henry, 43
Harper, William Rainey, 4, 225
Haydn, Joseph, 299n13
Hazard, Marshall C., 162
Heisenberg, Werner, 212, 239, 308n15
Hemingway Ernest, 190
Henderson, Lawrence Joseph, 101
Hiroshima, 199
Hitchcock, Edward, 3
Hocking, William Ernest, 43–44
Hodge, Charles: *What Is Darwinism?*, 43
Hood, Thomas, 203
Hoover, Herbert, 31
Hopkins, Ernest Martin, 71
Hornaday, William T., 73
Howe, Julia Ward, 191
Hoyle, Fred, 145
Hughes, Charles Evans, 67–68
Humanist Manifesto, 90, 277n73, 288n22
humanity, age of, 172
Hunter, George William: *A Civic Biology*, 50–51
Hutchinson, John, 294n46
Hutton, James, 242, 308n17
Huxley, Thomas Henry, 47, 86, 112, 208
Huygens, Christiaan, 97, 293n45

Icarus, 47
immaterial spirit, 173–74, 196, 215, 248–49
immortality, 73, 80, 100, 109, 128, 136–37, 159, 173, 193, 210–13, 248–49. See also *Life After Death*
Inchofer, Melchior, 128, 295n11

indeterminism in nature, 212, 239
Ingersoll, Robert, 65
Institute on Religion in an Age of Science, 238
intelligent design, 14, 36, 101–2, 140–41, 160, 190, 211–12, 219

James, Saint, 99
James, William, 216
Janeway, Frank, 72
Jastrow, Ignaz, 135
Jeans, James, 214, 243
Jeffrey, Edward Charles, 43
Jenner, Edward, 207
Jennings, Herbert Spencer, 43
Jerome, Saint, 21
Jesus Christ: atonement, 9, 35–36, 73, 207; attitude toward death, 229–30; authority of, 146; Christianity the main cause of moral improvement, 172; conception of God, 157; deity of Incarnation, 7, 9, 16, 83–84, 91, 100, 128, 191–92, 211; ministry of healing, 207; miracles of, 35, 83, 90, 92; as a modernist, 21, 150–53; preached love and fatherhood, 154, 157, 191, 196, 236; reveals the infinite personality, 142; Resurrection, 7, 9–10, 16, 33, 35, 43, 84, 100, 137, 213, 227, 232; as Son of God, 191; as the source of evolution, 24; as supreme moral exemplar, 23, 35, 250–51; virgin birth, 9, 16, 35, 84, 100
John, Saint, 236
Johnson, Francis Howe: *God in Evolution*, 93–94
Johnson, Phillip E., 102
Johnston, Howard Agnew, 79, 276n38
Jonah, xiii, 84
Jordan, David Starr, 222, 306n23
Joshua, 128, 155
Judd, John Wesley, 22, 276n38
Judson, Adoniram, 206

Kant, Immanuel, 163
Kay, George F., 73
Kearney, Carrie Belle, 69
Keas, Michael Newton, 11
Keen, William Williams, 55, 59, 73; *I Believe in God and Evolution*, 52
Kellogg, Vernon Lyman, 41, 59, 72; *Evolution: The Way of Man*, 46; *Headquarters Nights*, 12, 70, 236
Kelly, Howard Atwood, 69
Kelman, John, 29

Kelvin, William Thomson, Lord, 113, 148–49, 157, 159, 299n17
Kepler, Johannes, 174–75, 193
Kidd, Benjamin: *The Science of Power*, 18
King, Henry Churchill, 152; *Reconstruction in Theology*, 90
Kingsley, John Sterling, 85–86
Kirk, Albert East, 25
Kissinger, John R., 206–7
Kraatz, Walter C., 19
Krauss, Lawrence M., 102
Krutch, Joseph Wood: *The Modern Temper*, 230
Kuhn, Thomas S., 97

Lake, Kirsopp, 43
Lane, Alfred C., 70
Lane, Henry Higgins, 85; *Evolution and Christian Faith*, 38, 74, 83–84
Larson, Edward J., 5–6, 47
Lashley, Karl Spencer, 216
Lawrence, William, 29
Laws, Curtis Lee, 10
Lazear, Jesse William, 206–7
LeConte, Joseph: *Evolution and Its Relation to Religious Thought*, 91
Leeuwenhoek, Antoni van, 202
Leffmann, Henry, 75–76
Leuba, James, 15–16, 46
Levine, Lawrence, 12
*Life After Death*, 5, 63–64, 66, 100, 136, 210–11, 213–34, 236
Lillie, Frank R., 32, 83
Lincoln, William E., 73, 95
Linnaeus, Carl, 115, 181
Lister, Joseph, 207
Little, Clarence Cook, 218
Livingstone, David, 206
Loeb, Jacques, 26–27
Lorentz, Hendrik, 28
Lucretius, 42, 193
Lull, Richard Swan, 41, 43
Luther, Martin, 21, 121, 128
Lyell, Charles, 120

MacInnis, John Murdoch, 88–89
Macintosh, Douglas Clyde: *Why I Believe in Immortality*, 63
MacKenzie, J. D., 217, 306n16

Manning, William Thomas, 29, 61–62
Marsden, George M., 10–11
Marsh, Othniel Charles, 81
Mary, Saint, 65
materialism, 22, 99; difficult to identify proponents of atheistic evolution, 41–46; seen as declining, 22, 38–39, 100. *See also* Bryan, William Jennings; Compton, Arthur Holly; Conklin, Edwin Grant; Fosdick, Harry Emerson; Mather, Kirtley Fletcher; Mathews, Shailer; Millikan, Robert Andrews; Osborn, Henry Fairfield; Schmucker, Samuel Christian
Mather, Cotton, 3, 204, 235, 304n26
Mather, Increase, 235, 304n26
Mather, Kirtley Fletcher, 7, 19, 40, 43–44; conception of God, 71, 99–100, 110, 135, 237–38, 242, 245; concerns about social Darwinism, 70, 236; modernist faith of, 83, 99–100, 236–38; objections to creationism, 19; objections to "god of the gaps," 27; objections to materialism and reductionism, 100; *The Religion of a Geologist*, 5, 63–64, 69, 107, 235, 237–38; *Science in Search of God*, 191, 237; and Scopes trial, 78, 236–37
Mather, Richard, 235
Mathews, Edward B., 84
Mathews, Robert E., 66
Mathews, Shailer, 5, 19–20, 34–36, 56–67, 73–74, 77–79, 84–90, 94–95, 97, 107, 143–44, 152, 191–92, 213, 235–36; acceptance of hypotheses in science, 140; and Chautauqua Institution, 39, 177; conception of God, 99–102, 135–36, 141–42, 162, 238, 245; concerns about youth and education, 26; *Contributions of Science to Religion*, 60, 135, 161–62; *How Science Helps Our Faith*, 4, 37, 64–65, 69–70, 72–73, 123, 134, 137–42, 198, 238; *Immortality and the Cosmic Process*, 211; *Life After Death: From the Point of View of a Theologian*, 5, 136, 222–27; modernist faith of, 79, 85–86, 99, 134–35, 137, 235; objections to materialism and reductionism, 36–38, 46, 100, 110, 136, 223–25; on science and religion, 84–87, 138, 211; *Will Christ come again?*, 4, 86–89, 95
Matthew, William Diller, 19
Mauro, Philip, 88, 276n38
Maxwell, James Clerk, 148–49, 196
McCabe, Joseph, 45
McCann, Alfred Watterson, 45, 201, 276n38
McDougall, William, 46–47

McPherson, George Wilson, 89
Melanchthon, Philipp, 121, 293n44
Mencken, H. L., 76
Metcalf, Maynard Mayo, 46
Mercier, Louis J. A., 40–41, 43, 46
Merriam, John C., 29, 32, 150
Merton, Robert K., 98
Michelson, Albert, 29, 31–32
Micou, Richard Wilde, 79
Miller, Hugh: *The Testimony of the Rocks*, 71
Miller, Theodore A., 110
Millikan, Mabel, 28
Millikan, Robert Andrews, 5, 7, 22, 44, 46–47, 59–61, 72, 83, 90, 136, 190, 211; and academic freedom, 30; atheism does not really exist, 155, 157; conception of God, 100–101, 110, 135, 143–45, 155–58; concerns about youth and education, 30, 147–48; *Evolution in Science and Religion*, 99; involvement with and importance of local churches, 144, 147, 150; "A Joint Statement Upon the Relations of Science and Religion," 7, 28–34, 47, 59, 90, 98, 101, 143–44, 298n24, 298n31; modernist faith of, 83, 99, 144–45, 153–54; objections to materialism and reductionism, 100, 144, 156; on science and religion, 35, 143–44, 148–52; *A Scientist Confesses His Faith*, 4, 23, 33–34, 60, 65–66, 70–71, 73, 77, 143–44, 145–59; against sectarian religious differences, 35, 153
Milton, John, 23, 163, 175, 276n38, 300n32
Mitman, Gregg, 70
modernism: distinctives, 9; and miracles, 11, 90–92; about right living and not theology, 34–35; and science, 5, 7–8, 15, 36–37, 197
Moody, Dwight L., 132
Moore, Aubrey Lackington, 27
Moore, E. H., 32
Moran, John J., 207
More, Louis Trenchard, 41, 45
More, Paul Elmer, 40–41
Morgan, C. Lloyd: *Emergent Evolution*, 219
Morgan, G. Campbell, 89
Morgan, Thomas Hunt, 32, 115
Morris, William, 232, 307n43
Morrison, Robert, 206
Morton, Thomas Green, 207
Moses, 83, 155, 191
Mullen, Lincoln A., 11
Mullins, Edgar Young, 79

National Academy of Sciences, 29, 58, 59, 150; statement on evolution, 19–20, 31
National Research Council, 31, 59
Native Americans, 154, 186
Neanderthals, 119
Needham, James G., 75
Neff, Pat, 67
Newman, Horatio Hackett, 41, 68, 72
Newton, Isaac, 148–49, 174, 189, 191, 193, 196, 200, 204, 301n9
Nicene Creed, 9, 82, 85–86, 109
Nietzsche, Friedrich, 17
Noah, 23
Norris, Frank, 20
Noyes, Arthur Amos, 29–31, 46, 150
Noyes, William Albert, 150
Numbers, Ronald L., xiv, 18

Obama, Barack, 102
Orchard, William Edwin, 88
Osborn, Henry Fairfield, 22, 28–30, 32, 38, 41, 43, 46, 50–51, 60, 72, 150; aristogenesis, 47; editorial in *New York Times*, 3–4, 22, 49, 51, 126; objections to materialism and reductionism, 46–48
Osburn, Raymond Carroll, 66, 72

Pace, Ernest James, 16–17
pacifism, 34, 199
Paine, Thomas, 65
Palmer, Alice Freeman, 233
Palmer, George Herbert, 232–33
pantheism, 9, 34, 39, 42, 48, 81, 89, 99, 110, 288n20, 292n17
Parker, George Howard, 41, 43
Pasteur, Louis, 149, 174
Paton, John Gibson, 206
Patten, William, 74; concerns about social Darwinism, 70; *The Grand Strategy of Evolution*, 70; *Why I Teach Evolution*, 49, 71, 74, 105
Patterson, Alexander, 276n38
Paul, Saint, 12, 93, 146, 155, 232, 236
Peabody, Francis Greenwood, 230–31, 307n38
Peabody, Francis W., 231, 307n39
Peay, Austin, 18
Peter, Saint, 229
Pfeiffer, August, 294n9
Phillips, William D., 103

Pier, Arthur Stanwood, 43
Pope, Alexander, 171
Porter, James F., 37–38
postmillennialism, 86–87
prayer, 7, 9, 15, 100, 110, 137, 139, 199
preformationism, 118
premillennialism, 10–12, 86–88
Preston, May, 44
Price, George McCready, 19
process theism, 61, 178
Pupin, Michael Idvorsky, 7, 29, 31, 33, 60–62, 71–72, 78, 150, 278n81; conception of God, 191–92; *Creative Co-ordination*, 5, 61–62, 191, 192–96; *From Immigrant to Inventor*, 61, 190; *The New Reformation*, 62, 190, 237; not a modernist, 191; objections to materialism and reductionism, 62

Rachel, 198
Ranke, Leopold von, 135
Raulston, John T., 78
Rauschenbusch, Walter, 124
Rayleigh, John William Strutt, Baron, 148
Reed, Walter, 206–7
Reese, Herbert M., 35
Reighard, Jacob E., 74
Rice, Edward L., 20–22, 46, 278n81
Rice, William North, 41, 43, 52; *Christian Faith in an Age of Science*, 20
Riley, William Bell, 102; *The Menace of Modernism*, 13; *The Scientific Accuracy of the Sacred Scriptures*, 52
Rimmer, Harry, xiii
Roberts, Jon H., 11, 29
Robinson, John L.: *Evolution and Religion*, 75
Rockefeller, John David, Jr., 5, 56, 62, 64; *The Christian Church: What of Its Future?*, 55
Rockefeller, John David, Sr., 55, 210
Rockefeller Foundation, 54, 56, 58, 63, 93, 210, 235
Rodeheaver, Homer, 75
Rodeheaver, J. N., 74
Rogers, Fred T., 20, 67
Romanes, George, 117
Rosen, Edward E., 97
Royce, Josiah, 226
Russell, Bertrand, 97, 209, 230
Russell, Charles Taze, 289n33
Russell, Henry Norris, 32, 59, 306n22

Sanhedrin, 112
Sarton, George, 98
Schleiermacher, Friedrich, 12
Schmucker, Beale Melanchthon, 178
Schmucker, Samuel Christian, xiii, 5, 7, 48, 83, 100, 108; conception of God, 39, 100, 178–79, 188–89; *The Meaning of Evolution*, 61; objections to materialism and reductionism, 39, 100; *Through Science to God*, 4, 178, 179–89
Schmucker, Samuel Simon, 178
science falsely so-called, 11–12
Science Service, 44, 59–60, 72, 74. *See also* Slosson, Edwin Emery
Scopes, John, 18, 20, 38, 68, 78
Scopes trial, xiii, 3, 5, 7, 12, 18–20, 30, 46–47, 58, 61, 100, 101, 144, 190, 192; Darrow represented Scopes for free, 76; expert witnesses, 22, 38, 78, 238, 278n81; modernist concerns about, 26, 77–79, 236
Scott, William Berryman, 75
Scripps, Edward Willis, 44, 60
seminaries: teaching science and religion, 93–94
Shapley, Harlow, 44, 238
Shaw, Henry Wheeler (pseudonym of Josh Billings), 115
Shufeldt, Robert Wilson, 77
Sigerfoos, Charles P., 49, 70
Silliman, Benjamin, 3, 81
Simpson, James Young: *The Spiritual Interpretation of Nature*, 92
Simpson, Sir James Young, 92
skeptics, objections to religion. *See* Burbank, Luther; Eddy, J. Arthur; Floyd, William; McCabe, Joseph; Porter, James F.; Shufeldt, Robert Wilson
Skinner, J. E., 68
Slosson, Edwin Emery, 44–46, 59–60, 72. *See also* Science Service
Smallwood, William M., 70
Smith, Gerald Birney, 94–95; "Significant Movements in Modern Theology," 89–93, 288n23
Smith, Thomas Vernor, 211
Soares, Theodore Gerald, 144, 152
social gospel, 124, 135, 235
Society of Free Thinkers, 76
Spangler, Robert C., 75
Spencer, Herbert, 89

Sperry, Willard Learoyd, 234
Spong, John Shelby, 102
Steiglitz, Julius, 44
Stevenson, Ross, 33
Stewart, George Walter, 60, 72
Stewart, Lyman, 11
Stewart, Milton, 11
Straton, John Roach, 43, 45, 50–51, 69, 73
Strausbaugh, Perry Daniel, 75
Stone, John Timothy, 161
Strong, Augustus H., 79, 83
suicide, 110, 206, 209
*Sunday School Times*, 13, 16–17, 95
supernatural, 3, 7, 12, 61, 63; modernist rejection of, 100–101, 107, 109–10, 114–15, 118, 123, 161, 175, 237, 239–40

Tabrum, Arthur H.: *Religious Beliefs of Scientists*, 287n5
Taft, Lorado, 232
theistic evolution, 21, 37, 48, 81, 84, 86, 102, 108, 112, 131, 296n22; defined, 14; difficult for scientists to identify proponents of, 40–46. *See also* Bryan, William Jennings; Gray, Asa
theodicy, 134–35, 142, 158, 174–75, 295n2
Thompson, Silvanus P., 149
Thomson, J. Arthur, 41–42, 44, 46
Thoreau, Henry David, 228
Tilton, John T., 67
*Titanic*, RMS, 158
Torrey, John, 80
Torrey, Reuben Archer, 69, 89
Townes, Charles, 103
Trumbull, Charles Gallaudet, 17
Turner, William DeGarmo, 76

uniformity of nature, 239–40
universe: age of, 20, 299n8; great magnitude of, 139–40, 164–67, 198, 202–3; not the source of its own laws, 169–70; steady state, 145
University of Chicago Divinity School, xiii, 4–5, 38, 55–56, 93–94, 135–36, 213. *See also* American Institute of Sacred Literature
Ussher, James, 121

van Dyke, Henry, 53, 190
Venable, Francis P., 75
Voliva, Wilbur Glenn, 120

Wagner, Adolf, 135
Walcott, Charles Doolittle, 29, 59, 150
Ward, Henshaw, 41–42, 46
Warfield, Benjamin Breckinridge, 79, 83
Watson, James, 102
Watson, John B., 47
Weissmann, August, 236
Wellhausen, Julius, 12
Wesley, John: on Copernicus, 97, 121, 293n45; on evolution, 151, 297n22
Wharton, Edith, 190
Wheeler, William Morton, 43
White, Andrew Dickson, 6, 23, 37; *A History of the Warfare of Science with Theology in Christendom*, 86, 95–97, 99, 110, 197; influence on Conklin, 97, 99, 110–11; influence on Fosdick, 99, 197–99, 294n9, 295n11, 302n9, 303n12, 303n24, 304n27; influence on Mathews, 85; influence on Sarton, 98; influence on Smith, 95; misquoted Calvin and Wesley, 97; religious beliefs, 96, 98–99
Wieland, George Reber, 71–72
Wiggam, Albert Edward, 47, 190
Wiley, Harvey W., 34–35
Wilson, Woodrow, 61, 87, 108
Wise, Stephen Samuel, 29
Woelfkin, Cornelius, 56, 62
Woodruff, Lorande Loss, 41
Wordsworth, William, 48, 101, 202–3
World's Christian Fundamentals Association, 13, 102, 143, 292n23
Wright, George Frederick, 81, 93

yellow fever, 15, 206–7, 304n29
Yerkes, Robert M., 72
Young Men's Christian Association, 4, 64, 66–67, 70, 72

Zhang, Shizhang, 134
Zinsser, Hans, 231, 307n39

# Explore other books from HOPKINS PRESS

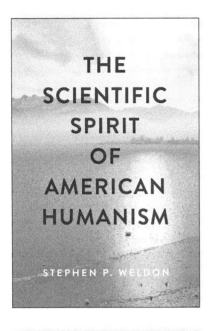

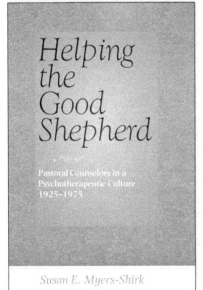

## JOHNS HOPKINS UNIVERSITY PRESS

PRESS.JHU.EDU

Milton Keynes UK
Ingram Content Group UK Ltd.
UKHW040954120924
448226UK00005B/20